CBS PROBLEMS and SOLUTIONS SERIES

Problems and Solutions in Signals and Systems

- This book is suitable for examinations of all Indian universities.
- Solutions of 250 problems are given in most simplified manner.
- All types of problems are included and the solutions given in detail.
- To explain the basics lucidly, computerised figures have been liberally added in the solutions. This volume contains more than 250 line diagrams.

CBS PROBLEMS and SOLUTIONS SERIES

Problems and Solutions in Signals and Systems

R GOPAL
Engineer

CBSPD

CBS Publishers & Distributors Pvt Ltd

New Delhi • Bengaluru • Chennai • Kochi • Kolkata • Lucknow • Mumbai
Hyderabad • Jharkhand • Nagpur • Patna • Pune • Uttarakhand

Problems and Solutions in Signals and Systems

Disclaimer

Science and technology are constantly changing fields. New research and experience broaden the scope of information and knowledge. The author has tried his best in giving information available to him while preparing the material for this book. Although, all efforts have been made to ensure optimum accuracy of the material, yet it is quite possible some errors might have been left uncorrected. The publisher, printer and the author will not be held responsible for any inadvertent errors or inaccuracies.

ISBN: 978-81-239-1145-8

First Edition: 2004

Reprint:2006, 2008, 2011, 2015, **2024**

Published by Satish Kumar Jain and Produced by Varun Jain for

CBS Publishers & Distributors Pvt Ltd

4819/XI Prahlad Street, 24 Ansari Road, Daryaganj, New Delhi 110 002, India.4819/XI Prahlad Street, 24 Ansari Road, Daryaganj, New Delhi 110 002, India.

Ph: 23289259, 23266861 Website: www.cbspd.com

e-mail: delhi@cbspd.com

Corporate Office: 204 FIE, Industrial Area, Patparganj, Delhi 110 092

Ph: 011-4934 4934 Fax: 011-4934 4935 e-mail: publishing@cbspd.com; publicity@cbspd.com

Branches

- **Bengaluru:** Seema House 2975, 17th Cross, K.R. Road, Banasankari 2nd Stage, Bengaluru 560 070, Karnataka
  Ph: +91-80-26771678/79 Fax: +91-80-26771680 e-mail: bangalore@cbspd.com
- **Chennai:** 7, Subbaraya Street, Shenoy Nagar, Chennai 600 030, Tamil Nadu, India
  Ph: +91-44-26680620/26681266 Fax: +91-44-42032115 e-mail: chennai@cbspd.com
- **Kochi:** 42/1325, 1326, Power House Road, Opp KSEB, Power House, Ernakulam 682 018, Kochi, Kerala, India
  Ph: +91-484-4059061-65, 67 Fax: +91-484-4059065 e-mail: kochi@cbspd.com
- **Kolkata:** 147, Hind Ceramics Compound, 1st Floor, Nilgunj Road, Belghoria, Kolkata-700056, West Bengal, India
  Ph: +033-25633055, 033-25633056 e-mail: kolkata@cbspd.com
- **Lucknow:** Basement, Khushnuma Complex, 7 Meerabai Marg (Behind Jawahar Bhawan), Lucknow-226001, UP, India
  Ph: +91-522-4000032 e-mail: tiwari.lucknow@cbspd.com
- **Mumbai:** PWD Shed, Gala no 25/26, Ramchandra Bhatt Marg, Next to JJ Hospital Gate no. 2, Opp. Union Bank of India Noorbaug, Mumbai-400009, Maharashtra, India
  Ph: 022-66661880/89 e-mail: mumbai@cbspd.com

Representatives

- **Hyderabad** 0-9885175004
- **Patna** 0-9334159340
- **Jharkhand** 0-9811541605
- **Pune** 0-9664372571
- **Nagpur** 0-8692091830
- **Uttarakhand** 0-9716462459

Printed at: SRK Graphics, Delhi, India

Preface

Signals and Systems is one of the fundamental subjects for the students of electronics and telecommunication engineering. The author feels it a great pleasure in presenting this intricate subject through typical problems and their solutions.

This book covers all important and major topics taught at the undergraduate level in an engineering college. All the problems have been discussed at length, and detailed line diagrams given extensively to help the reader understand the basics and grasp the logic. This book is suitable for preparing for all university examinations and also to face the competitions. This feature is expected to help the reader in visualising the patterns of problems and the extent and level of difficulty in solving the questions likely to be encountered in the examination, and thus tackling the problems with confidence and success.

Although every care has been taken to ensure accuracy, yet some errors might have crept in. The author will be grateful if these are brought to his notice so that these could be rectified in the subsequent printings and editions of the book. Readers are requested to send their suggestions for further improvement and revision to me through the publishers at the following e-mail address: cbspubs@del3.vsnl.net.in.

R. GOPAL

Contents

Chapter 1

Introduction to Signals

Problem 1.1: Find the even and odd components of each of the following signals:

(a) $x(t) = \cos(t) + \sin(t) + \sin(t)\cos(t)$

(b) $x(t) = 1 + t + 3t^2 + 5t^3 + 9t^4$

(c) $x(t) = 1 + t\cos(t) + t^2 \sin(t) + t^3 \sin(t)\cos(t)$

(d) $x(t) = (1 + t^3)\cos^3(10t)$

Solution:

For even part,

$$x_e(-t) = x_e(t) \text{ for all } t$$

and for odd part

$$x_o(t) = x_o(t) \text{ for all } t$$

(*a*)
$$x_e(t) = \cos t$$
$$x_o(t) = \sin t + \sin(t)\cdot \cos(t)$$

(*b*)
$$x_e(t) = 1 + 3t^2 + 9t^4$$
$$x_o(t) = t + 5t^3$$

(*c*)
$$x_e(t) = 1 + t^3 \sin t \cos t$$
$$x_o(t) = t\cos t + t^2 \sin t$$

(*d*)
$$x_e(t) = \cos^3(10t)$$
$$x_o(t) = t^3 \cos^3(10t)$$

Problem 1.2: Determine whether the following signals are periodic. If they are periodic, find the fundamental period.

(a) $x(t) = (\cos(2\pi t))^2$

(b) $x(t) = \sum_{k=-5}^{5} w(t-2k)$ for $w(t)$ depicted in Fig. P. 1.2 (a)

(c) $x(t) = \sum_{k=-\infty}^{\infty} w(t-3k)$ for $w(t)$ depicted in Fig. P. 1.2 (a)

(d) $x[n] = (-1)^n$

(e) $x[n] = (-1)^{n^2}$

(f) $x[n]$ depicted in Fig. P. 1.2 (b).

(g) $x(t)$ depicted in Fig. P. 1.2 (c).

(h) $x[n] = \cos(2n)$

(i) $x[n] = \cos(2\pi n)$

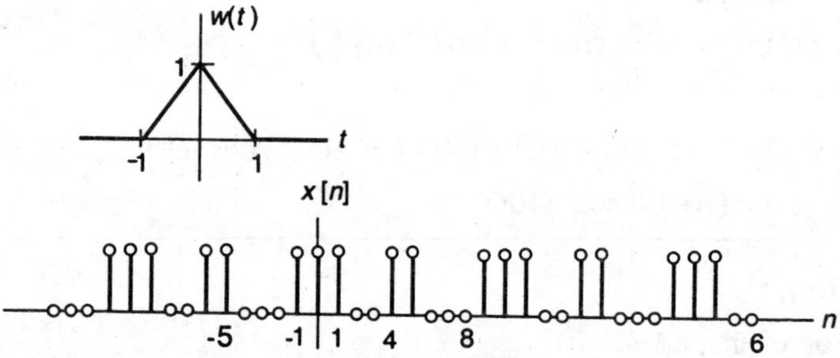

Fig. P. 1.2 (a)–(b)

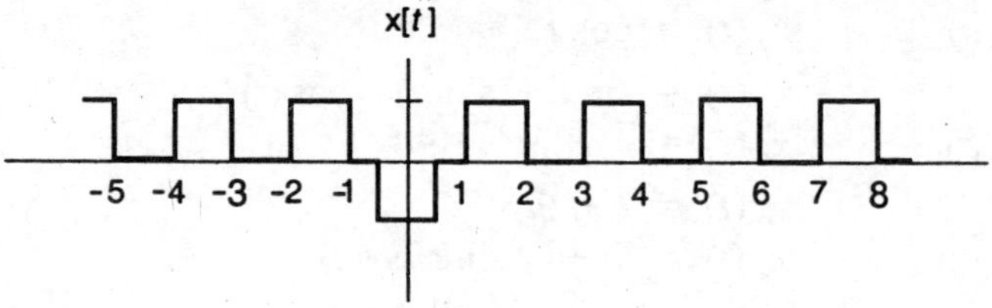

Fig. P. 1.2 (c)

Solution :

(a) $$x(t) = [\cos(2\pi t)]^2 = \frac{1+\cos(4\pi t)}{2}$$

For periodic signal,

$$x(t) = x(t + T) \text{ for all } t$$

Where T is a positive constant

The smallest value of T is the fundamental period,

i.e., $T = 0.5$

(*b*) $$x(t) = \sum_{K=-5}^{5} w(t - 2k) \text{ is depicted next}$$

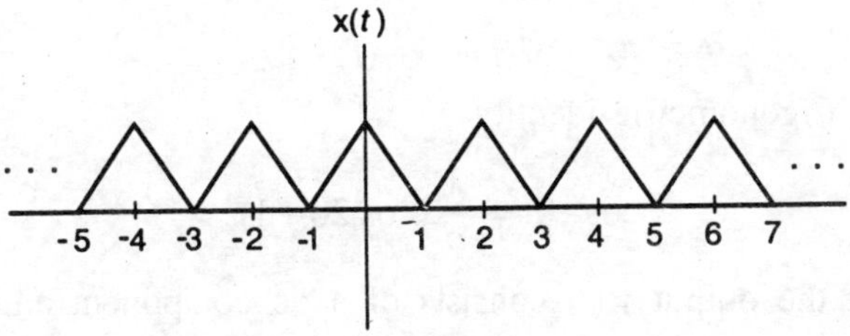

Fig. 1.2 (d)

Here, $x(t) = x(t + 2)$ for all t

Then the fundamental period $T = 2$

(c) $x(t)$ is non-periodic signal because fundamental period is infinite here.

(*d*) $$x[n] = (-1)^n$$

For a discrete signal to be periodic,

$$x[n] = x[n + N]$$

For all integers N, n

Where N is a positive integer

Here, $N = 2$ (fundamental period)

(*e*) $$x[n] = (-1)^{n^2}$$

$x[n]$ periodic signal and fundamental period is 1.

(*f*) For $x[n]$ to be periodic,

$$x[n] = x(n + 10)$$

the fundamental period is $N = 10$

(*g*) $x(t) \neq x(t + T)$ and thus $x(t)$ is non-periodic

(*h*) Here $x[n] = \cos(2n) \neq x[n + N] = x[2(n + N)]$

for any positive integer N

and, thus $x[n]$ is non-periodic

(i) $x[n] = \cos(2\pi n)$

$x[n] = x[n+1]$ for any n

and the fundamental period is $N = 1$

Problem 1.3: The sinusoidal signal

$$x(t) = 3\cos(200t + \pi/6)$$

is passed through a square-law device defined by the input-output relation

$$y(t) = x^2(t)$$

Using the trigonometric identity

$$\cos^2\theta = \frac{1}{2}(\cos 2\theta + 1)$$

show that the output $y(t)$ consists of a dc component and a sinusoidal component.

(a) Specify the dc component.

(b) Specify the amplitude and fundamental frequency of the sinusoidal component in the output $y(t)$.

Solution:

$$y(t) = x^2(t) = \left[3\cos\left(200t + \frac{\pi}{6}\right)\right]^2$$

$$y(t) = \frac{9}{2}\left[1 + \cos\left(400t + \frac{\pi}{3}\right)\right]$$

(*a*) The d.c. component is $\frac{9}{2}$

(*b*) The amplitude $= \frac{9}{2}$

and fundamental frequency $= \frac{\pi}{200}$

Problem 1.4: Categorize each of the following signals as an energy or power signal, and find the energy or power of the signal.

(a) $x(t) = \begin{cases} t, & 0 \le t \le 1 \\ 2-t, & 1 \le t \le 2 \\ 0, & \text{otherwise} \end{cases}$

(b) $x[n] = \begin{cases} n, & 0 \le n \le 5 \\ 10-n, & 5 \le n \le 10 \\ 0, & \text{otherwise} \end{cases}$

(c) $x(t) = 5\cos(\pi t) + \sin(5\pi t), \quad -\infty < t < \infty$

(d) $x(t) = \begin{cases} 5\cos(\pi t) & -1 \le t \le 1 \\ 0, & \text{otherwise} \end{cases}$

(e) $x[n] = \begin{cases} \sin(\pi n) & -4 \le n \le 4 \\ 0, & \text{otherwise} \end{cases}$

Solution:

(*a*) $$E = \underset{T\to\infty}{Lt} \int_{-T/2}^{T/2} x^2(t)\,dt = \int_{-\infty}^{\infty} x^2(t)\,dt$$

$$= \int_0^1 t^2\,dt + \int_1^2 (2-t)^2\,dt = \frac{1}{3} + \frac{1}{3} = \frac{2}{3}$$

Since $0 < E < \infty$, $x(t)$ is an energy signal

(*b*) $$E = \sum_{n=-\infty}^{\infty} x^2[n] = \sum_{n=0}^{5} n^2 + \sum_{n=5}^{10} (10-n)^2 = 110$$

Since $0 < E < \infty$, $x[n]$ is an energy signal

(*c*) $$x(t) = 5\cos(\pi t) + \sin(5\pi t)$$

$$E = \int_{-\infty}^{\infty} x^2(t)\,dt$$

$$= \int_{-\infty}^{\infty} [25\cos^2(\pi t) + \sin^2(5\pi t) + 10\sin(5\pi t)\cdot\cos\pi t]\,dt = \infty$$

$x(t)$ can't be an energy signal

$$P = \underset{T\to\infty}{Lt} \frac{1}{T} \int_{-T/2}^{T/2} x^2(t)\, dt = \frac{1}{T} \int_{-T/2}^{T/2} [5 \cos(\pi t) + \sin(5\pi t)]^2 \cdot dt$$

Where T is the fundamental period equal to 2.

$$= \frac{1}{2} \int_{-1}^{1} [25 \cos^2 \pi t + \sin^2 5\pi t + 5 \sin(6\pi t) + \sin(4\pi t)]\, dt$$

$$= \frac{1}{2} \int_{-1}^{1} \left[\frac{25}{2}(1 + \cos 2\pi t) + \frac{(1 - \cos 2\pi t)}{2} + 5 \sin(6\pi t) + \sin(9\pi t) \right] dt = \frac{13}{2}$$

Since $0 < P < \infty$, P is a power signal

(*d*) $$x(t) = \begin{cases} 5\cos(\pi t), & -1 \le t \le 1 \\ 0, & \text{otherwise} \end{cases}$$

$$E = \int_{-\infty}^{\infty} x^2(t)\, dt = \int_{-\infty}^{\infty} 25 \cos^2(\pi t)\, dt = \frac{25}{2} \int_{-1}^{1} (1 + \cos 2\pi t)\, dt$$

$$= \frac{25}{2} \left\{ t + \frac{\sin 2\pi t}{2\pi} \Big|_{-1}^{1} \right\} = 25$$

Since $0 < E < \infty$, $x(t)$ is an energy signal

$$P = \frac{1}{T} \int_{-T/2}^{T/2} x^2(t)\, dt \quad \text{here, } T = 2$$

$$= \frac{1}{2} \int_{-1}^{1} 25 \cos^2(\pi t)\, dt = \frac{25}{2} \int_{-1}^{1} \{1 + \cos(2\pi t)\}\, dt = \frac{25}{2}$$

Since $0 < P < \infty$, $x(t)$ is an power signal also

(*e*) $$E = \sum_{n=-4}^{4} \sin^2(\pi n) = \sum_{n=-4}^{4} \left(\frac{1 - \cos 2\pi n}{2} \right) = 4$$

Since $0 < E < \infty$, $x[n]$ is an energy signal

$$P = \frac{1}{N} \sum_{n=0}^{N-1} x^2[n]$$

Where N is the fundamental period here, $N = 2$

$$P = \frac{1}{2}\sum_{n=0}^{1} \sin^2(\pi n) = \frac{\sin^2 \pi}{2} = 0$$

$x(n)$ cannot be a power sigma

Problem 1.5: Consider the sinusoidal signal

$$x(t) = A\cos(\omega t + \phi)$$

Determine the average power of $x(t)$.

Solution:

$$P = \frac{1}{T}\int_{-T/2}^{T/2} x^2(t)\, dt$$

Where T is the fundamental period equal to

$$T = \frac{2\pi}{\omega}$$

Now, $$P = \frac{\omega}{2\pi}\int_{-\pi/\omega}^{\pi/\omega} A^2\cos^2(\omega t + \phi)\, dt = \frac{\omega}{2\pi}\int_{-\pi/\omega}^{\pi/\omega} [1 + \cos 2(\omega t + \phi)\, dt$$

$$= \frac{\omega A^2}{4\pi}\left[t + \frac{\sin 2(\omega t + \phi)}{2\omega}\Bigg|_{-\frac{\pi}{\omega}}^{\frac{\pi}{\omega}}\right] = \frac{\omega A^2}{4\pi}\cdot\left(\frac{2\pi}{\omega}\right) = \frac{A^2}{2}$$

Problem 1.6: The angular frequency Ω of the sinusoidal signal

$$x[n] = A\cos(\Omega n + \phi)$$

satisfies the condition for $x[n]$ to be periodic. Determine the average power of $x[n]$.

Solution:

Average power, $$P = \frac{1}{N}\sum_{n=0}^{N-1} x^2[n]$$

Here,. $$N = \frac{2\pi}{\Omega} \text{ (fundamental period)}$$

Then, $$P = \frac{\Omega}{2\pi} \sum_{n=0}^{\left(\frac{2\pi}{\Omega}-1\right)} \frac{A^2}{2} \{\cos(\Omega n + \phi) + 1\}$$

$$= \frac{A^2\Omega}{4\pi} \{1 + \cos 2\phi + 1 + \cos[2(2\pi - \Omega) + 2\phi]\}$$

$$= \frac{A^2\Omega}{4\pi} \{1 + \cos 2\phi + 1 + \cos(2\phi - 2\Omega)\}$$

Problem 1.7: The raised-cosine pulse $x(t)$ shown in Fig. P. 1.7 is defined as

$$x(t) = \begin{cases} \frac{1}{2}[\cos(\omega t) + 1], & -\pi/\omega \le t \le \pi/\omega \\ 0, & \text{otherwise} \end{cases}$$

Determine the total energy of $x(t)$.

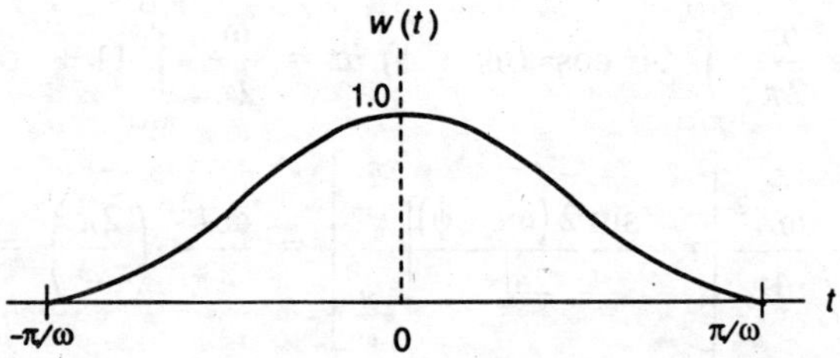

Fig. P. 1.7

Solution:

$$E = \int_{-\infty}^{\infty} x^2(t)\, dt = \int_{-\pi/\omega}^{\pi/\omega} \frac{1}{2} [\cos \omega t + 1]^2\, dt$$

$$= \frac{1}{2} \int_{-\pi/\omega}^{\pi/\omega} [1 + \cos^2 \omega t + 2 \cos \omega t]\, dt$$

$$= \frac{1}{2} \int_{-\pi/\omega}^{\pi/\omega} \left[1 + \frac{1 + \cos 2\omega t}{2} + 2 \cos \omega t\right] dt$$

$$= \frac{1}{2} \int_{-\pi/\omega}^{\pi/\omega} \left[t + \frac{\sin 2\omega t}{4\omega} + \frac{t}{2} + 2 \sin \omega t \right] dt = \frac{1}{2} \frac{3\pi}{\omega} = \frac{3\pi}{2\omega} \text{ Ans.}$$

Problem 1.8: The trapezoidal pulse $x(t)$ shown in Fig. P. 1.8 is defined by

$$x(t) = \begin{cases} 5 - t, & 4 \le t \le 5 \\ 1, & -4 \le t \le 4 \\ t + 5 & -5 \le t \le -4 \\ 0, & \text{otherwise} \end{cases}$$

Determine the total energy of $x(t)$.

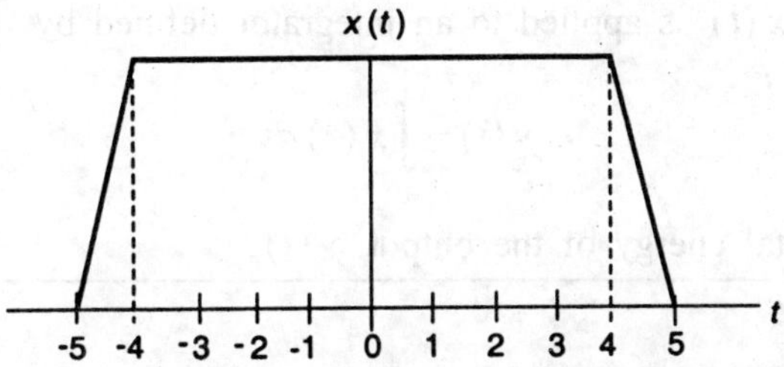

Fig. P. 1.8

Solution:

$$E = \int_{-\infty}^{\infty} x^2(t)\, dt = +\int_{-5}^{-4} (t+5)^2\, dt + \int_{-4}^{4} dt + \int_{4}^{5} (5-t)^2\, dt$$

$$= \left. \frac{(t+5)^3}{3} \right|_{-5}^{-4} + t\Big|_{-4}^{4} - \left. \frac{(5-t)^3}{3} \right|_{4}^{5} = \frac{1}{3} + 8 + \frac{1}{3} = \frac{26}{3} \text{ units}$$

Problem 1.9: The trapezoidal pulse $x(t)$ of Fig. P. 1.8 is applied to a differentiator, defined by

$$y(t) = \frac{d}{dt} x(t)$$

(a) Determine the resulting output $y(t)$ of the differentiator.

(b) Determine the total energy of $y(t)$.

Solution:

(a)
$$y(t) = \begin{cases} -1, & 4 \le t \le 5 \\ 0, & -4 \le t \le 4 \\ 1, & -5 \le t \le -4 \\ 0, & \text{otherwise} \end{cases}$$

(b)
$$E = \int_{-\infty}^{\infty} x^2(t)\, dt = \int_{-5}^{-4} 1 \cdot dt + \int_{4}^{5} -1 \cdot dt = 1 - 1 = 0$$

Problem 1.10: A rectangular pulse $x(t)$ is defined by

$$x(t) = \begin{cases} A, & 0 \le t \le T \\ 0, & \text{otherwise} \end{cases}$$

The pulse $x(t)$ is applied to an integrator defined by

$$y(t) = \int_0^t x(\tau)\, d\tau$$

Find the total energy of the output $y(t)$.

Solution:

$$x(t) = \begin{cases} A, \; 0 \le t \le T \\ 0, \; \text{otherwise} \end{cases}$$

$$y(t) = \int_0^t x(\tau)\, d\tau = \int_0^t A\, d\tau = At, \; 0 < t \le T$$

and, the total energy, $E = \int_{-\infty}^{\infty} y^2(t)\, dt = \int_0^T A^2 t^2\, dt = \left.\frac{A^2 t^3}{3}\right|_0^T = \frac{A^2 T^3}{3}$

Problem 1.11: A triangular pulse signal $x(t)$ is depicted in Fig. P. 1.11. Sketch each of the following signals derived from $x(t)$:

(a) $x(3t)$

(b) $x(3t+2)$

(c) $x(-2t-1)$

(d) $x(2(t+2))$

(e) $x(2(t-2))$

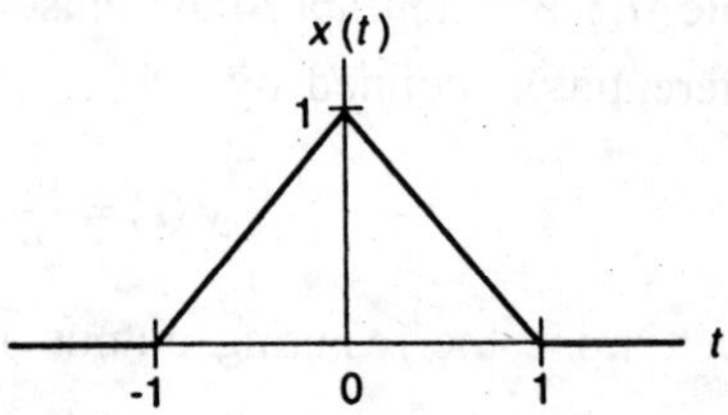

Fig. P. 1.11

Solution:

(*a*) $x(3t)$

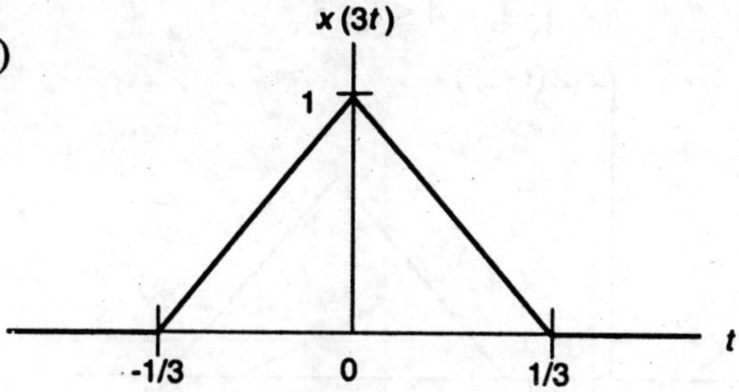

Fig. P. 1.11 (a)

(*b*)

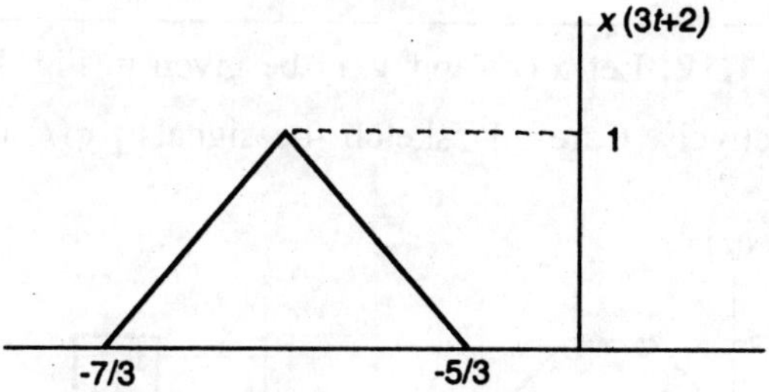

Fig. P. 1.11 (b)

(*c*)

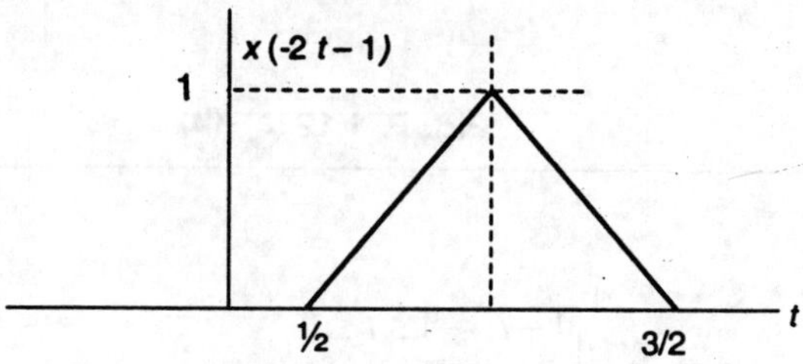

Fig. P. 1.11 (c)

(*d*)

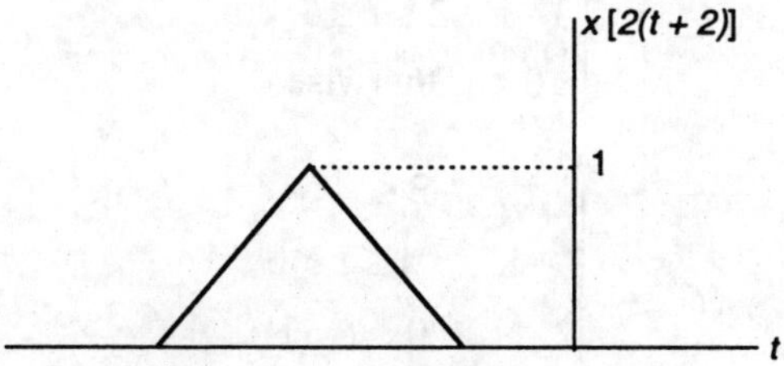

Fig. P. 1.11 (d)

(e)

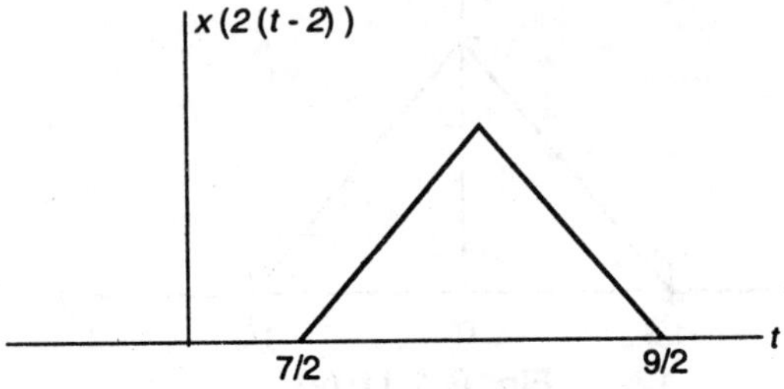

Fig. P. 1.11 (e)

Problem 1.12: Let $x(t)$ and $y(t)$ be given in Fig. P. 1.12 (a) and (b), respectively. Carefully sketch the signals: $x(t)\, y(t-1)$

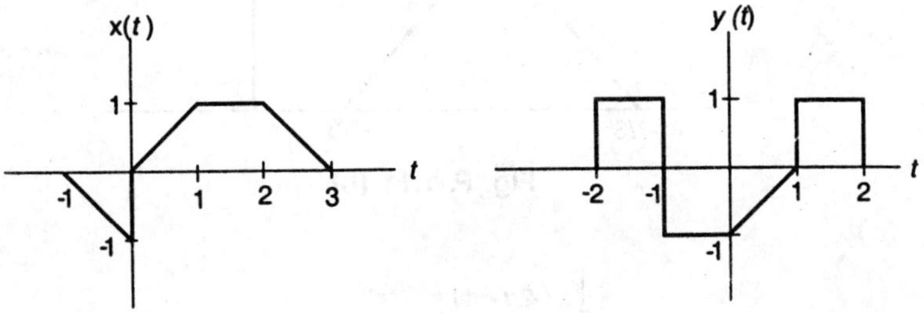

Fig. P. 1.12 (a)–(b)

Solution:

$$x(t) = \begin{cases} 1-t, & -1 \le t \le 0 \\ t, & 0 \le t < 1 \\ 1, & 1 \le t < 2 \\ 3-t, & 2 \le t \le 3 \\ 0, & \text{otherwise} \end{cases}$$

$$y(t) = \begin{cases} 1, & -2 \le t < -1 \\ -1, & -1 \le t < 0 \\ t-1, & 0 \le t < 1 \\ 1 & 1 \le t \le 2 \end{cases}$$

$$\text{Here } y(t-1) = \begin{cases} 1, & -1 \le t < 0 \\ 1, & 0 \le t < 1 \\ t-2, & 1 \le t < 2 \\ 1, & 2 \le t \le 3 \\ 0, & \text{otherwise} \end{cases}$$

$$\text{Now,} \quad g(t) = x(t)\, y(t-1) = \begin{cases} 1-t, & -1 \le t < 0 \\ t, & 0 \le t < 1 \\ t-2, & 1 \le t < 2 \\ 3-t, & 2 \le t \le 3 \\ 0, & \text{otherwise} \end{cases}$$

The sketch of $g(t)$ is shown below

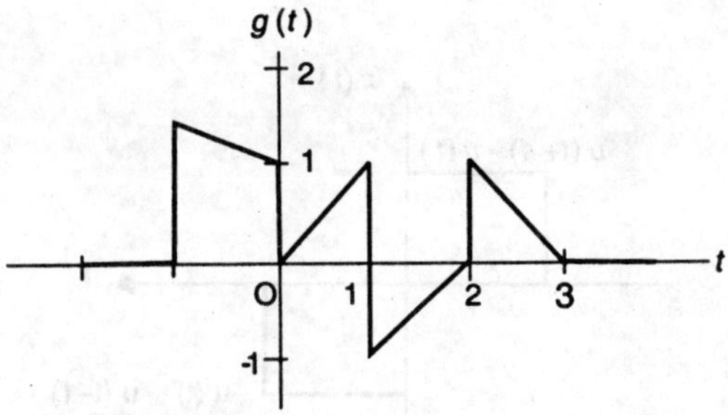

Fig. P. 1.12 (c)

Problem 1.13: Sketch the waveforms of the following signals:

(a) $x(t) = u(t) - u(t-2)$

(b) $x(t) = u(t+1) - 2u(t) + u(t-1)$

(c) $x(t) = -u(t+3) + 2u(t+1) - 2u(t-1) + u(t-3)$

(d) $y(t) = r(t+1) - r(t) + r(t-2)$

(e) $y(t) = r(t+2) - r(t+1) - r(t-1) + r(t-2)$

Solution:

(a) $x(t) = u(t) - u(t-2)$

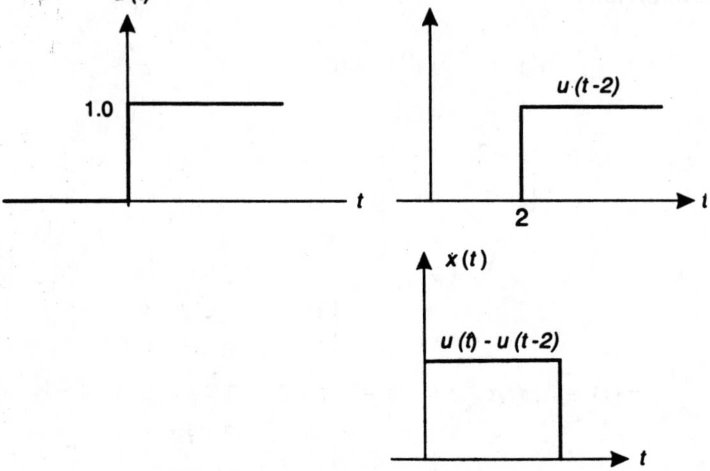

Fig. P. 1.13 (a)

(b) $x(t) = u(t+1) - 2u(t) + u(t-1)$

or, $x(t) = u(t+1) - u(t) - u(t) + u(t-1)$

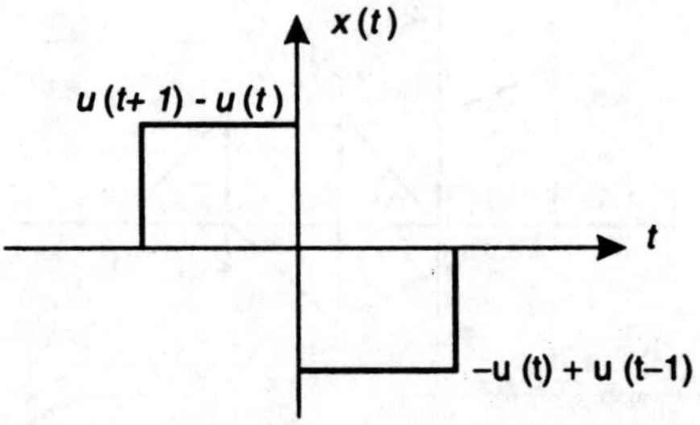

Fig. P. 1.13 (b)

(c) $x(t) = -u(t+3) + 2u(t+1) - 2u(t-1) + u(t-3)$

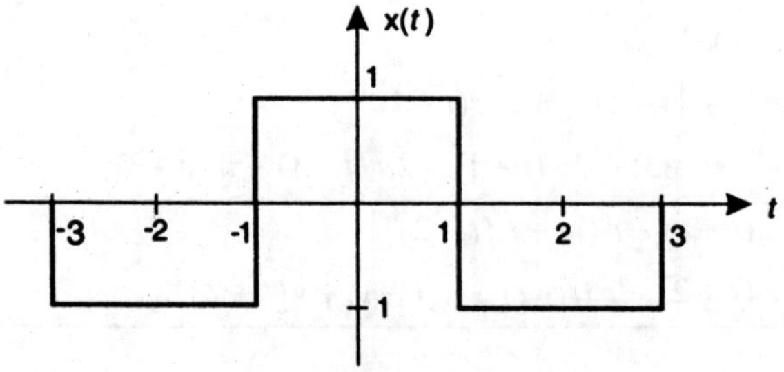

Fig. P. 1.13 (c)

(d) $y(t) = r(t+1) - r(t) + r(t-2)$

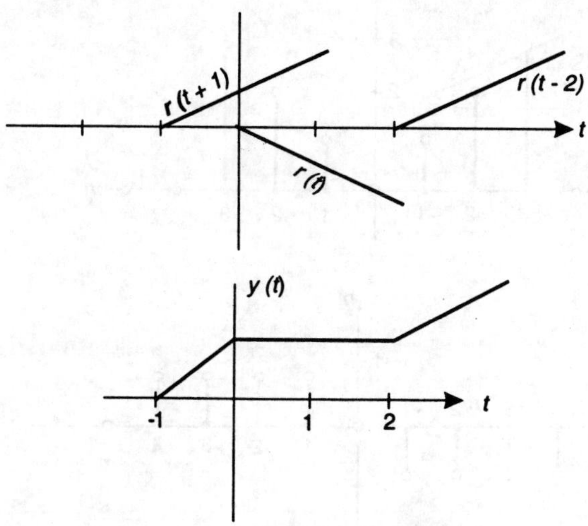

Fig. P. 1.13 (d)

(e) $y(t) = r(t+2) - r(t+1) - r(t-1) + r(t-2)$

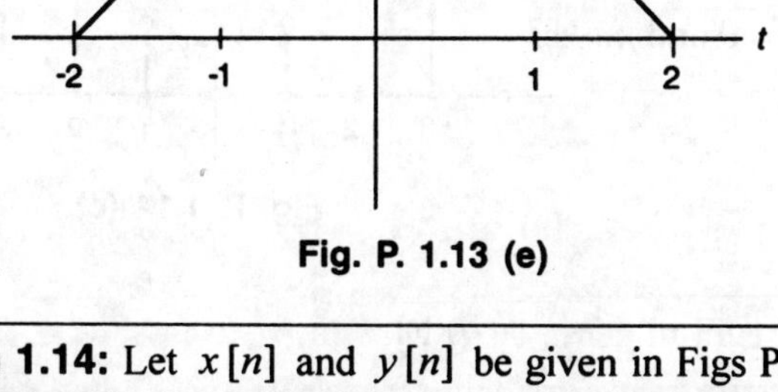

Fig. P. 1.13 (e)

Problem 1.14: Let $x[n]$ and $y[n]$ be given in Figs P. 1.14 (a) and (b), respectively. Carefully sketch the following signals:

(a) $x[2n]$ (b) $x[3n-1]$

(c) $y[1-n]$ (d) $y[2-2n]$

(e) $x[n-2] + y[n+2]$ (f) $x[2n] + y[n-4]$

(g) $x[n+2]\, y[n-2]$ (h) $x[-n]\, y[-n]$

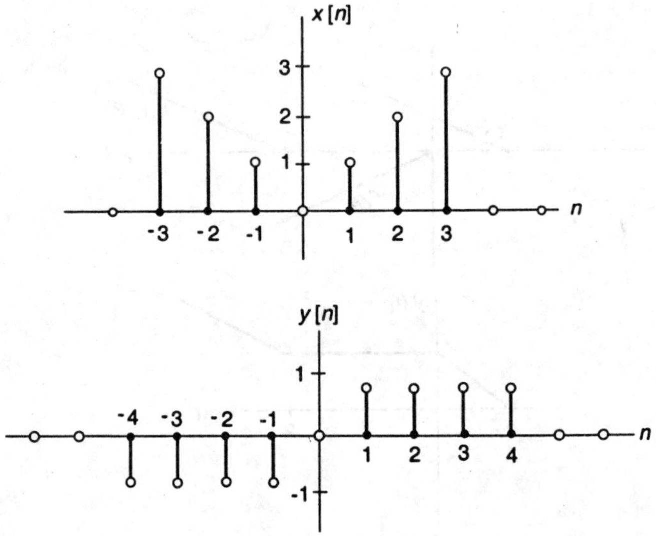

Figs P. 1.14 (a) and (b)

Solution:

$$x[n] = \begin{cases} 1, & n = \pm 1 \\ 2, & n = \pm 2 \\ 3, & n = \pm 3 \\ 0, & n = 0 \text{ and } n > \pm 3 \end{cases}$$

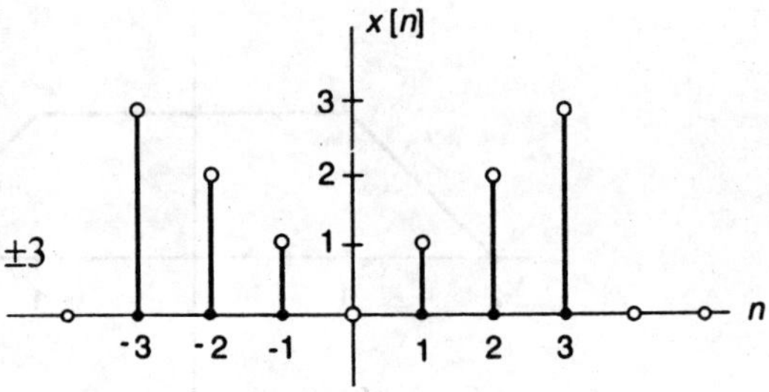

Fig. P. 1.14 (c)

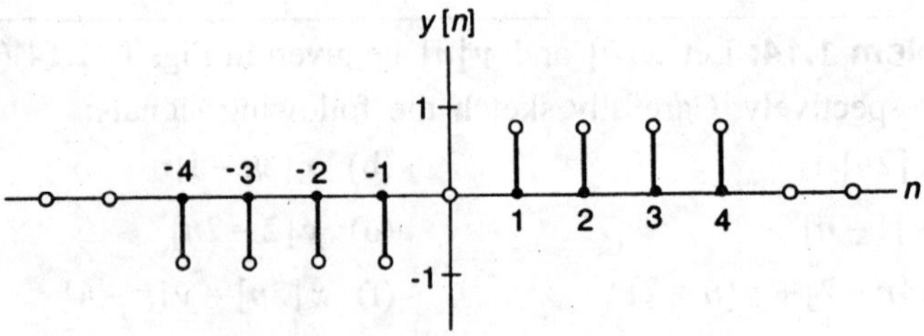

Fig. P. 1.14 (d)

$$y\text{-}[n] = \begin{cases} 1, & n = 1, 2, 3, 4 \\ -1, & n = -1, -2, -3, -4 \\ 0, & n = 0 \text{ and } n > \pm 4 \end{cases}$$

(*a*) $x[2n]$

From Fig. P. 1.14 (*c*):

$$x[2n] = \begin{cases} 2, & n = \pm 1 \\ 0, & n = 0 \text{ and } n > \pm 1 \end{cases}$$

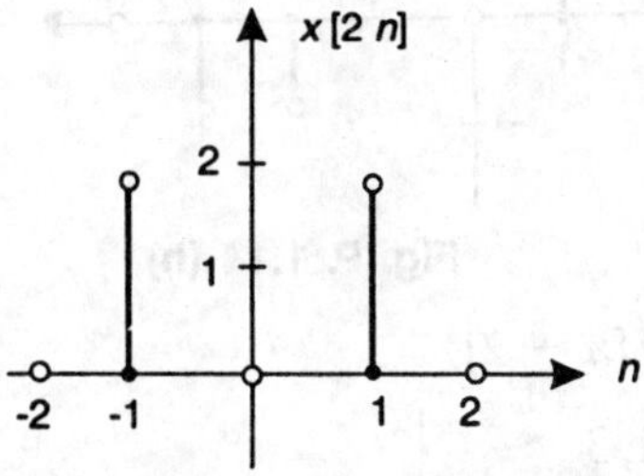

Fig. P. 1.14 (e)

(*b*) $x[3n - 1]$

From Fig. P. 1.14 (*c*):

$$x[3n - 1] = \begin{cases} 1, & n = 1 \\ 0, & n \neq 1 \text{ and } 0 \\ -1 & n = 0 \end{cases}$$

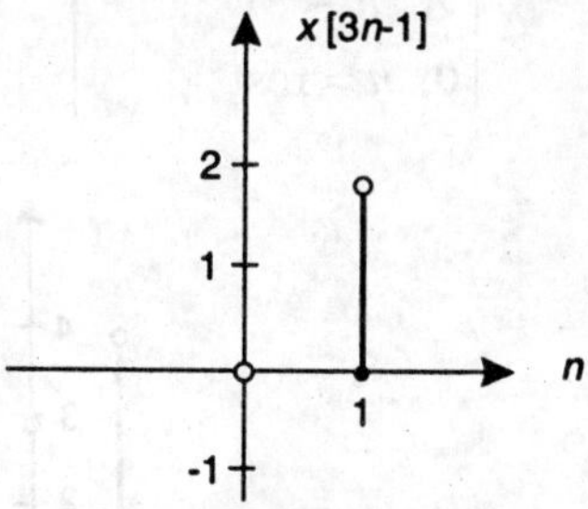

Fig. P. 1.14 (f)

(*c*) $y[1 - n]$

From Fig. P. 1.14 (*d*):

$$y[1 - n] = \begin{cases} 1, & n = 0, -1, -2, -3 \\ -1, & n = 2, 3, 4, 5 \\ 0, & n = \text{rest} \end{cases}$$

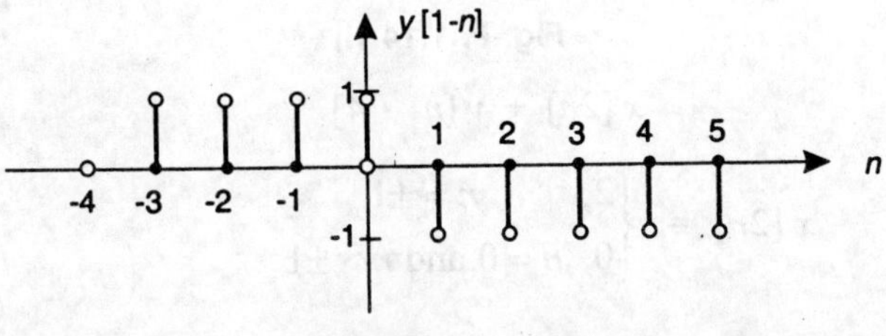

Fig. P. 1.14 (g)

(*d*) $y[2 - 2n]$

From Fig. P. 1.14 (*d*):

$$y[2-2n] = \begin{cases} 1, & n = 0, -1 \\ -1, & n = 2, 3 \\ 0, & n = 1, \text{ rest} \end{cases}$$

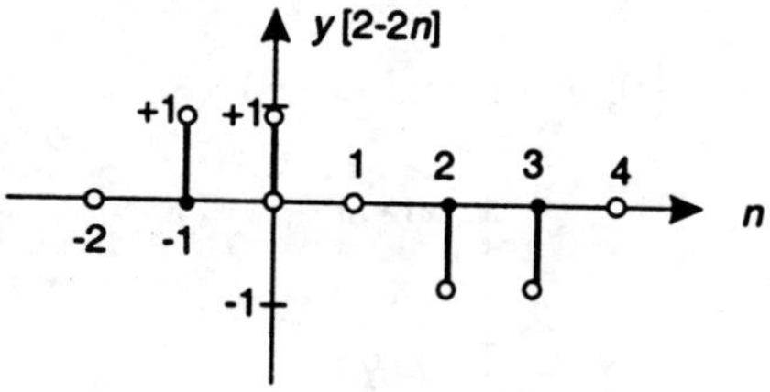

Fig. P. 1.14 (h)

(*e*) $x[n - 2] + y[n + 2]$

$$x[n-2] = \begin{cases} 1, & n = 1, 3 \\ 2, & n = 0, 4 \\ 3, & n = -1 \\ 0, & n = \text{rest} \end{cases}$$

$$y[n+2] = \begin{cases} 1, & n = 0, -1, 1, 2 \\ -1, & n = -3, -4, -5, -6 \\ 0 & n = -2 \text{ and rest} \end{cases}$$

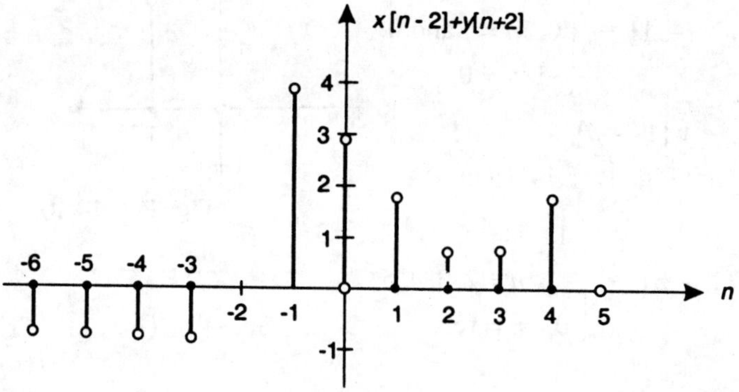

Fig. P. 1.14 (i)

(*f*) $x[2n] + y[n - 4]$

$$x[2n] = \begin{cases} 2, & n = \pm 1 \\ 0, & n = 0 \text{ and } n > \pm 1 \end{cases}$$

$$y[n-4] = \begin{cases} 1, & n = 5, 6, 7, 8 \\ -1, & n = 0, 1, 2, 3 \\ 0 & n = \text{rest} \end{cases}$$

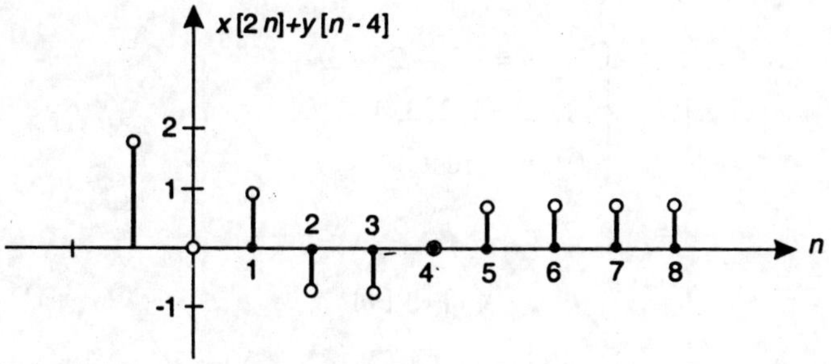

Fig. P. 1.14 (j)

(g) $x[n+2]\,y[n-2]$

$$x[n+2] = \begin{cases} 1, & n = -1, -3 \\ 2, & n = 0 \\ 3, & n = 1, -5 \\ 0, & n = \text{rest} \end{cases}$$

$$y[n-2] = \begin{cases} 1, & n = 3, 4, 5, 6 \\ -1, & n = 1, 0, -1, -2 \\ 0, & n = \text{rest} \end{cases}$$

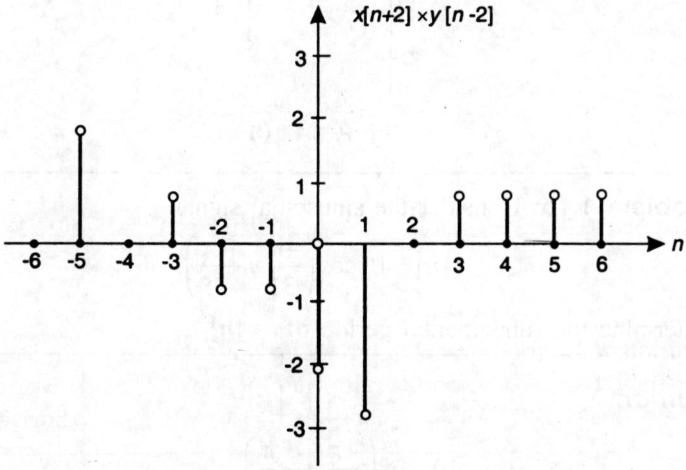

Fig. P. 1.14 (k)

(*h*) $x[-n]\ y[-n]$

$$x[-n] = \begin{cases} 1, & n = \pm 1 \\ 2, & n = \pm 2 \\ 3, & n = \pm 3 \\ 0, & n = \text{rest} \end{cases}$$

$$y[-n] = \begin{cases} 1, & n = -1, -2, -3, -4 \\ -1, & n = 1, 2, 3, 4 \\ 0 & n = \text{rest} \end{cases}$$

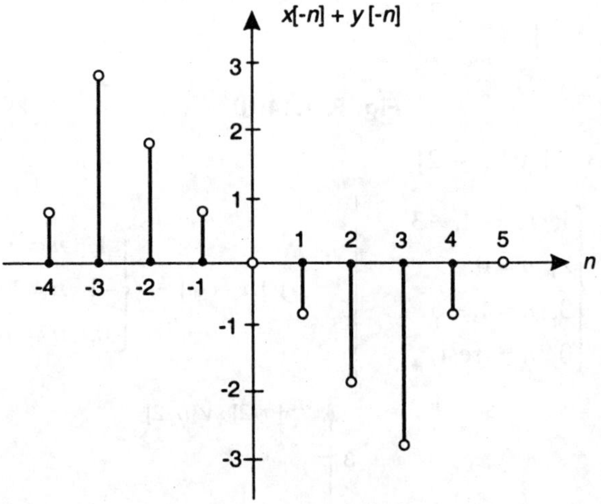

Fig. P. 1.14 (l)

Problem 1.15: Consider the sinusoidal signal

$$x[n] = 10\cos\left(\frac{4\pi}{31}n + \frac{\pi}{5}\right)$$

Determine the fundamental period of $x[n]$

Solution:

$$x[n] = 10\cos\left(\frac{4\pi}{31}n + \frac{\pi}{5}\right)$$

Here $\quad \Omega = \frac{4\pi}{31}$, Ω is called fundamental frequency of $x[n]$

So, $\quad \Omega = \frac{2\pi m}{N}$

Where, N is the smallest value of integer known as fundamental period of the discrete time signal $x[n]$

So, $\quad \frac{2\pi m}{N} = \frac{4\pi}{31}$

Or, $\quad N = \frac{31m}{2}$

Since N is the smallest integer, hence $m = 2$

So that, $\quad N = \frac{31}{2} \times 2 = 31$ be an integer

So, fundamental period of $x[n] = 31$

> **Problem 1.16:** The sinusoidal signal $x[n]$ has fundamental period $N = 10$ samples. Determine the smallest angular frequency Ω for which $x[n]$ is periodic.

Solution:

Angular frequency

$$\Omega = \frac{2\pi m}{N}$$

Let $\quad m = 1$

$$\Omega = \frac{2\pi}{10} = \frac{\pi}{5} \text{ radians/cycle}$$

> **Problem 1.17:** Determine whether the following signals are periodic. If they are periodic, find the fundamental periods.
>
> (*a*) $\quad x[n] = \cos\left(\frac{8}{15}\pi n\right)$
>
> (*b*) $\quad x[n] = \cos\left(\frac{7}{15}\pi n\right)$

(c)	$x(t) = \cos(2t) + \sin(3t)$
(d)	$x(t) = \sum_{k=-\infty}^{\infty} (-1)^k \, \delta(t - 2k)$
(e)	$x[n] = \sum_{k=-\infty}^{\infty} \{\delta[n - 3k] + \delta[n - k^2]$
(f)	$x(t) = \cos(t)\, u(t)$
(g)	$x[n] = \cos\left(\frac{1}{5}\pi n\right) \sin\left(\frac{1}{3}\pi n\right)$

Solution:

(a) It is periodic. Its fundamental period is 15.
(b) It is periodic. Its fundamental period is 30.

(c) period of $\cos 2t = \frac{2\pi}{2} = \pi$

and of $\sin 3t = \frac{2\pi}{3}$

T (period) $= n_1 T_1 = n_2 T_2$, for period $\frac{T_1}{T_2}$ should be rational.

i.e., $\frac{T_1}{T_2} = \frac{3}{2} = \frac{\text{Integer}}{\text{Integer}}$, so above

$\cos 2t + \sin 3t$ is periodic. Its fundamental period is

$$T = n_1 T_1 = n_2 T_2$$

$\therefore$ $$n_1 T_1 = n_2 T_2$$

Or, $$\frac{n_1}{n_2} = \frac{T_2}{T_1} = \frac{2}{3}$$

So, $$n_1 = 2$$

and $$n_2 = 3 \text{ (Smallest integer)}$$

$\therefore$ $$T = 2 \times \pi = 2\pi$$

So, "2π" is its fundamental period.

(d) It is periodic. Its fundamental period is 4. [This fundamental period is due to $(-1)^k$ otherwise it will be 2.]

(e) It is not periodic. Because impulse function is not repeating at the same interval.

(f) It is a period function. Its fundamental period is 2π.

$$x(t) = \begin{cases} \cos(t), & t \geq 0 \\ 0, & t < 0 \end{cases}$$

(g) $$\cos\left(\frac{1}{5}\pi n\right) \cdot \sin\left(\frac{1}{3}\pi n\right) = \frac{1}{2}\left[\sin\left(\frac{2\pi n}{15}\right) + \sin\left(\frac{8\pi n}{15}\right)\right]$$

Period of $\sin\left(\frac{2\pi n}{15}\right) = 15$ and of $\sin\left(\frac{8\pi n}{15}\right) = 15$

$$\therefore \quad T_1 = 15$$
$$T_2 = 15$$

$$\therefore \quad \frac{T_1}{T_2} = \frac{15}{15} = 1 \text{ is a rational number}$$

hence given function is a periodic function and its fundamental period is $T = n_1 T_1 = n_2 T_2$

$$\frac{n_1}{n_2} = \frac{T_2}{T_1} = 1$$

So, $n_1 = n_2 = 1$ as smallest integer

$\therefore$ $T = 1 \times T_1 = 1 \times 15$

$T = 15$ is the fundamental period

Problem 1.18: A complex sinusoidal signal $x[t]$ has the following components

$$\text{Re}\{x(t)\} = x_R(t) = A\cos(\omega t + \phi)$$
$$\text{Im}\{x(t)\} = x_I(t) = A\sin(\omega t + \phi)$$

The amplitude of $x(t)$ is defined by the square root of $x_R^2(t) + x_I^2(t)$. Show that this amplitude equals A, independent of the phase angle ϕ.

Solution:

$$x(t) = \sqrt{x_R^2(t) + x_I^2(t)} = \sqrt{A^2\cos^2(\omega t + \phi) + A^2\sin^2(\omega t + \phi)}$$

$$= A\sqrt{\cos^2(\omega t + \phi) + \sin^2(\omega t + \phi)} = A\sqrt{1} \quad (\because \sin^2\theta + \cos^2\theta = 1)$$

$$x(t) = A$$

Thus $x(t)$ is independent of the phase angle ϕ.

Problem 1.19: Consider the complex-valued exponential signal

$$x(t) = Ae^{\alpha t + j\omega t}, \ \alpha > 0$$

Evaluate the real and imaginary components of $x(t)$.

Solution:

$$x(t) = Ae^{\alpha t + j\omega t} = Ae^{\alpha t}(e^{j\omega t})$$
$$= Ae^{\alpha t}(\cos \omega t + j \sin t) = Ae^{\alpha t}\cos \omega t + jAe^{\alpha t}\sin \omega t$$

Real part of $x(t) = x_R(t) = Ae^{\alpha t}\cos \omega t$

Imaginary part of $x(t) = x_I(t) = Ae^{\alpha t}\sin \omega t$

Problem 1.20: Consider the continuous-time signal

$$x(t) = \begin{cases} \frac{1}{T}, & -\frac{T}{2} \le t \le \frac{T}{2} \\ 0, & t \ge \frac{T}{2} \\ 0, & t < -\frac{T}{2} \end{cases}$$

Whichis applied to a differentiator. Show that the output of the differentiator approaches the unit impulse $\delta(t)$ as T approaches zero.

Solution:

When $x(t)$ is applied to a differentiator then its output:

$$y(t) = \begin{cases} \frac{1}{T}, & -\frac{T}{2} \le t \le \frac{T}{2} \\ 0, & t \ge \frac{T}{2} \\ 0, & t < -\frac{T}{2} \end{cases}$$

It can be graphically represented as:

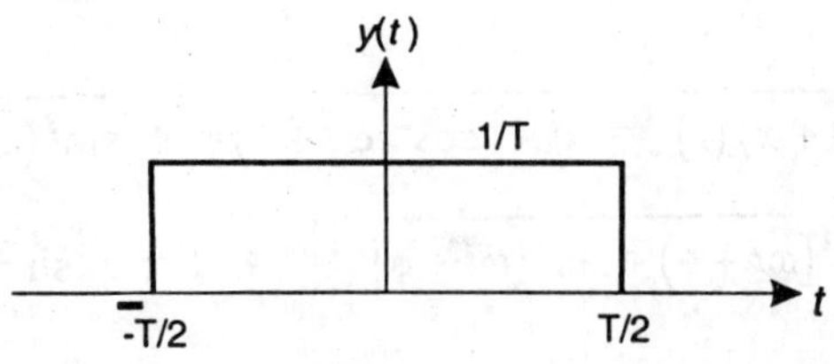

Fig. P. 1.20 (a)

Let $y_T(t)$ is a pulse of even function with duration T and unit area

Now $\underset{T \to 0}{Lt}\ y_T(t) = 1 = \delta(t)$

which is an unit impulse function $\delta(t)$ as T approaches zero

Problem 1.21: The derivative of impulse function $\delta(t)$ is referred to as a doublet. It is denoted by $\delta'(t)$. Show that $\delta'(t)$ satisfies the sitting property.

$$\int_{-\infty}^{\infty} \delta'(t - t_0)\, f(t)\, dt = f'(t_0)$$

where $f'(t_0) = \dfrac{d}{dt} f(t)\big|_{t = t_0}$

Assume that the function $f(t)$ has a continuous derivative at time $t = t_0$.

Solution:

Unit impulse function $\delta(t)$ is denoted as:

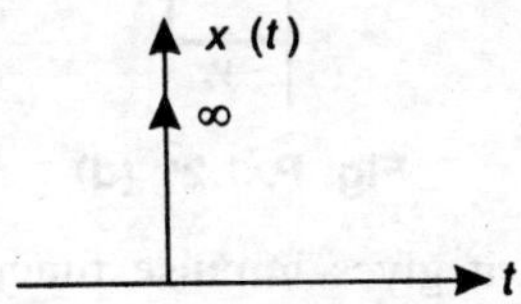

Fig. P. 1.21 (a)

and as in the form of shifting, i.e. $\delta(t - a)$

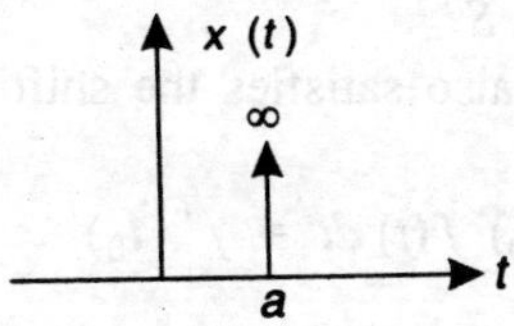

Fig. P. 1.21 (b)

The derivative of unit impulse function $\delta'(t - a)$ is defined as:

$$\delta'(t - a) = \begin{cases} +\infty \text{ and } -\infty & \text{for } t = a \\ 0 & \text{for } t \neq a \end{cases}$$

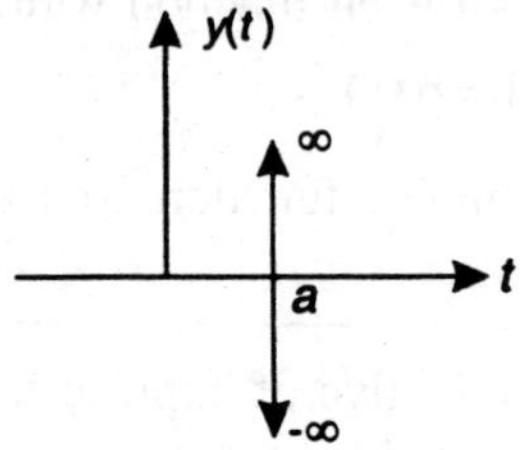

Fig. P. 1.21 (c)

Since derivative of impulse function denoted as impulse function in both direction rather than unit direction (as in impulse function). So its derivation is also known as doublet function. It can be shown as:

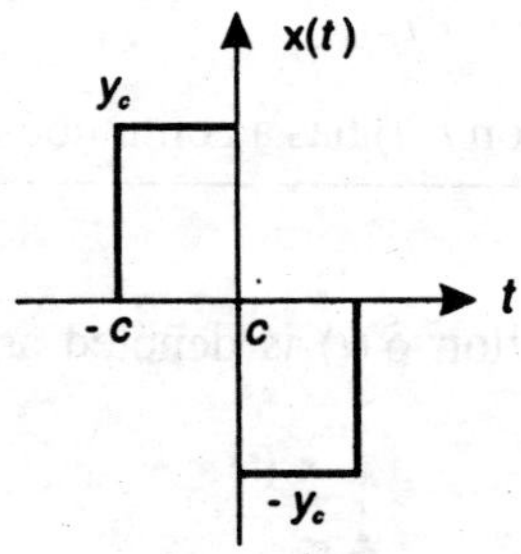

Fig. P. 1.21 (d)

When $c \to 0$ then it gives impulse function in two direction. So it is called doublet function.

Laplace transform of doublet function is given by S

i.e. $\mathcal{L}\ \delta'(t) = S$

Similarly for triplent function

$\mathcal{L}\ \delta''(t) = S^2$

Doublet function also satisfies the shifting property:

$$\int_{-\infty}^{\infty} \delta'(t - t_0)\, f(t)\, dt = f'(t_0) \qquad \ldots(1)$$

proof:

$$\because \quad \int_{-\infty}^{\infty} \phi(t)\, \delta(t)\, dt = \phi(0) \qquad \ldots(2)$$

$$\text{and} \quad \int_{-\infty}^{\infty} \phi(t)\, \delta(t - t_0)\, dt = \phi(t_0) \qquad \ldots(3)$$

Using the above equation (2) and (3) in equation (1), we get:

$$\int_{-\infty}^{\infty} f(t)\ \delta(t - t_0)\ dt = f'(t_0)$$

Where
$$f'(t_0) = \frac{d}{dt} f(t)\Big|_{t=t_0}$$

Problem 1.22: A system consits of several subsystems connected as shown in fig. Find the operator H relating $x(t)$ to $y(t)$ for the subsystem operators given by:

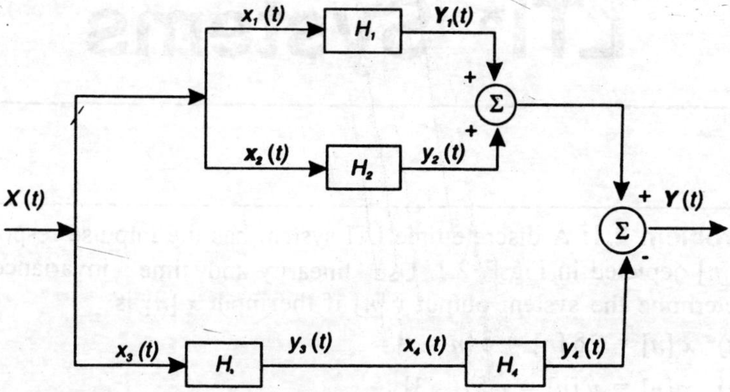

Fig. P. 1.22

$H_1 : y_1(t) = x_1(t)\, x_1(t-1)$ $\qquad$ $H_2 : y_2(t) = |x_2(t)|$

$H_3 : y_3(t) = 1 + 2x_3(t)$ $\qquad$ $H_y : y_4(t) = \cos(x_4(t))$

Solution:

$$x(t) = x_1(t) + x_2(t) + x_3(t)$$

$$y(t) = y_1(t) + y_2(t) - y_4(t) = x(t)\, x_1(t - 1) + |x_2(t)| - \cos[x_4(t)]$$

$$\because \quad y_3(t) = 1 + 2\, x_3(t) = x_4(t)$$

$$\therefore \quad y_4(t) = \cos[1 + 2\, x_3(t)]$$

$$\therefore \quad y(t) = x_1(t)\, x_1(t - 1) + |x_2(t)| - \cos[1 + 2\, x_3(t)]$$

operator
$$H = \frac{y(t)}{x(t)}$$

Or,
$$H = \frac{x_1(t)\, x_1(t-1) + |x_2(t)| - \cos[1 + 2x_3(t)]}{x_1(t) + x_2(t)\, x_3(t)} \quad \textbf{Ans.}$$

Chapter 2

Time-Domain Representation of LTI* Systems

Problem 2.1: A discrete-time LTI system has the impulse response $h[n]$ depicted in Fig P. 2.1. Use linearity and time invariance to determine the system output $y[n]$ if the input $x[n]$ is

(*a*) $x[n] = 2\delta[n] - \delta[n - i]$

(*b*) $x[n] = u[n] - u[n - 3]$

(*c*) $x[n]$ as given in Fig P. 2.1 (b)

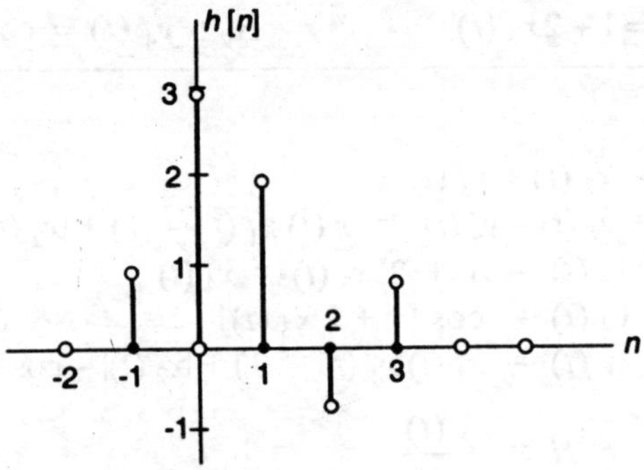

Fig. P. 2.1

---

* Linear Time-Invariant

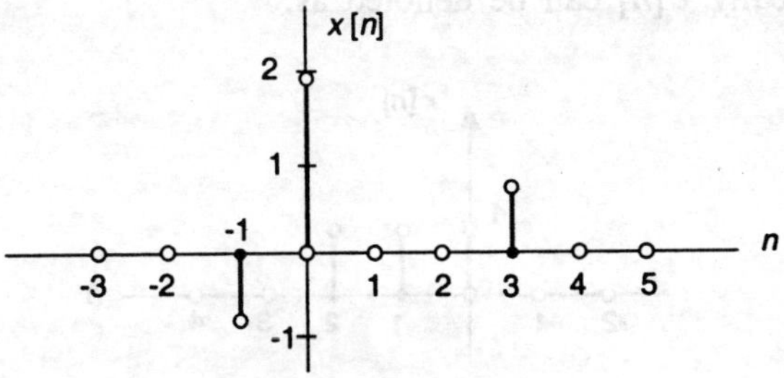

Fig. P. 2.1 (a)

Solution:

$$h[n] = \begin{cases} 1, & n = -1, 3 \\ -1, & n = 2 \\ 3, & n = 0 \\ 2, & n = 1 \\ 0, & \text{otherwise} \end{cases}$$

(*a*) $x[n] = 2\delta[n] - \delta[n-1]$

$$y[n] = \sum_{K=-\infty}^{\infty} x[k]\, h[n-k]$$

$\therefore$ $y[n] = 2h[n] - h[n-1]$

$$\therefore \quad y[n] = \begin{cases} 0, & n \le -2 \\ 2, & n = -1 \\ 5, & n = 0 \\ 1, & n = 1 \\ -4, & n = 2 \\ 1, & n = 3 \\ -1, & n = 4 \\ 0, & n \ge 5 \end{cases}$$

Hints:

For $n = 0$

$$y[n] = 2\,h[0] - h[0-1] = 2\,h[0] - h[-1]$$
$$= 2 \times 3 - (1) = 6 - 1 = 5$$

Similarly, all

(*b*) $x[n] = u[n] - u[n-3]$

graphically $x[n]$ can be denoted as:

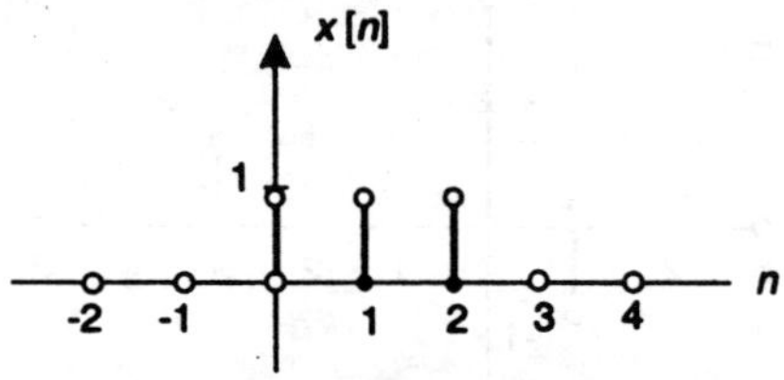

Fig. P. 2.1 (b)

This unit step function can be easily expressed in sum of impulse function.

i.e. $x[n] = \delta[n] + \delta[n-1] + \delta[n-2]$

$\therefore$ $y[n] = h[n] + h[n-1] + h[n-2]$

Similarly $y[n]$ can be expressed as: $$y[n] = \begin{cases} 0, & n \le -2 \\ 1, & n = -1 \\ 4, & n = 0 \\ 6, & n = 1 \\ 4, & n = 2 \\ 2, & n = 3 \\ 0, & n = 4 \\ 1, & n = 5 \\ 0, & n \ge 6 \end{cases}$$

(*c*) $x[n]$ can be represented as the sum of impulsive function of Fig. P. 2.1(*b*) as: $x[n] = -\delta[n+1] + 2\delta[n] + \delta[n-3]$

$\therefore$ $y[n] = -h[n+1] + 2h[n] + h[n-3]$

Similarly $y[n]$ can be expressed as: $$y[n] = \begin{cases} 0, & n \le -3 \\ 1, & n = -2 \\ -1, & n = -1 \\ 4, & n = 0 \\ 5, & n = 1 \\ -2 & n = 2 \\ 5, & n = 3 \\ 2, & n = 4 \\ -1, & n = 5 \\ 1, & n = 6 \\ 0, & n \ge 7 \end{cases}$$

Problem 2.2: Evaluate the discrete-time convolution sums given below:

(*a*) $y[n] = u[n] * u[n-3]$

(*b*) $y[n] = 2^n u[-n+2] * u[n-3]$

(*c*) $y[n] = \left(\frac{1}{2}\right)^n u[n-2] * u[n]$

(*d*) $y[n] = \cos\left(\frac{1}{2}\pi n\right) u[n] * u[n-1]$

(*e*) $y[n] = (u[n+10] - 2u[n+5] + u[n-6]) * u[n-2]$

(*f*) $y[n] = (u[n+10] - 2u[n+5] + u[n-6]) * \cos\left(\frac{1}{2}\pi n\right)$

Solution:

(*a*) $y[n] = u[n] * u[n-3]$

$u[n] =$

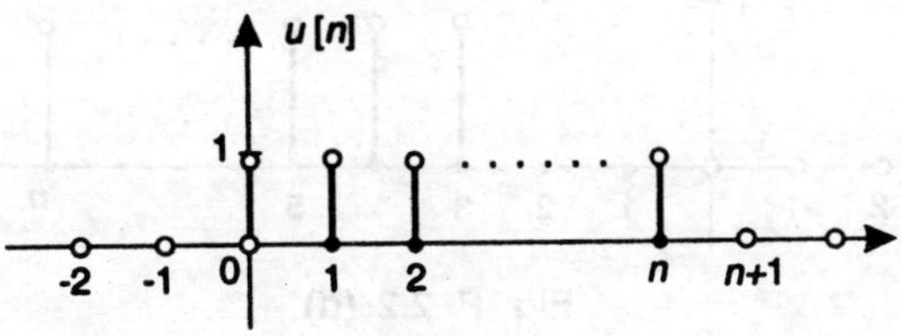

Fig. P. 2.2 (a)

$u[n-3] =$

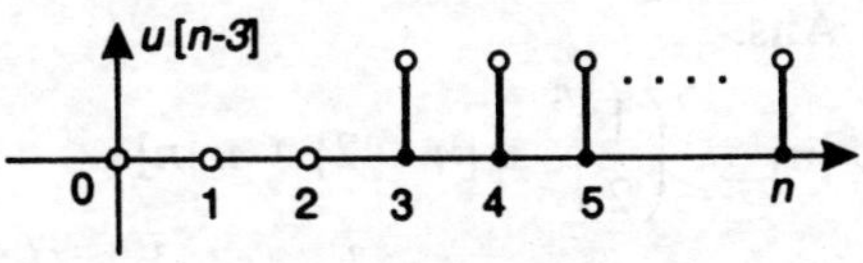

Fig. P. 2.2 (b)

$$y[n] = u[n] * u[n-3]$$

$$\because y[n] = \ldots + y[-4]\,\delta[n+4] + y[-3]\,\delta[n+3] + y[-2]\,\delta[n+2] + y[-1]\,\delta[n+1] + y[0]\,\delta[n] + y[1]\,\delta[n-1] + \ldots\ldots$$

$$= \sum_{K=-\infty}^{\infty} x[k]\,\delta[n-k]$$

$\therefore y[n] = 0 + 0 + \ldots\ 0 + 1.0 + 1.0 + 1.0 + 1.1 + 1.1 + \ldots\ldots +$ 1.1 upto n

$= 0 + 0 + 0 + 1 + 1 + 1 + \ldots\ldots + 1$ upto n

$$y[n] = n - 2 \quad \textbf{Ans.}$$

$$[\text{as } y[0]\ \delta[n] = y[1]\ \delta[n-1] = y[2]\ \delta[n-2] = 0]$$

(*b*) $$y[n] = 2^n\, u[-n+2] * u[n-3]$$

$2^n\, u[-n+2]$ can be graphically represented as:

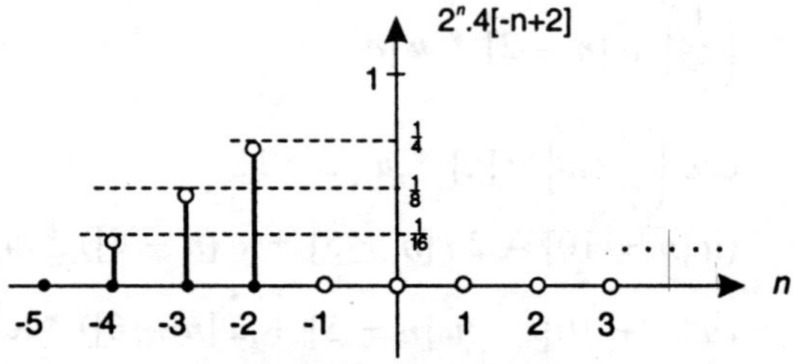

Fig. P. 2.2 (c)

and $u[n-3]$ can be graphically represented as:

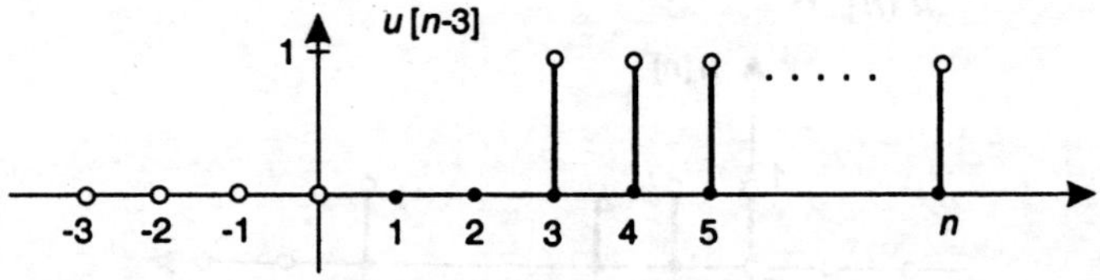

Fig. P. 2.2 (d)

$$y[n] = \sum_{K=-\infty}^{\infty} x[k]\ \delta[n-k] = 0 + 0 + 0 + \ldots\ldots + 0 \text{ upto } n$$

$$y[n] = 0 \ \textbf{Ans.}$$

(*c*) $$y[n] = \left(\frac{1}{2}\right)^n u[n-2] * u[n]$$

$$y[n] = \sum_{K=-\infty}^{\infty} x[k]\ \delta[n-k] = \left(\frac{1}{2}\right)^2 + \left(\frac{1}{2}\right)^3 + \left(\frac{1}{2}\right)^4 + \ldots\ldots + \left(\frac{1}{2}\right)^n$$

Or, $$y[n] = \left(\frac{1-\left(\frac{1}{2}\right)^n}{\left(1-\frac{1}{2}\right)}\right) - \frac{1}{2} = 2\left(1-\left(\frac{1}{2}\right)^n\right) - \frac{1}{2} \quad \textbf{Ans.}$$

When n is very-very large, then $\left(\frac{1}{2}\right)^n = 0$

$$y[n] = 2 - \frac{1}{2} = \frac{3}{2}$$

(*d*) $$y[n] = \cos\left(\frac{1}{2}\pi n\right) u[n] * u[n-1]$$

$\cos\left(\frac{1}{2}\pi n\right) \cdot u[n]$ can be represented as:

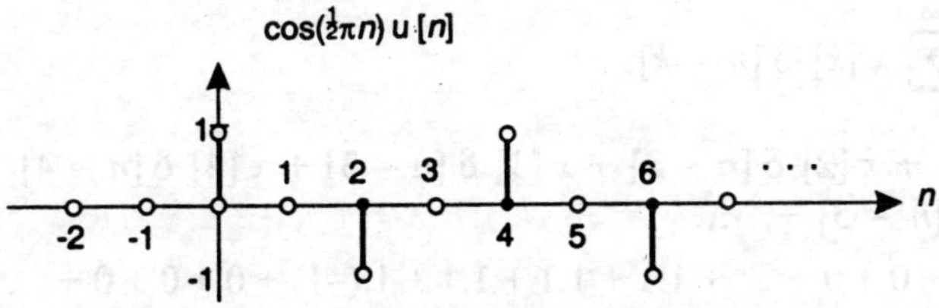

Fig. P. 2.2 (e)

and $u[n-1]$ can be represented as:

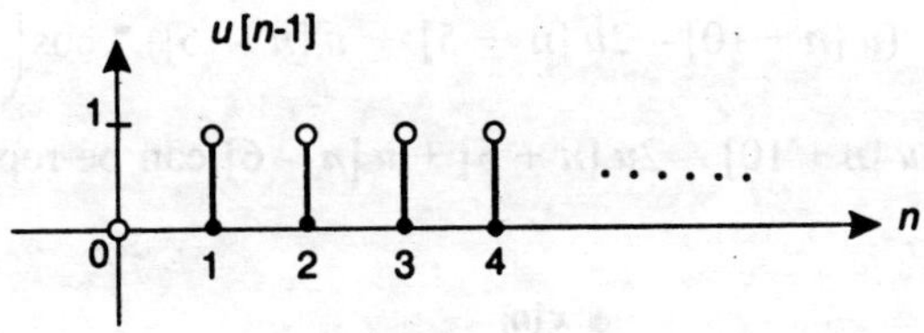

Fig. P. 2.2 (f)

$$y[n] = \sum_{K=-\infty}^{\infty} y[k]\, \delta[n-k]$$

$y[n] = 0 + 0 + (-1) + 0 + (1) + 0 + (-1) + 0 + (1) + \ldots\ldots$ upto n

$y[n] = 0$

(*i*) $y[n] = (u[n+10] - 2\,u[n+5] + u[n-6]) * u[n-2]$

$(u[n+10] - 2u[n+5] + u[n-6])$ can be represented as:

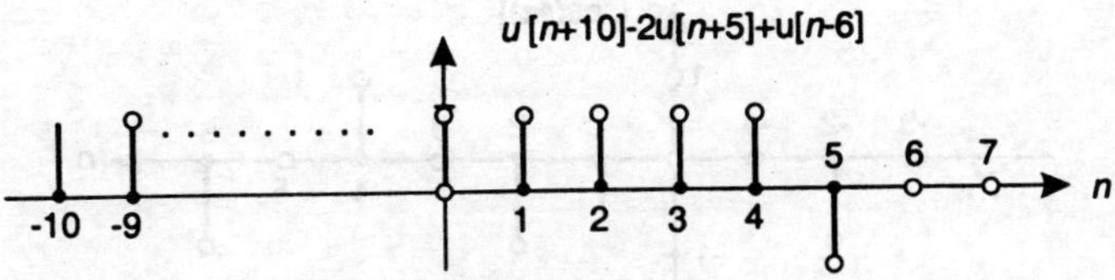

Fig. P. 2.2 (g)

and $u[n-2]$ can be represented as:

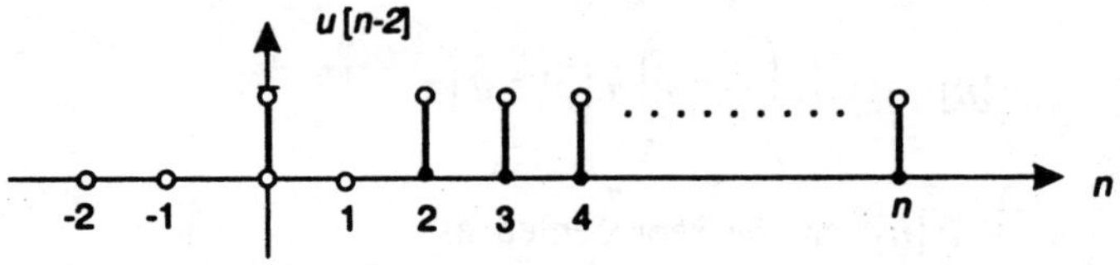

Fig. P. 2.2 (h)

$$y[n] = \sum_{K=-\infty}^{\infty} x[k]\,\delta[n-k]$$

$$= \ldots + x[2]\,\delta[n-2] + x[3]\,\delta[n-3] + x[4]\,\delta[n-4] + x[5]\,\delta[n-5] + \ldots\ldots$$

$$= 0 + 0 + 0 + \ldots + 1.1 + 1.1 + 1.1 + 1.(-1) + 0 + 0 + 0 + \ldots \text{ upto } n$$

$$= 1 + 1 + 1 - 1 = 2$$

$\therefore \quad y[n] = 2$ **Ans.**

(*k*) $y[n] = (u[n+10] - 2u[n+5] + u[n-6]) * \cos\left(\frac{1}{2}\pi n\right)$

Let $x[n] = u[n+10] - 2u[n+5] + u[n-6]$ can be represented as:

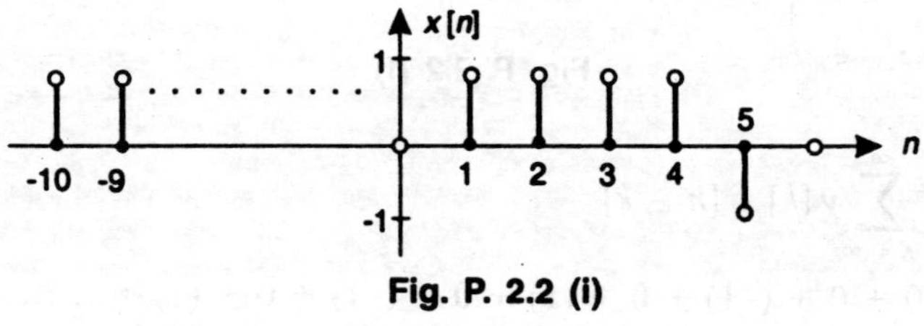

Fig. P. 2.2 (i)

Let $\quad \delta[n] = \cos\left(\frac{1}{2}\pi n\right)$ can be represented as

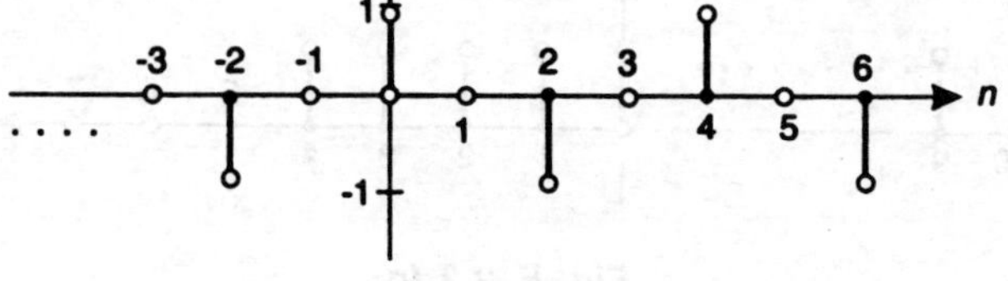

Fig. P. 2.2 (j)

$$\therefore \quad y[n] = \sum_{K=-\infty}^{\infty} x[k]\, \delta[n-k]$$

$$= \ldots + x[-10]\,\delta[n+10] + x[-9]\,\delta[n+9] + \ldots + x[5]\,\delta[n-5] + \ldots$$

$$= 0 + 0 + 0 + -1.1 + 0 + 1.1 + 0 - 1.1 + 0 + \ 1.1 + 0 + -1.1 + 0 + 1.1 + 0 + -1.1 + 0 + 1.1 + 0 + 0 + 0 + \ldots$$

$$= -1 + 1 - 1 + 1 - 1 + 1 - 1 + 1$$

$$= -4 + 4 = 0$$

$\therefore \quad y[n] = 0$ **Ans.**

Problem 2.3: Consider the discrete-time signals depicted in figures below. Evaluate the convolution sums indicated below.

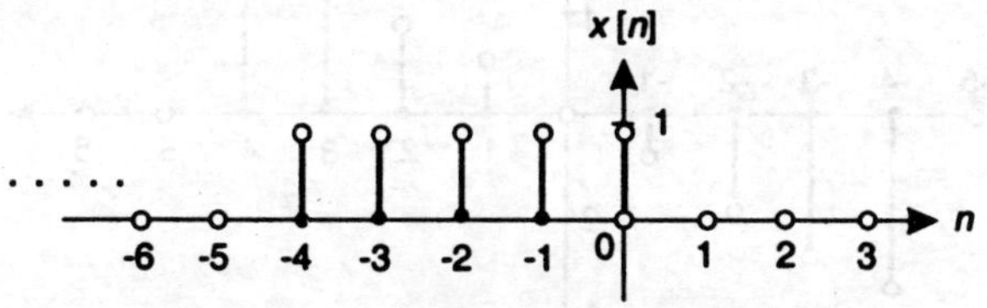

Fig. P. 2.3

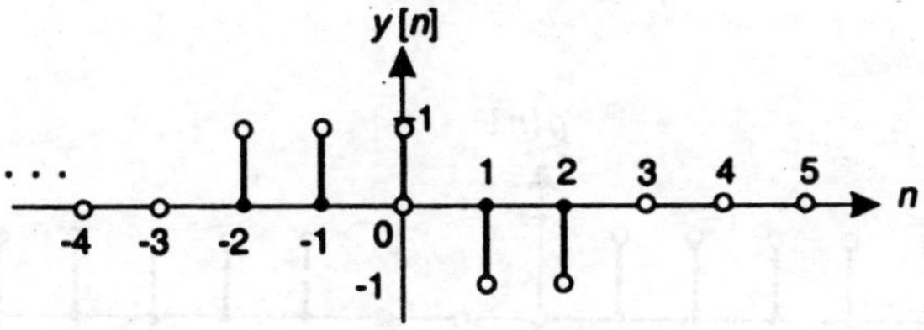

Fig. P. 2.3 (a)

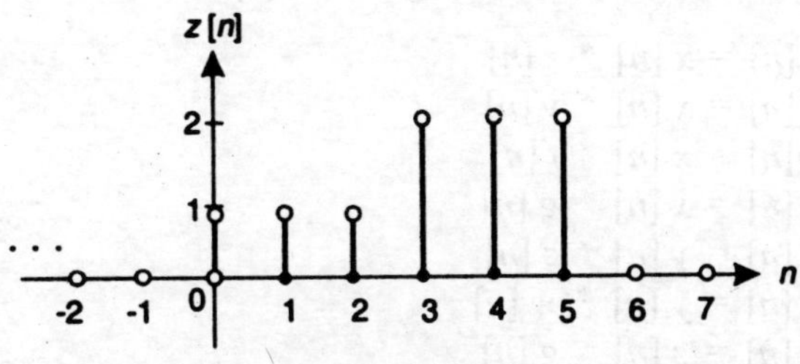

Fig. P. 2.3 (b)

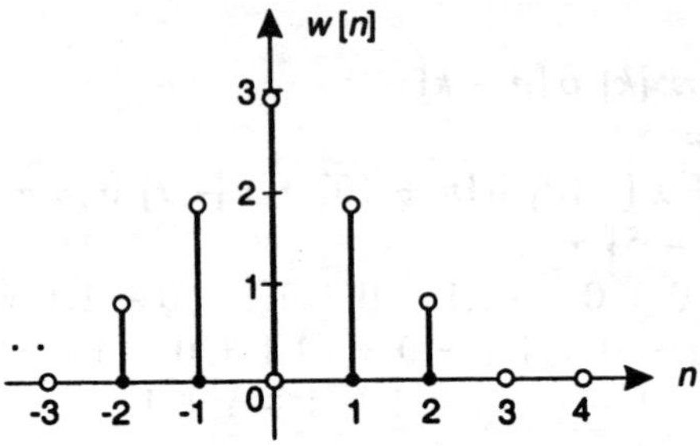

Fig. P. 2.3 (c)

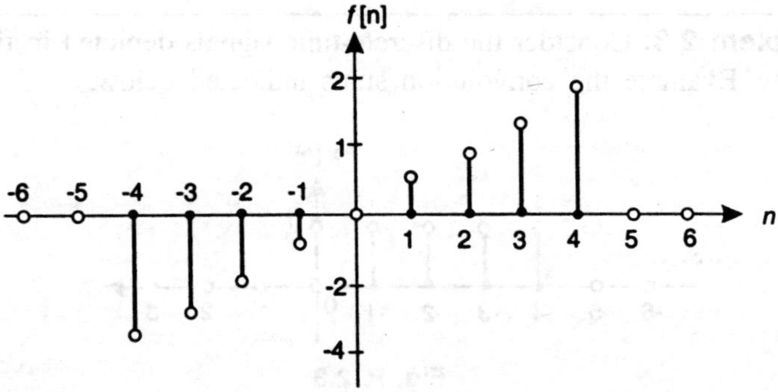

Fig. P. 2.3 (d)

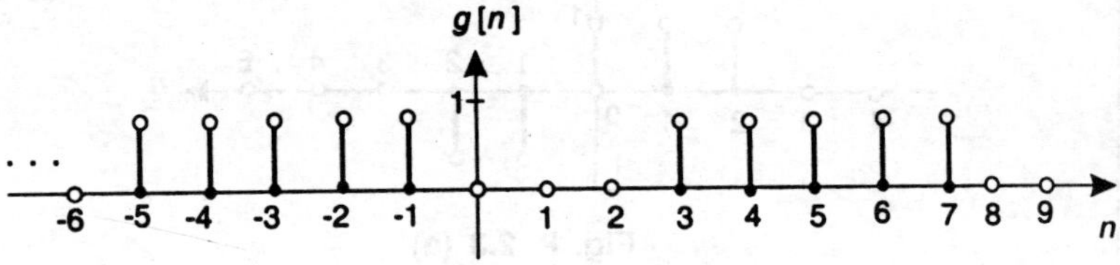

Fig. P. 2.3 (e)

(a) $m[n] = x[n] * z[n]$
(b) $m[n] = x[n] * y[n]$
(c) $m[n] = x[n] * f[n]$
(d) $m[n] = x[n] * g[n]$
(e) $m[n] = y[n] * z[n]$
(f) $m[n] = y[n] * w[n]$
(g) $m[n] = z[n] * g[n]$
(h) $m[n] = f[n] * g[n]$

Solution:

(*a*) $m[n] = x[n] * z[n]$

$\because m[n] = \ldots + x[-4]\ \delta[n+4] + x[-3]\ \delta[n+3] + x[-2]\ \delta[n+2]$
$+ x[-1]\ \delta[n+1] + x[0]\ \delta[n] + x[1]\ \delta[n-1] + x[2]$
$\delta[n-2] + \ldots\ldots$
$= 0 + 0 + 0 + \ldots\ldots + 0 + 0 + 0 + 0 + 1.1 + 0 + 0 + 0 + \ldots\ldots$
$= 1$

So $\therefore$ $m[n] = 1$ **Ans.**

(*b*) $m[n] = x[n] * y[n]$

$$\because m[n] = \sum_{K=-\infty}^{\infty} x[k]\ \delta[n-k]$$

$k = -\infty$

$= \ldots + x[-2]\ \delta[n+2] + x[-1]\ \delta[n+1] + x[0]\ \delta[n] + x[1]$
$\delta[n-1] + x[2]\ \delta[n-2] + \ldots$
$= 0 + 0 + 0 + \ldots + 1 \times 1 + 1 \times 1 + 1 \times 1 + 0 + 0 + 0 + 0 \ldots$
$= 1 + 1 + 1 = 3$

So, $m[n] = 3$ **Ans.**

(*c*) $m[n] = x[n] * f[n]$

$$\because m[n] = \sum_{K=-\infty}^{\infty} x[k]\ \delta[n-k]$$

$= 0 + 0 + 0 + \ldots + (-4 \times 1) + (-3 \times 1) + (-2 \times 1) +$
$(-1 \times 1) + 1 \times 0 + 0 + 0 + 0 + \ldots$
$= -4 - 3 - 2 - 1 = -10$

$\therefore$ $m[n] = -10$ **Ans.**

(*d*) $m[n] = x[n] * g[n]$

$$\because m[n] = \sum_{K=-\infty}^{\infty} x[n]\ \delta[n-k]$$

$= 0 + 0 + 0 + \ldots + (+1 \times 0) + 1 \times 1 + 1 \times 1 + 1 \times 1 +$
$1 \times 1 + 1 \times 0 + 0 \times 0 + 0 \times 0 + 0 \times 1 + 0 \times 1 + \ldots +$
$0 + 0 \ldots$
$= 1 + 1 + 1 + 1 = 4$

$\therefore$ $m[n] = 4$ **Ans.**

(*e*) $m[n] = y[n] * z[n]$

$$\because\ m[n] = \sum_{K=-\infty}^{\infty} y[n]\,\delta[n-k]$$

$= 0 + 0 + 0 + \ldots + 1 \times 0 + 1 \times 0 + 1 \times 1 + 1 \times -1 + 1 \times -1 + 2 \times 0 + 2 \times 0 + 2 \times 0 + 0 + 0 + 0 + \ldots$

$= 1 - 1 - 1 = -1$

$\therefore\ m[n] = -1$ **Ans.**

(*f*) $m[n] = y[n] * w[n]$

$$\because\ m[n] = \sum_{K=-\infty}^{\infty} y[n]\,\delta[n-k]$$

$= 0 + 0 + 0 + \ldots + 1 \times 1 + 1 \times 2 + 1 \times 3 + (-1) \times 2 + (-1) \times 1 + 0 + 0 + 0 + \ldots$

$= 1 + 2 + 3 - 2 - 1 = 3$

$\therefore\ m[n] = 3$ **Ans.**

(*g*) $m[n] = z[n] * g[n]$

$$\because\ m[n] = \sum_{K=-\infty}^{\infty} z[n]\,\delta[n-k]$$

$= 0 + 0 + 0 + \ldots + 1 \times 0 + 1 \times 0 + 1 \times 0 + 1 \times 0 + 1 \times 0 + 1 \times 0 + 1 \times 0 + 1 \times 0 + 2 \times 1 + 2 \times 1 + 2 \times 1 + 0 + 0 + 0 + \ldots$

$= 2 + 2 + 2 = 6$

$\therefore\ m[n] = 6$ **Ans.**

(*h*) $m[n] = f[n] * g[n]$

$$\because\ m[n] = \sum_{K=-\infty}^{\infty} f[n]\,\delta[n-k]$$

$= 0 + 0 + 0 + \ldots + (-4) \times 1 + (-3) \times 1 + (-2) \times 1 + (-1) \times 1 + 0 + 0 + 0 + 1.5 \times 1 + 2 \times 1 + 0 + 0 + \ldots$

$= -4 - 3 - 2 - 1 + 1.5 + 2 = -6.5$

$\therefore\ m[n] = -6.5$ **Ans.**

Problem 2.4: Evaluate the continuous time convolution integration given below:

(*a*) $y(t) = u(t+1) * u(t-2)$

(*b*) $y(t) = e^{-2t}\,u(t) * u(t+2)$

(c) $y(t) = \cos(\pi t)\,(u(t+1) - u(t-3) * u(t)$
(d) $y(t) = (u(t+2) - u(t-1) * u(-t+2) = 0$

Solution:

(a) $y(t) = u(t+1) * u(t-2)$

Let $x(t) = u(t+1)$

$h(t) = u(t-2)$

$\therefore \quad x(\tau) = u(\tau - 1)$

$h(t-\tau) = u(t-\tau-2) = u[(t-2)-\tau]$

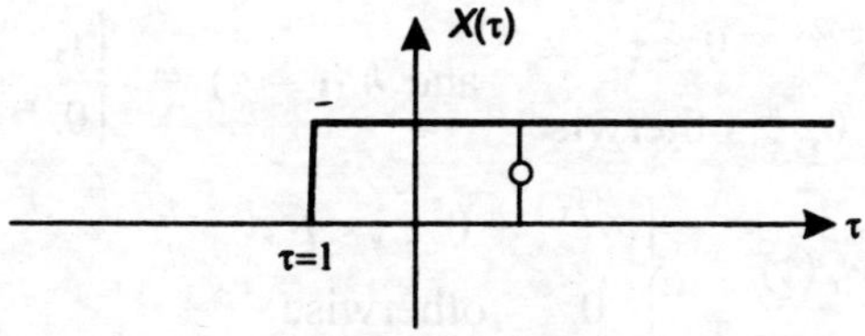

Fig. P. 2.4 (a)

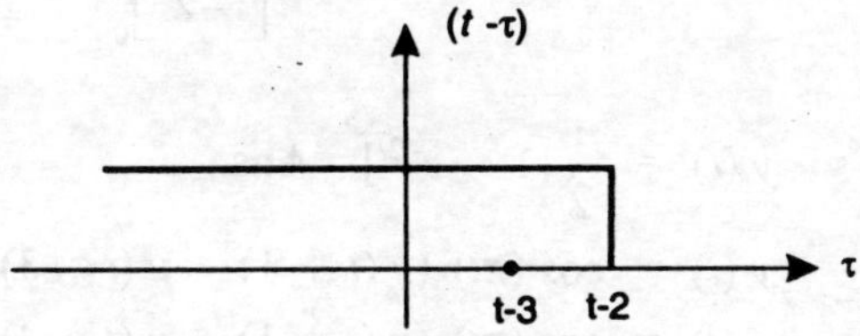

Fig. P. 2.4 (b)

For

$t < -1,\; w_t(\tau) = 0$

For $t > 3$

$$w_t(\tau) = \begin{cases} 1, & -1 < \tau < t-2 \\ 0, & \text{otherwise} \end{cases}$$

$$\therefore \; y(t) = \int_{-1}^{t-2} 1.1\, d\tau = [\tau]_{-1}^{t-2} = [t-2-(-1)] = [t-2+1]$$

$y(t) = t - 1$ **Ans.**

(b) $y(t) = e^{-2t}\,u(t) * u(t+2)$

Let $x(t) = e^{-2t}\,u(t)$

$$\therefore \quad x(\tau) = e^{-2\tau} u(\tau)$$

$$\text{and let} \quad h(t) = u(t+2)$$

$$\therefore \quad h(t-\tau) = u(t-\tau+2)$$

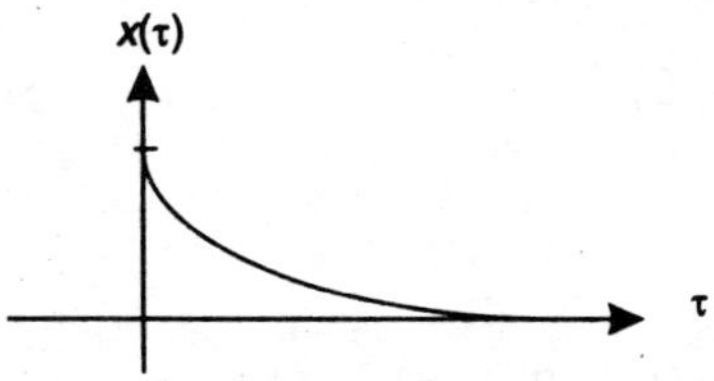

Fig. P. 2.4 (c)

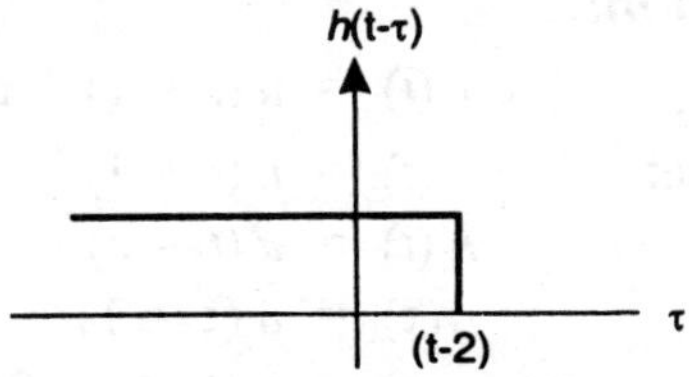

Fig. P. 2.4 (d)

$$x(\tau) = \begin{cases} e^{-2\tau} & 0 < \tau < \infty \\ 0, & \text{otherwise} \end{cases} \quad \text{and } h(t-\tau) = \begin{cases} 1, & \tau < t+2 \\ 0, & \text{otherwise} \end{cases}$$

$$\therefore \quad w_t(\tau) = \begin{cases} e^{-2\tau}, & 0 < \tau < t \\ 0, & \text{otherwise} \end{cases}$$

$$\therefore \quad y(t) = \int_0^t e^{-2\tau} \cdot 1 \cdot d\tau = \left[\frac{e^{-2\tau}}{-2}\right]_0^t = \frac{1}{2}(1 - e^{-2t})$$

$$\therefore \quad y(t) = \frac{1}{2}[1 - e^{-2t}] \quad \textbf{Ans.}$$

(*c*) $y(t) = \cos(\pi t)\,(u(t+1) - u(t-3)) * u(t)$

Let $x(t) = \cos(\pi t)\,(u(t+1) - u(t-3))$

$$x(\tau) = \cos(\pi\tau)\,(u(\tau+1) - u(\tau-3))$$

$$x(\tau) = \begin{cases} \cos(\pi\tau), & -1 < \tau < 3 \\ 0, & \text{otherwise} \end{cases}$$

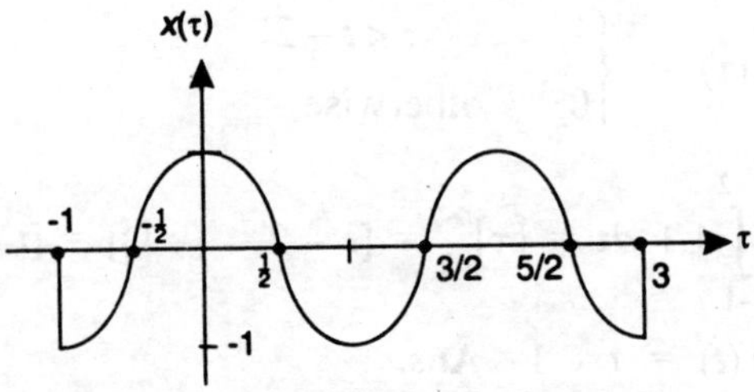

Fig. P. 2.4 (e)

Let $h(t) = u(t)$

$$h(t-\tau) = u(t-\tau)$$

$$h(t-\tau) = \begin{cases} 1, & \tau < t \\ 0, & \text{otherwise} \end{cases}$$

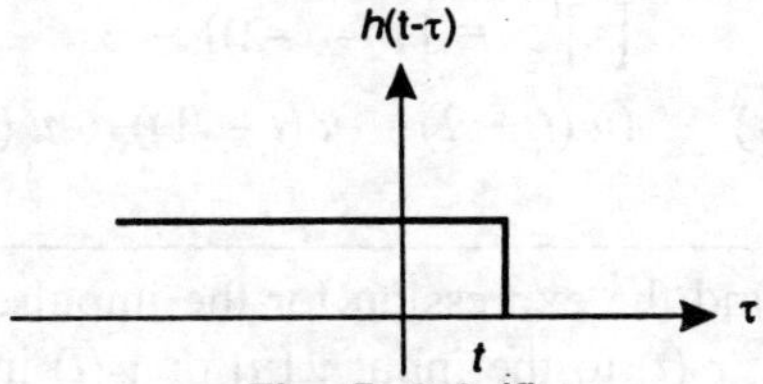

Fig. P. 2.4 (f)

$$\therefore y(t) = \int_{-\infty}^{\infty} x(\tau)\, h(t-\tau)\, d\tau = \int_{-1}^{3} \cos(\pi\tau) \cdot 1 \cdot d\tau + 0 = \left[\sin(\pi\tau)\right]_{-1}^{3} = 0$$

$$\therefore \quad y(t) = \cos(\pi t)\,(u(t+1) - u(t-3)) * (t) = 0 \quad \textbf{Ans.}$$

(*d*) $\quad y(t) = (u(t+2) - u(t-1)) * u(-t+2)$

Let $\quad x(t) = u(t+2) - u(t-1)$

$$x(\tau) = u(\tau+2) - u(\tau-1)$$

$$x(\tau) = \begin{cases} 1, & -2 < \tau < 1 \\ 0, & \text{otherwise} \end{cases}$$

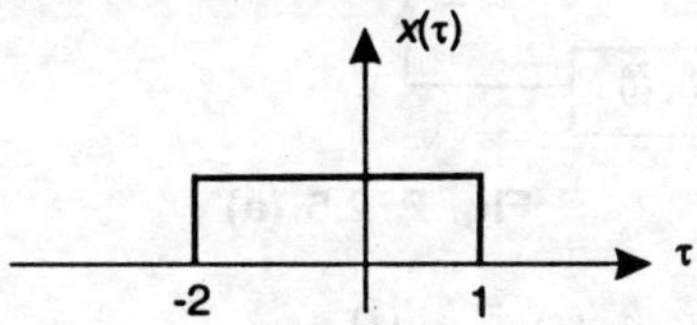

Fig. P. 2.4 (g)

Let $\quad h(t) = u(-t+2)$

$$h(t-\tau) = u(-t-\tau+2) = u(2-t-\tau)$$

$$h(t-\tau) = \begin{cases} 1, & \tau < 2-t \\ 0, & \text{otherwise} \end{cases}$$

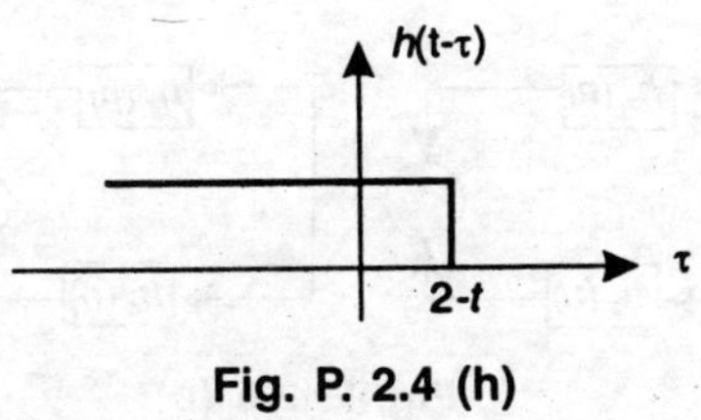

Fig. P. 2.4 (h)

$$\therefore \quad y(t) = \int_{-\infty}^{\infty} x(\tau)\, h(t-\tau)\, d\tau = \int_{-1}^{1} 1 \cdot 1 \cdot d\tau$$

$$= [\tau]_{-2}^{1} = [1 - (-2)] = 3$$

$$\therefore \quad y(t) = (u(t+2) - u(t-1)) * u(-t+2) = 3\cdot$$

Problem 2.5: Find the expression for the impulse response relating the input $x[n]$ or $x(t)$ to the input $y[n]$ or $y(t)$ in terms of impulse response of each subsystem for LT1 systems depicted in

Solution: (*a*)

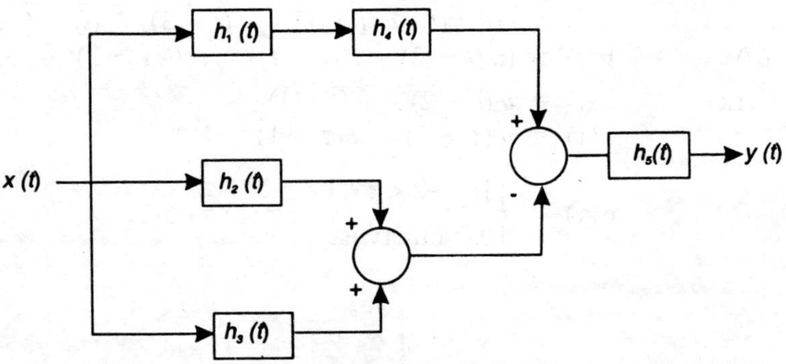

Fig. P. 2.5 (a)

$$h_{14}(t) = h_1(t) * h_4(t)$$

$$h_{23}(t) = h_2(t) + h_3(t)$$

$$h(t) = \frac{y(t)}{x(t)} = (-(h_2(t) + h_3(t)) + h_1(t) * h_4(t)) * h_5(t)$$

or, $h(t) = (h_1(t) * h_4(t) - (h_2(t) + h_3(t))) * h_5(t)$ **Ans.**

(*b*)

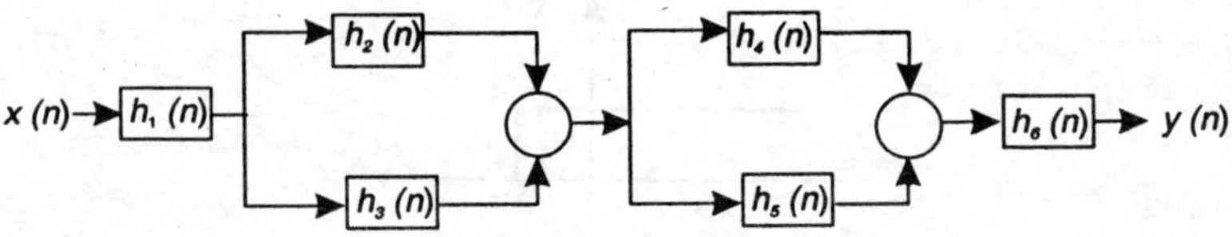

Fig. P. 2.5 (b)

$$h_{23}[n] = h_2[n] + h_3[n]$$

$$h_{45}[n] = -h_4[n] + h_5[n]$$

$$\therefore\ x[n] \rightarrow \boxed{h_1[n]} - \boxed{h_{23}[n]} - \boxed{h_{45}[n]} - \boxed{h_6[n]} \rightarrow y[n]$$

Or, $h[n] = [[h_1[n] * h_{23}[n]] * h_{45}[n] * h_6[n]$

Or, $h[n] = [[h_1[n] * [h_2[n] + h_3[n]]] * [h_5[n] - h_4[n]] * h_6[n]$

(*c*)

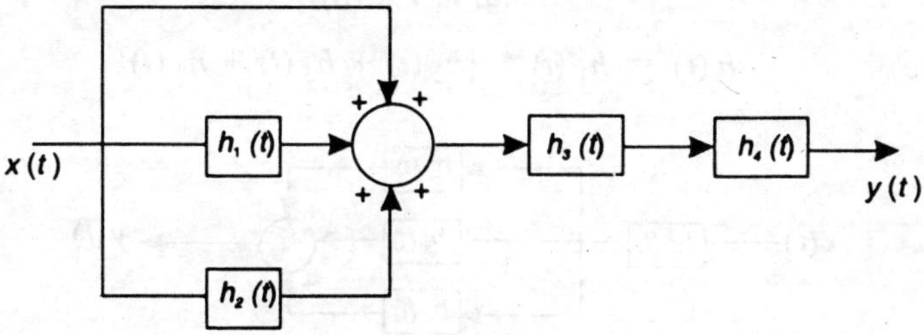

Fig. P. 2.5 (c)

$$h[n] = [[-h_1(t) + h_2(t) + 1] * h_3(t)] * h_4(t)$$

Problem 2.6: Let $h_1(t)$, $h_2(t)$, $h_3(t)$ and $h_4(t)$ be impulse responses of LT1 systems. Construct a system with impulse response $h(t)$ using $h_1(t)$, $h_2(t)$, $h_3(t)$ and $h_4(t)$ as subsystems. Draw the interconnection of systems require to obtain:

(*a*) $h(t) = h_1(t) + \{h_2(t) + h_3(t)\} * h_4(t)$

(*b*) $h(t) = h_1(t) * h_2(t) + h_3(t) * h_4(t)$

(*c*) $h(t) = h_1(t) * \{h_2(t) + h_3(t) + h_4(t)\}$

Solution:

(*a*) $$h(t) = h_1(t) + [h_2(t) + h_3(t)] * h_4(t)$$

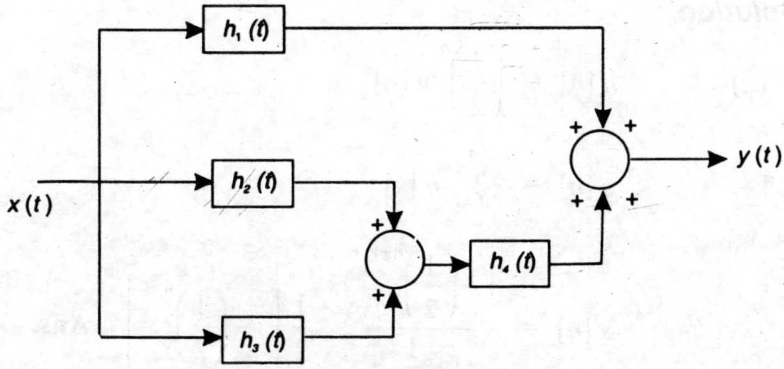

Fig. P. 2.6 (a)

(*b*) $$h(t) = h_1(t) * h_2(t) + h_3(t) * h_4(t)$$

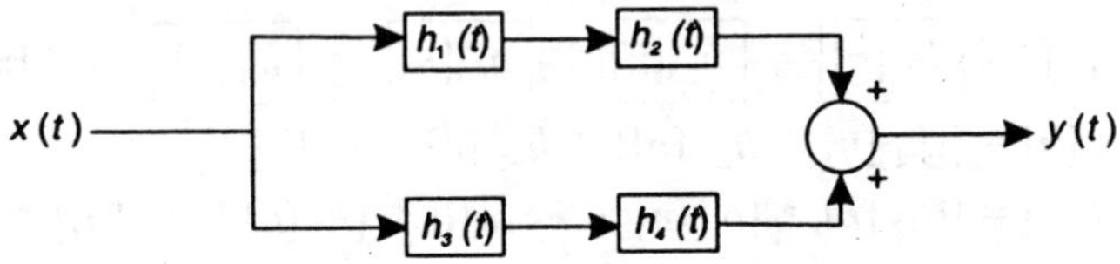

Fig. P. 2.6 (b)

(*c*) $$h(t) = h_1(t) * \{h_2(t) + h_3(t) + h_4(t)\}$$

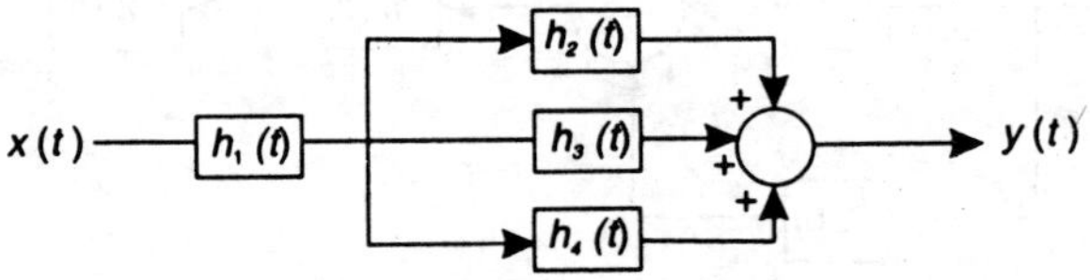

Fig. P. 2.6 (c)

Problem 2.7: Evaluate the step response for the LTI system represented by the following impulse response

(*a*) $h[n] = \left(\frac{1}{2}\right)^n u[n]$

(*b*) $h[n] = (-1)^n \{u(n+2) - u(n-3)\}$

(*c*) $h[n] = u[n]$

(*d*) $h(t) = e^{-|t|}$

(*e*) $h(t) = \delta(t) - \delta(t-1)$

(*f*) $h(t) = u(t+1) - u(t-1)$

(*g*) $h(t) = u(t+1) - u(t-1)$

Solution:

(*a*) $$h[n] = \left(\frac{1}{2}\right)^n u[n]$$

$$s[n] = \sum_{K=-\infty}^{n} h[k]$$

$$\therefore \quad s[n] = \frac{1-\left(\frac{1}{2}\right)^{n+1}}{1-\frac{1}{2}} = \frac{1}{2}\left[1-\left(\frac{1}{2}\right)^{n+1}\right] \quad \textbf{Ans.}$$

(*b*) $h[n] = (-1^n)\{u(n+2) - u(n-3)\}$

$$s[n] = \sum_{K=-\infty}^{n} h[k]$$

graphically $h[n]$ can be represented as:

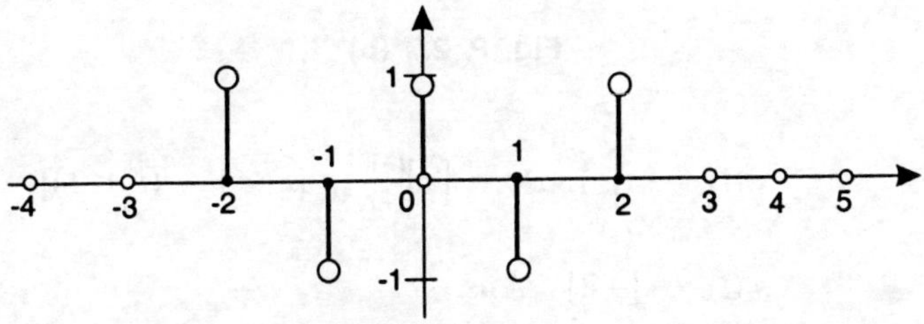

Fig. P. 2.7 (a)

$\therefore$ $s[n] = +1 - 1 + 1 - 1 + 1$

$s[n] = 1$ **Ans.**

(*c*) $h[n] = u[n]$

$$s[n] = \sum_{K=-\infty}^{n} h[k]$$

$\therefore$ $s[n] = 1 + 1 + 1 + \ldots$ upto n

$s[n] = n$ **Ans.**

(*d*) $h(t) = e^{-|t|}$

$$s(t) = \int_{-\infty}^{t} e^{-|\tau|}\, d\tau = \left[\frac{e^{-|e|}}{-1}\right]_{-\infty}^{t}$$

$$= -1\,[e^{-t} - e^{-\infty}] = -1\,[e^{-t} - 0]$$

$s(t) = -e^{-t}$ **Ans.**

(*e*) $h(t) = \delta(t) - \delta(t-1)$

$$s(t) = \int_{-\infty}^{t} (\delta(\tau) - \delta(\tau - 1)\, d\tau$$

$$= [u(t) - u(t-1)]$$ **Ans.**

(*f*) $h(t) = u(t+1) - u(t-1)$

$h(t)$ can be graphically represented as:

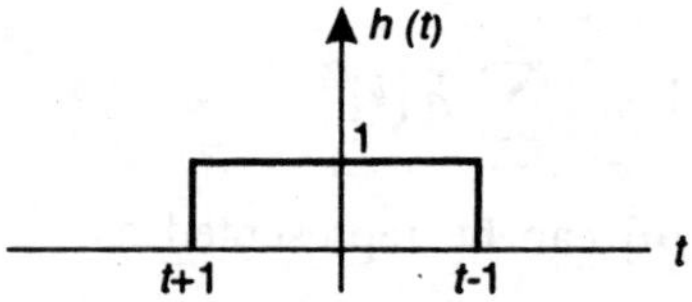

Fig. P. 2.7 (b)

$$\therefore \qquad s(t) = \int_{t+1}^{t-1} 1 \cdot d\tau = [\tau]_{t+1}^{t-1} = [t - 1 - (t + 1)]$$

$$s(t) = [-2] \quad \textbf{Ans.}$$

(*g*) $$h(t) = tu(t)$$

$$h(t) = \begin{cases} 0, & \text{for } t < 0 \\ t & \text{for } t \geq 0 \end{cases}$$

$$\therefore \qquad s(t) = \int_0^t \tau \, d\tau = \left[\frac{\tau^2}{2}\right]_0^t$$

$$s(t) = \left[\frac{t^2}{2}\right]$$

> **Problem 2.8:** Evaluate the frequency response for the LTI system represented by the following impulse response:
>
> (*a*) $h[n] = \left(\frac{1}{2}\right)^n u[n]$
>
> (*b*) $h[n] = \delta[n] - \delta[n-1]$
> (*c*) $h[n] = (-1)^n\{u[n+2] - u[n-3]\}$
> (*d*) $h[t] = e^{-|t|}$
> (*e*) $h[t] = -\delta(t+1) + \delta(t) - \delta(t-1)$
> (*f*) $h[t] = \cos(\pi t)\ \{u(t+3) - u(t-3)\}$
> (*g*) $h(t) = e^{2t}\, u(-t)$

Solution:

(*a*) $$h[n] = \left(\frac{1}{2}\right)^n u[n]$$

$$h[n] = \begin{cases} 0, & t < 0 \\ \left(\frac{1}{2}\right)^n, & t \geq 0 \end{cases}$$

Frequency response:

$$\therefore \quad h[e^{j\Omega}] = \sum_{K=-\infty}^{\infty} h[k]\, e^{-j\Omega k}$$

$$h[e^{j\Omega}] = 1 + \frac{1}{2} e^{-j\Omega} + \frac{1}{4} e^{-2j\Omega} + \frac{1}{8} e^{-3j\Omega} + \ldots\ldots + \left(\frac{1}{2}\right)^n e^{-nj\Omega} \quad \textbf{Ans.}$$

(*b*) $$h[n] = \delta[n] - \delta[n-1]$$

$$h[e^{j\Omega}] = \sum_{K=-\infty}^{\infty} h[k]\, e^{-j\Omega k}$$

$$h[e^{j\Omega}] = 1 - e^{-j\Omega} \quad \textbf{Ans.}$$

(*c*) $$h[n] = (-1)^n \{u[n+2] - u[n-3]$$

$h[n]$ can be graphically represented as:

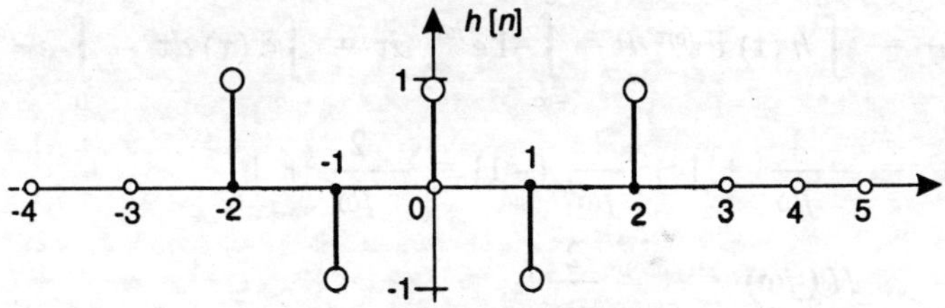

Fig. P. 2.8 (a)

$$\therefore \quad h[e^{j\Omega}] = \sum_{K=-\infty}^{\infty} h[k]\, e^{-j\Omega k}$$

$$= e^{2j\Omega} - e^{j\Omega} + 1 - e^{-j\Omega} + e^{-2j\Omega}$$

$$= e^{j\Omega}(e^{j\Omega} - 1) + 1 - e^{j\Omega}(1 - e^{-j\Omega})$$

or, $$h[e^{j\Omega}] = 2\frac{\left(e^{2j\Omega} + e^{-2j\Omega}\right)}{2} + 1 - \frac{\left(e^{j\Omega} + e^{-j\Omega}\right) \times 2}{2}$$

$$= 2\cos 2\Omega + 1 - 2\cos\Omega$$

$$h[e^{j\Omega}] = 1 + 2(\cos 2\Omega - \cos\Omega) \quad \textbf{Ans.}$$

(*d*) $$h(t) = \bar{e}^{|t|}$$

$$H(j\omega) = \int_{-\infty}^{\infty} h(\tau)\, \bar{e}^{j\omega\tau}\, d\tau = \int_{-\infty}^{\infty} e^{|\tau|}\, \bar{e}^{j\omega\tau}\, d\tau$$

$$= \int_{-\infty}^{0} e^{+\tau}\, \bar{e}^{j\omega\tau}\, d\tau + \int_{0}^{\infty} e^{-\tau}\, \bar{e}^{j\omega\tau}\, d\tau \quad \left[as|\tau| \text{ is} \begin{cases} = +ve, & t > 0 \\ = -ve, & t < 0 \end{cases} \right]$$

$$= \int_{-\infty}^{0} e^{(1-j\omega)\tau}\, d\tau + \int_{0}^{\infty} e^{-(1+j\omega)\tau}\, \bar{e}^{j\omega\tau}\, d\tau$$

$$= \left.\frac{e^{(1-j\omega)\tau}}{(1-j\omega)}\right|_{-\infty}^{0} + \left.\frac{\bar{e}^{(1+j\omega)\tau}}{-(1+j\omega)}\right|_{0}^{\infty}$$

$$= \frac{1}{(1-j\omega)}[1-0] + \frac{1}{-(1+j\omega)}[0-1] = \frac{1}{(1-j\omega)} + \frac{1}{(1+j\omega)}$$

$$H(j\omega) = \frac{1}{1+\omega^2}$$

(e) $$h(t) = -\delta(t+1) + \delta(t) - \delta(t-1)$$

$$H(j\omega) = \int_{-\infty}^{\infty} h(\tau)\, \bar{e}^{j\omega\tau} d\tau = \int_{-\infty}^{0} -1\, e^{j\omega\tau}\, d\tau + \int_{-\infty}^{t} \delta(\tau)\, d\tau - \int_{0}^{\infty} e^{-j\omega\tau}\, d\tau$$

$$= -\frac{1}{j\omega} + 1 + \frac{1}{j\omega}[-1] = -\frac{2}{j\omega} + 1$$

$$H(j\omega) = 1 - \frac{2}{j\omega}$$

(f) $$h(t) = \cos(\pi t)\, \{u(t+3) - u(t-3)\}$$

$$h(t) = \begin{cases} \cos \pi t, & -3 < t < +3 \\ 0, & \text{otherwise} \end{cases}$$

$$\therefore \quad H(j\omega) = \int_{-\infty}^{\infty} h(\tau)\, \bar{e}^{j\omega\tau}\, d\tau = \int_{-3}^{3} h(\tau)\, \bar{e}^{j\omega\tau}\, d\tau$$

$$\therefore H(j\omega) = \int_{-3}^{3} \cos(\pi\tau) \cdot \bar{e}^{j\omega\tau}\, d\tau$$

$$= \left[\frac{\cos(\pi\tau)\, e^{-j\omega\tau}}{-2j\omega}\right]_{-3}^{3} + \left[\frac{\sin(\pi\tau)\, e^{-j\omega\tau}}{2j\omega}\right]_{-3}^{3}$$

$$= \frac{1}{-2j\omega}\left[\cos 3\pi\, e^{-3j\omega} - \cos(-3\pi)\, e^{+3j\omega}\right] +$$

$$[\sin 3\pi \cdot e^{-3j\omega} - \sin(-3\pi) \cdot e^{+3j\omega}]$$

$$= \frac{1}{-2j\omega}\,[-e^{-3j\omega} + e^{3j\omega}] + 0 = -\frac{1}{j\omega}\left[\frac{e^{3j\omega} - e^{-3j\omega}}{2}\right]$$

$$H(j\omega) = -\frac{1}{j\omega}\,\sin 3\Omega \quad \textbf{Ans.}$$

(g) $$h(t) = e^{2t}\, u(-t)$$

$$\because \quad h(t) = \begin{cases} 0, & t > 0 \\ e^{2t}, & t \le 0 \end{cases}$$

$$\therefore \quad H(j\omega) = \int_{-\infty}^{0} e^{2\tau}\,.\,\bar{e}^{\,j\omega\tau}\, d\tau = \int_{-\infty}^{0} e^{(2-j\omega)\tau}\, d\tau$$

$$= \left[\frac{e^{(2-j\omega)\tau}}{2-j\omega}\right]_{-\infty}^{0} = \frac{1}{2-j\omega}\,[1-0]$$

$$H(j\omega) = \frac{1}{(2-j\omega)} \quad \textbf{Ans.}$$

Problem 2.9: Write a differential equation description relating the output to the input of the circuit given below.

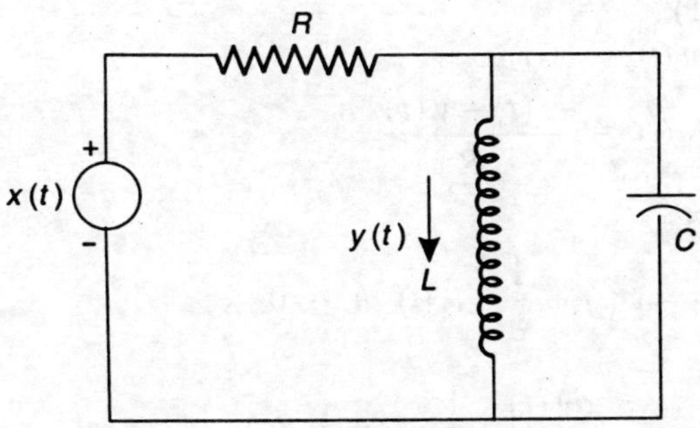

Fig. P. 2.9

Solution:

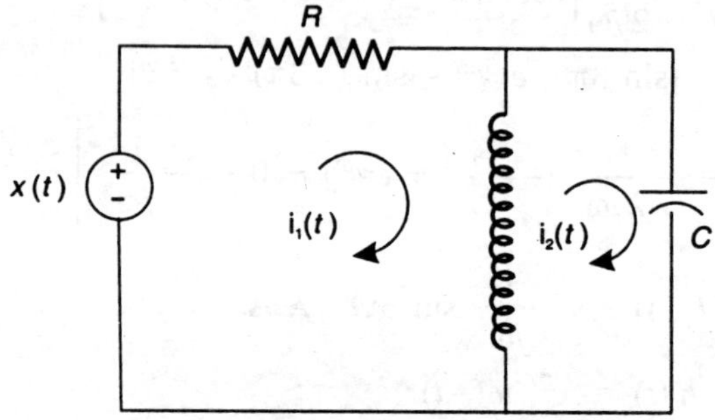

Fig. P. 2.9 (a)

Now taking KVL in loop 1 and loop 2

$$Ri_1(t) + L\frac{d}{dt}[i_1(t) - i_2(t)] = x(t) \quad ...(1)$$

$$L\frac{d}{dt}[i_2(t) - i_1(t)] + \frac{1}{c}\int_{-\infty}^{t} i_2(t)\,dt = 0 \quad ...(2)$$

$$\because \quad y(t) = L\frac{d}{dt}[i_1(t) - i_2(t)]$$

Taking integration

$$\int_{-\infty}^{t} y(t) = L[i_1(t) - i_2(t)]$$

or, $$\frac{1}{L}\int_{-\infty}^{t} y(t) = [i_1(t) - i_2(t)] \quad ...(3)$$

From eq (1)

$$Ri_1(t) + y(t) = x(t)$$

$$\therefore \quad i_1(t) = \frac{x(t) - y(t)}{R} \quad ...(4)$$

From eq (2)

$$-y(t) + \frac{1}{c}\int_{-\infty}^{t} i_2(t)\,dt = 0$$

or, $$-\frac{dy(t)}{dt} + \frac{1}{c}\,i_2(t) = 0$$

or, $$i_2(t) = +c\frac{dy(t)}{dt} \quad \ldots(5)$$

putting eq (4) and (5) in eq (3)

$$\frac{1}{L}\int_{-\infty}^{t} y(t) = \frac{x(t)-y(t)}{R} - \left(+c\frac{dy(t)}{dt}\right)$$

or, $$\frac{R}{L}\int_{-\infty}^{t} y(t) = x(t) - y(t) - Rc\,\frac{dy(t)}{dt}$$

Now, differentiating both sides

$$\frac{R}{L}y(t) = \frac{d}{dt}x(t) - \frac{dy(t)}{dt} - Rc\frac{d^2y(t)}{dt^2}$$

or, $$Rc\frac{d^2y(t)}{dt^2} + \frac{dy(t)}{dt} + \frac{R}{L}y(t) = \frac{d}{dt}x(t)$$

or, $$\frac{d^2y(t)}{dt^2} + \frac{1}{Rc}\frac{dy(t)}{dt} + \frac{1}{Lc}y(t) = +\frac{1}{Rc}\frac{d}{dt}x(t)$$ **Ans.**

Problem 2.10: Determine the natural response for the systems described by the following differential equations:

(*a*) $5\frac{d}{dt}y(t) + 10\,y(t) = 2\,x(t),\ y(0) = 3$

(*b*) $\frac{d^2}{dt^2}y(t) + 5\frac{d}{dt}y(t) + 6\,y(t) =$

$$2xt + \frac{d}{dt}x(t),\ y(0) = 2, \frac{d}{dt}y(t)\Big|_{t=0} = 1$$

(*c*) $\frac{d^2}{dt^2}y(t) + 3\frac{d}{dt}y(t) + 2\,y(t) =$

$$x(t) + \frac{d}{dt}x(t),\ y(0) = 0,\ \frac{d}{dt}y(t)\Big|_{t=0} = 1$$

(*d*) $\frac{d^2}{at^2}y(t) + 2\frac{d}{dt}y(t) + y(t) =$

$$\frac{d}{dt}x(t), y(0) = 0, \frac{d}{dt}y(t)\Big|_{t=0} = 1$$

(e) $\frac{d^2}{dt} y(t) + 2\frac{d}{dt} y(t) + 2y(t) =$

$\frac{d}{dt} x(t), y(0) = 1, \frac{d}{dt} y(t)\Big|_{t=0} = 1$

Solution:

(a) $$5\frac{d}{dt} y(t) + 10\, y(t) = 2x(t),\ y(0) = 3$$

Let $\frac{d}{dt} = m$

$\therefore$ $(5m + 10)\, y(t) = 0$

Or, $5m + 10 = 0$

Or, $m = 2$

$\therefore$ $y^n(t) = c_1 e^{mt}$

$y(0) = c_1 = 3$

$\therefore$ $y^n(t) = 3e^{-2t}$ **Ans.**

(b) $$\frac{d^2}{dt^2} y(t) + 5\frac{d}{dt} y(t) + 6\, y(t) = 2x(t) + \frac{d}{dt} x(t)$$

$$y(0) = 2,\ \frac{d}{dt} y(t)\Big|_{t=0} = 1$$

above equation can be written as:

$(m^2 + 5m + 6)\, y(t) = 0$

or, $m^2 + 5m + 6 = 0$

or, $m_1, m_2 = -2, -3$

$\because$ $y^n(t) = c_1 e^{m_1 t} + c_2 e^{m_2 t}$ ...(1)

$y(0) = c_1 + c_2 = 2$ ...(2)

$\frac{dy(t)}{dt}\Big|_{t=0} = m_1 c_1 + m_2 c_2 = 1$ ...(3)

or, $-2c_1 - 3c_2 = 1$ ...(4)

From (2) and (4) eqn

$$\begin{aligned} -2c_1 - 3c_2 &= 1 \\ 2c_1 + 2c_2 &= 4 \\ \hline -c_2 &= 5 \end{aligned}$$

$$\text{or,} \quad c_2 = -5$$

$$\therefore \quad c_1 = 7$$

$$\therefore \quad y^n(t) = 7e^{-2t} - 5e^{-3t} \quad \textbf{Ans.}$$

(c) $\dfrac{d^2}{dt^2}y(t) + 3\dfrac{d}{dt}y(t) + 2y(t) = x(t) + \dfrac{d}{dt}x(t)$

$$y(0) = 0, \frac{d}{dt}y(t)\Big|_{t=0} = 1$$

Above eqn can be written as:

$$(m^2 + 3m + 2)\,y(t) = 0$$

$$\text{or,} \quad m^2 + 3m + 2 = 0$$

$$m_1, m_2 = -1, -2$$

$$\because \quad y^n(t) = c_1 e^{m_1 t} + c_2 e^{m_2 t}$$

$$y(0) = c_1 + c_2 = 0$$

$$\frac{dy(t)}{dt}\Bigg|_{t=0} = c_1 m_1 + c_2 m_2 = 1$$

$$\text{or,} \quad -c_1 - 2c_2 = 1$$

$$c_1 + c_2 = 0$$

$$\text{or,} \quad -c_2 = 1$$

$$\text{or} \quad c_2 = -1 \text{ and } c_1 = 1$$

$$\therefore \quad y^n(t) = e^{-t} - e^{-2t} \quad \textbf{Ans.}$$

(d) $\dfrac{d^2}{dt^2}y(t) + 2\dfrac{d}{dt}y(t) + y(t) = \dfrac{d}{dt}x(t)$

$$y(0) = 1, \frac{d}{dt}y(t)\Big|_{t=0} = 1$$

$$(m^2 + 2m + 1)\,y(t) = 0$$

$$\text{or,} \quad m^2 + 2m + 1 = 0$$

$$\text{or,} \quad m_1\ m_2 = -1, -1$$

Since both the roots are similar and real

$$\therefore \quad y^n(t) = (c_1 + c_2 t)\,e^{m_1 t}$$

$$y(0) = c_1 = 1$$

$$\frac{dy(t)}{dt}\Bigg|_{t=0} = c_1 + c_2 = -1 \qquad [\because m_1 = m_2 = -1]$$

$$c_2 = -2$$

$$y^n(t) = (1 - 2t)\,e^{-t} \quad \textbf{Ans.}$$

(*f*) $\frac{d^2}{dt^2} y(t) + 2\frac{d}{dt} y(t) + y(t) = \frac{d}{dt} x(t),\ y(0) = 1, \frac{d}{dt} y(t)\big|_{t=0} = 0$

$$(m^2 + 2m + 2)\, y(t) = 0$$

for natural response

$$\therefore \qquad m^2 + 2m + 2 = 0$$

$$m_1 = -1 + \sqrt{-1}, \qquad m_2 = -1 - \sqrt{-1}$$

When two roots are imaginary

then $\quad y^n(t) = e^{\alpha t}[c_1 \cos Bt + c_2 \sin Bt]$

Now, $\quad y(0) = c_1 = 1$

$$\left.\frac{dy(t)}{dt}\right|_{t=0} = \alpha c_1 + Bc_2 = 0 = -c_1 + 2c_2 = 0$$

or, $\quad 2c_2 = c_1 = 1$

$$\therefore \qquad c_2 = \frac{1}{2}$$

$$\therefore \qquad y^n(t) = e^{-t}\left[\cos 2t + \frac{1}{2} \sin 2t\right]$$

Problem 2.11: Find differential-equation descriptions for the three systems depicted in Fig. P2.11.

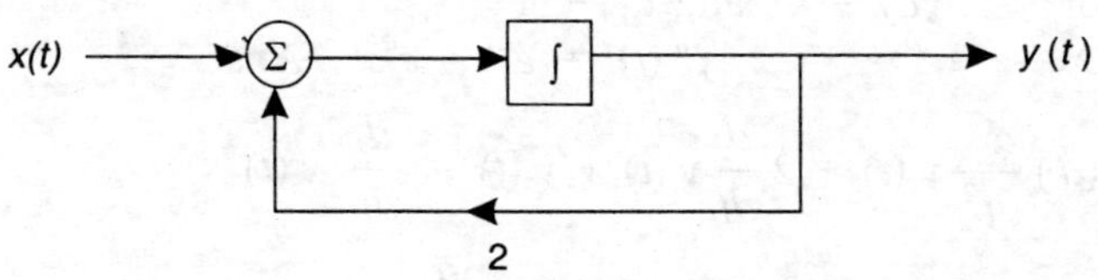

(a)

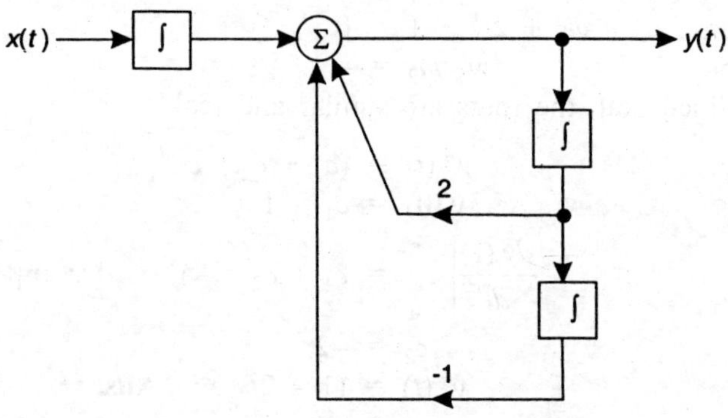

(b)

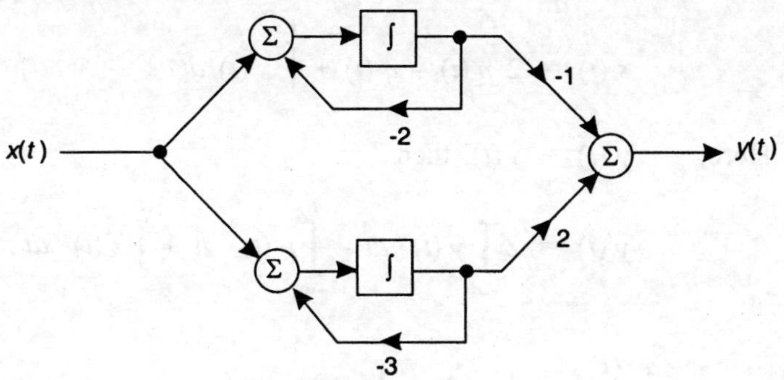

Fig. 2.11 (c)

Solution:

(*a*)

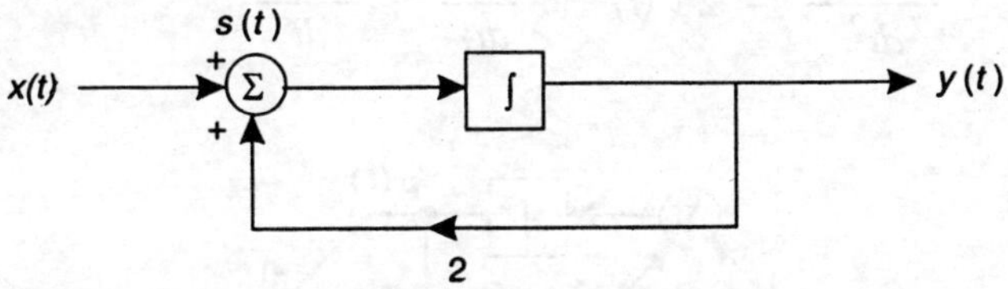

Fig. 2.11 (d)

At the summing point,

$$s(t) = [x(t) + 2y(t)] \text{ and, } y(t) = \int_{-\infty}^{t} s(t) \cdot dt$$

Differentiating both sides, we have,

$$\text{or,} \qquad \frac{dy(t)}{dt} = 2x(t) + 2y(t)$$

$$\text{or,} \qquad \frac{dy(t)}{dt} - 2y(t) = 2x(t)$$

(*b*)

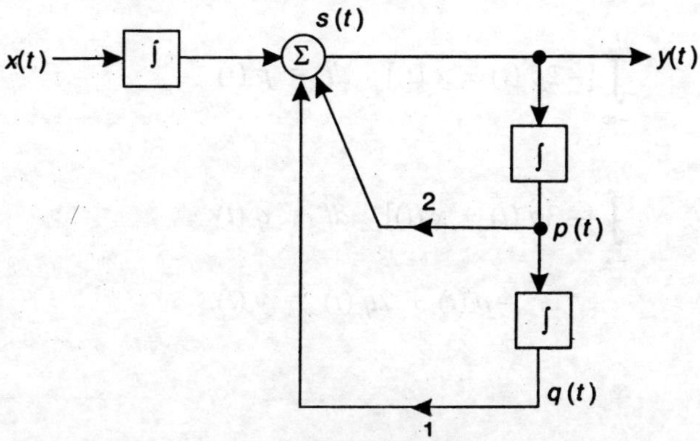

Fig. 2.11 (e)

The sum $s(t)$ can be expressed as,

$$s(t) = 2\,p(t) - q(t) + \int_{-\infty}^{t} x(t)\, dt$$

Here, $y(t) = s(t)$, then

$$y(t) = 2\int_{-\infty}^{t} y(t)\cdot dt - \int_{-\infty}^{t} p(t)\cdot dt + \int_{-\infty}^{t} x(t)\cdot dt$$

or $$\frac{dy(t)}{dt} = 2\,y(t) - \int_{-\infty}^{t} y(t)\cdot dt + x(t)$$

or, $$\frac{d^2y(t)}{dt^2} - 2\,y(t) + \frac{dy(t)}{dt} = \frac{dx(t)}{dt}$$

(*c*)

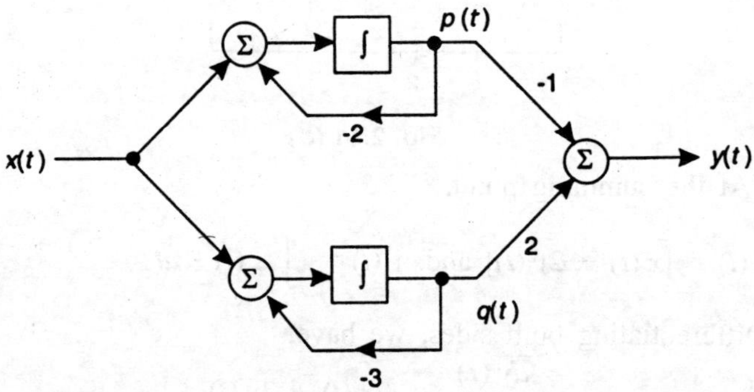

Fig. 2.27 (f)

From the figure redrawn we have,

$$\int_{-\infty}^{t} [-2p(t) + x(t)]\cdot dt = p(t) \qquad ...(1)$$

$$\int_{-\infty}^{t} [-3q(t) + x(t)]\cdot dt = q(t) \qquad ...(2)$$

and, $$-p(t) + 2q(t) = y(t) \qquad ...(3)$$

Chapter 3

Fourier Series Representation of LTI Systems

Problem 3.1: Determine the Fourier series representations for the following signals:

(*a*) Each $x(t)$ illustrated in Figures P. 3.1 (a)–(f).

(*b*) $x(t)$ periodic with period 2 and

$$x(t) = e^{-t} \text{ for } -1 < t < 1$$

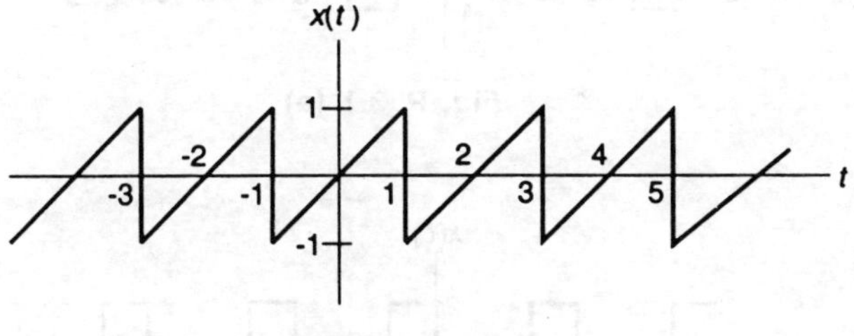

Fig. P. 3.1 (a)

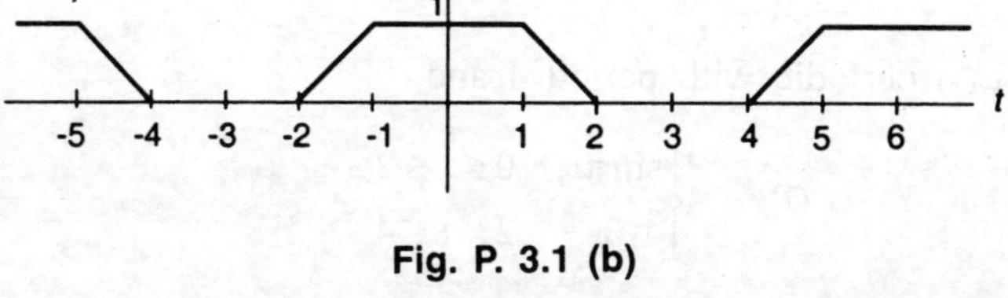

Fig. P. 3.1 (b)

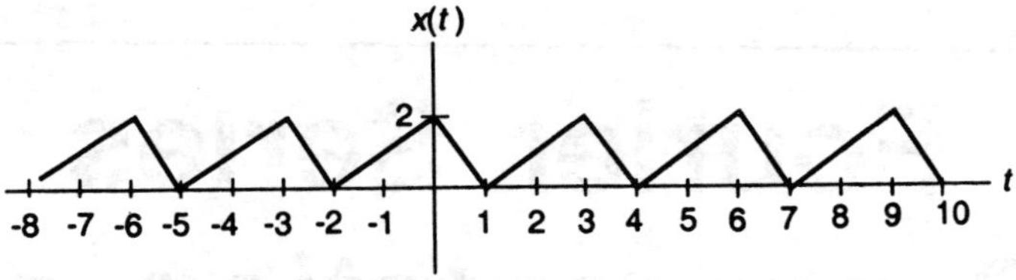

Fig. P. 3.1 (c)

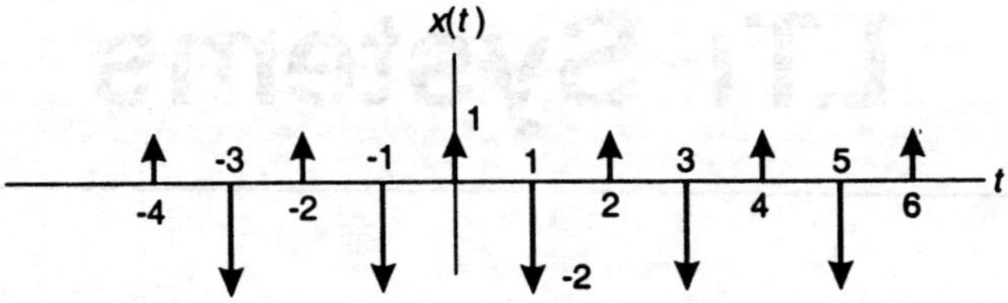

Fig. P. 3.1 (d)

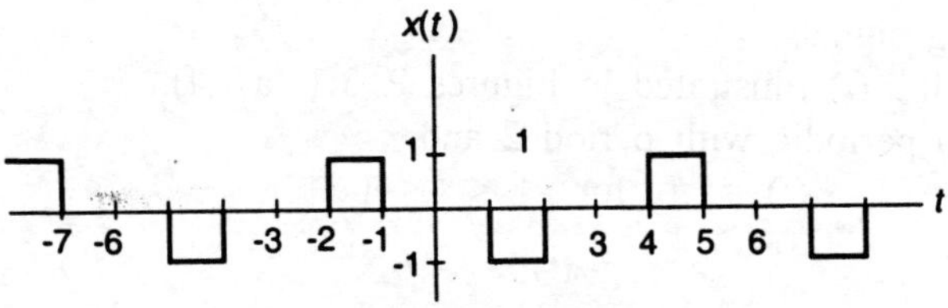

Fig. P. 3.1 (e)

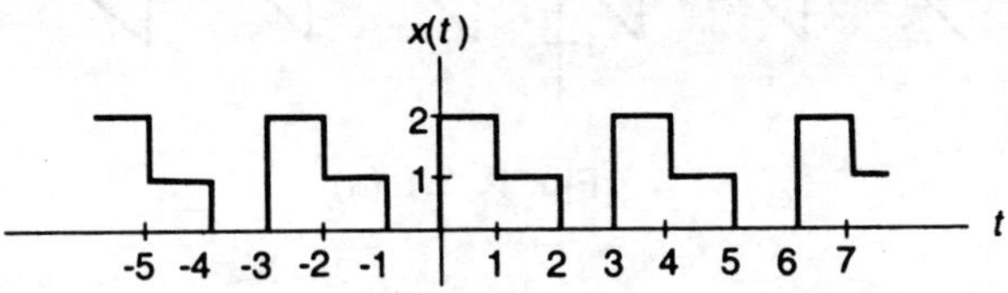

Fig. P. 3.1 (f)

(*c*) $x(t)$ periodic with period 4 and

$$x(t) = \begin{cases} \sin \pi t, & 0 \le t \le 2 \\ 0, & 2 < t \le 4 \end{cases}$$

Solution :

(*a*) (*i*) $T = 1, \quad a_o = 0,$

and $a_k = \dfrac{j(-1)k}{k\pi} \quad k \neq 0$

(*ii*) Here

$$x(t) = \begin{cases} t+2 & -2 < t < -1 \\ 1 & -1 < t < 1 \\ 2-t & 1 < t < 2 \end{cases}$$

$$T = 6 \qquad a_o = \frac{1}{2} \text{ and}$$

$$a_k = \begin{cases} 0 & k\text{-even} \\ \dfrac{6}{\pi^2 k^2} \sin\left(\dfrac{\pi k}{2}\right) \sin\left(\dfrac{\pi k}{6}\right) & k\text{-odd} \end{cases}$$

(*iii*) $T = 3,\ a_o = 1$ and

$$a_k = \frac{3j}{2\pi^2 k^2} \left[e^{j2\frac{\pi}{3}} \sin\left(2\pi \frac{k}{3}\right) + 2e^{jk\frac{\pi}{3}} \sin\left(k\frac{\pi}{3}\right) \right] \quad k \neq 0$$

(*iv*) $T = 2,\ a_o = -\dfrac{1}{2},$

$$a_k = \frac{1}{2} - (-1)^k \quad k \neq 0$$

(*v*) $T = 6,\ \omega_o = \dfrac{\pi}{3}$ and

$$a_k = \frac{\cos\left(2\pi \dfrac{k}{3}\right) - \cos\left(k\dfrac{\pi}{3}\right)}{jk\dfrac{\pi}{3}}$$

Note that

$$a_o = 0 \text{ and } a_k \text{ even} = 0$$

(*vi*) $T = 4, \quad \omega_o = \frac{\pi}{2}, \quad a_o = \frac{3}{4}$ and

$$a_k = \frac{e^{-jk\frac{\pi}{2}} \sin\left(k\frac{\pi}{2}\right) + e^{-j\pi\frac{k}{4}} \sin\left(k\frac{\pi}{4}\right)}{k\pi}$$

(*b*) $T = 2$

$$a_k = \frac{-j^k}{2(1+jk\pi)}(e - e^{-1}) \text{ for all } k$$

(*c*) $T = 3, \quad \omega_o = 2\frac{\pi}{3}, \quad a_o = 1$ and

$$a_k = \frac{2e^{-j\pi\frac{k}{3}}}{\pi k} \sin\left(2\pi\frac{k}{3}\right) + \frac{e^{-j\pi k}}{\pi k} \sin(\pi k)$$

Problem 3.2: In each of the following, we specify the Fourier series coefficients of a continuous time signal that is periodic with period 4. Determine the signal $x(t)$ in each case.

(*a*) $a_k = \begin{cases} 0, & k = 0 \\ (j)^k \dfrac{\sin k\frac{\pi}{4}}{k\pi} & \text{otherwise} \end{cases}$

(*b*) $a_k = (-1)^k \dfrac{\sin k\frac{\pi}{8}}{2k\pi}$

(*c*) $a_k = \begin{cases} jk, & |k|<3 \\ 0, & \text{otherwise} \end{cases}$

(*d*) $a_k = \begin{cases} 1, & k \text{ even} \\ 2, & k \text{ odd} \end{cases}$

Solution:

(*a*) Let us consider a signal $y(t)$ with *fs* coefficients

$$b_k = \frac{\sin\left(k\frac{\pi}{4}\right)}{k\pi}$$

We know that,

$y(t)$ must be periodic square wave, its period is

$$y(t) = \begin{cases} 1 & |t| < \frac{1}{2} \\ 0 & \frac{1}{2} < |t| < 2 \end{cases}$$

Now, Note that

$b_o = \frac{1}{4}$, let us define signal $\tau(t) = -\frac{1}{4}$, whose only non-zero Fourier series coefficient is $c_o = -\frac{1}{4}$, the signal $p(t) = y(t) + z(t)$ will have Fourier series coefficients

$$d_k = a_k + c_k = \begin{cases} 0 & k = 0 \\ \frac{\sin\left(k\frac{\pi}{4}\right)}{k\pi} & \text{otherwise} \end{cases}$$

Now, note that

$$a_k = d_k e^{j\left(\frac{\pi}{2}\right)k},$$

Therefore, the signal

$$x(t) = p(t-1),$$

which is as shown in figure below.

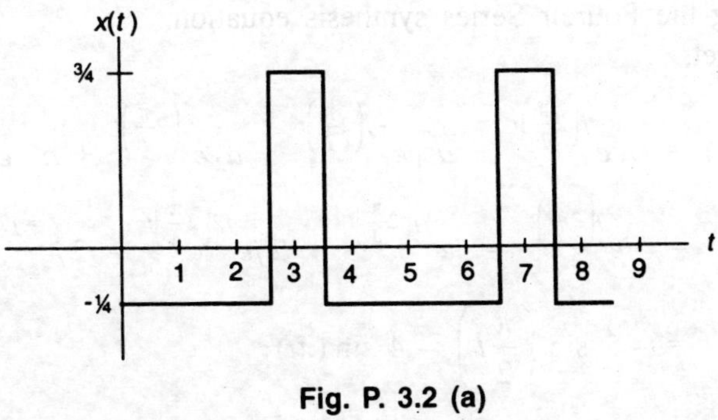

Fig. P. 3.2 (a)

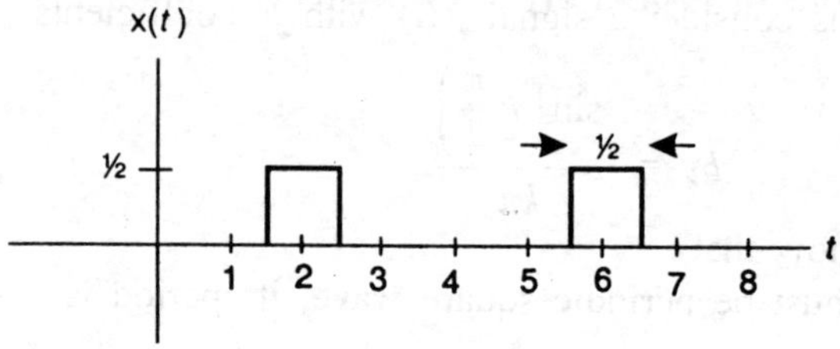

Fig. P. 3.2 (b)

(*b*) First let us consider a signal $y(t)$ with Fourier series coefficients

$$b_k = \frac{\sin\left(k\frac{\pi}{8}\right)}{2k\pi}$$

We know that $y(t)$ must be a periodic square wave which over one period is

$$y(t) = \begin{cases} \frac{1}{2} & |t| < \frac{1}{4} \\ 0 & \frac{1}{4} < |t| < 2 \end{cases}$$

Now, note that

$$a_k = b_k e^{jk\pi}$$

Therefore, the signal

$$x(t) = y(t+2),$$

which is shown in above fig. 3.2 (b)

(*c*) The only non zero Foureir Series coefficients are

$$a_1 = a_{-1}^* = j \text{ and}$$

$$a_2 = a_{-2}^* = 2j$$

Using the Foureir Series synthesis equation,
We get,

$$x(t) = a_1 e^{j\left(2\frac{\pi}{T}\right)t} + a_{-1} e^{-j\left(2\frac{\pi}{T}\right)t} + a_2 e^{j\left(2\frac{\pi}{T}\right)t} + a_{-2} e^{j2\left(2\frac{\pi}{T}\right)t}$$

$$= je^{j\left(2\frac{\pi}{T}\right)t} - je^{-j\left(2\frac{\pi}{4}\right)t} + 2je^{j2\left(2\frac{\pi}{4}\right)t} - 2je^{-j2\left(2\frac{\pi}{4}\right)t}$$

$$= -2\sin\left(\frac{\pi}{2}t\right) - 4\sin(\pi t)$$

(*d*) The Fourier Series coefficients *ak* may be written as the sum of the two sorts of Fourier Series coefficients b_k and c_k,

Where $b_k = 1$ for all k
and

$$c_k = \begin{cases} 1 & k\text{ - Odd} \\ 0 & k\text{ - even} \end{cases}$$

The Foureir Series coefficients b_k corresponds to the signal

$$y(t) = \sum_{K=-\infty}^{\infty} \delta(t - 4k)$$

and the Foureir Series coefficients c_k corresponds to the signal

$$\tau(t) = \sum_{K=-\infty}^{\infty} e^{j\left(\frac{\pi}{2}\right)t} \delta(t - 2k)$$

Therefore,

$$x(t) = y(t) + p(t) = \sum_{K=-\infty}^{\infty} \delta(t - 4k) + \sum_{K=-\infty}^{\infty} e^{j\left(-\frac{k}{2}\right)} \delta(t - 2k)$$

Problem 3.3: Let

$$x(t) = \begin{cases} t, & 0 \le t \le 1 \\ 2 - t, & 1 \le t \le 2 \end{cases}$$

be a periodic signal with fundamental period $T = 2$ and Fourier coefficients a_k.

(*a*) Determine the value of a_0.

(*b*) Determine the Fourier series representation of $dx(t)/dt$.

(*c*) Use the result of part (*b*) and the differentiation property of the continuous-time Fourier series to help determine the Fourier series coefficients of $\dot{x}(t)$.

Solution:

(*a*) We know that $a_0 = \frac{1}{2}\int_0^1 t\,dt + \frac{1}{2}\int_1^2 (2 - t)\,dt = \frac{1}{2}$

(*b*) The signal

$$g(t) = \frac{dx(t)}{dt} \quad \text{as shown in figure below}$$

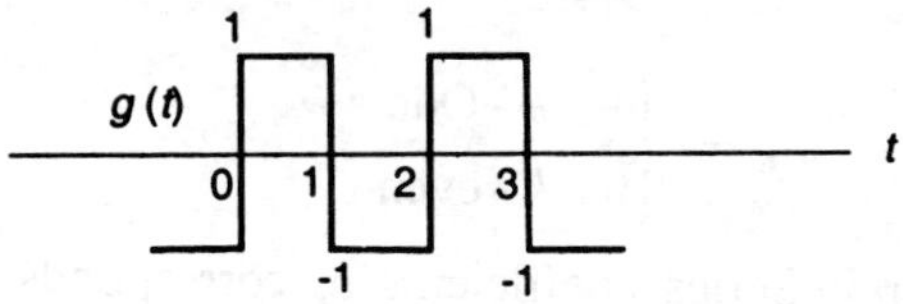

Fig. P. 3.3 (a)

The Fourier series coefficients b_k of $g(t)$ may be found as follows

$$b_0 = \frac{1}{2}\int_0^1 dt - \frac{1}{2}\int_1^2 dt = 0$$

and

$$b_k = \frac{1}{2}\int_0^1 e^{-j\pi kt} + \frac{1}{2}\int_1^2 e^{-jk\pi t} \cdot dt = \frac{1}{j\pi k}\left[1 - e^{-j\pi k}\right]$$

(*c*) We know that

$$g(t) = \frac{dx(t)}{dt} \overset{Fs}{\longleftrightarrow} b_k = jk\pi a_k$$

Therefore

$$a_k = \frac{1}{jk\pi} b_k = \frac{-1}{\pi^2 k^2}\left\{1 - e^{-j\pi k}\right\}$$

Problem 3.4: Consider the following three continuous-time signals with a fundamental period of $T = \frac{1}{2}$:

$$x(t) = \cos(4\pi t).$$
$$y(t) = \sin(4\pi t).$$
$$z(t) = x(t)\, y(t).$$

(*a*) Determine the Fourier series coefficients of $x(t)$

(*b*) Determine the Fourier series coefficients of $y(t)$

(*c*) Use the results of parts (*a*) and (*b*), alongwith the multiplication property of the continuous-time Fourier series, to determine the Fourier series coefficients of $z(t) = x(t)\, y(t)$.

(*d*) Determine the Fourier series coefficients of $z(t)$ through direct expansion of $z(t)$ in trigonometric form, and compare your result with that of part (*c*).

Solution:

(*a*) The non zero Fourier series coefficients of $x(t)$ are

$$a_1 = a_{-1} = \frac{1}{2}$$

(*b*) The non zero Fourier series coefficients of $x(t)$ are

$$b_1 = b_{-1} = \frac{1}{2}j$$

(*c*) Using the multplication theorem property,

We know that

$$\tau(t) = x(t)\,y(t) \xleftrightarrow{Fs} c_k = \sum_{l=-\infty}^{\infty} a_1 b_{k-l}$$

Therefore

$$c_k = a_k \cdot b_k = \frac{1}{4j}\delta[k-2] - \frac{1}{4j}\delta[k+2]$$

This implies that the non zero Fourier transform coefficients of $\tau(t)$ are $\frac{1}{4j}$

(*d*) We know that

$$\tau(t) = \sin(4t)\cos(4t) = \frac{1}{2}\sin(8t)$$

Therefore,

The non zero Fourier series coefficients of $\tau(t)$ are

$$c_2 = c_{-2} = \left(\frac{1}{4j}\right)$$

Problem 3.5: Let $x(t)$ be a periodic signal whose Fourier series coefficients are

$$a_k = \begin{cases} 2, & k = 0 \\ j\left(\frac{1}{2}\right)^{|k|}, & \text{otherwise} \end{cases}$$

Use Fourier series properties to answer the following questions:
(*a*) Is $z(t)$ real?
(*b*) Is $x(t)$ even?
(*c*) Is $dx(t)/dt$ even?

Solution:

(*a*) If $x(t)$ is real, then

$$x(t) = x^*(t)$$

This implies that for

$$x(t) = a_k = a_{-k}$$

Since this is not true in this case $x(t)$ is not real.

(*b*) If $x(t)$ is even, then

$$x(t) = x(-t) \text{ and } a_k = a_{-k}$$

Since this is not true for this case

$x(t)$ is even

(*c*) We also know that

$$g(t) = \frac{dx(t)}{dt} \xleftrightarrow{Fs} b_k = jk\frac{2\pi}{T_0}a_k$$

Therefore,

$$b_k = \begin{cases} 0 & k = 0 \\ -k\left(\frac{1}{2}\right)^{|k|}\left(2\frac{\pi}{T_0}\right) & \text{otherwise} \end{cases}$$

Since b_k is not even, $g(t)$ is not even.

Problem 3.6: A discrete-time periodic signal $x[n]$ is real valued and has a fundamental period $N = 5$. The non-zero Fourier series coefficients for $x[n]$ are

$$a_0 = 2,\ a_2 = a_{-2}^* = 2e^{j\pi/6},\ a_4 = a_{-4}^* = e^{j\frac{\pi}{3}}.$$

Express $x[n]$ in the form

$$x[n] = A_0 + \sum_{k=1}^{\infty} A_k \sin(\omega_k n + \phi_k)$$

Solution:

Using the Fourier series Synthesis

$$x[n] = a_0 + a_2 e^{j2\left(2\frac{\pi}{N}\right)n} + a_{-2} e^{-j2\left(2\frac{\pi}{N}\right)n} + a_4 e^{j4\left(2\frac{\pi}{N}\right)n} + a_{-4} e^{-j4\left(2\frac{\pi}{N}\right)n}$$

$$= 2 + 2e^{j\frac{\pi}{6}} e^{j\left(4\frac{\pi}{5}\right)n} + 2e^{-j\frac{\pi}{6}} e^{-j\left(4\frac{\pi}{5}\right)n} + e^{j\frac{\pi}{3}} e^{j\left(8\frac{\pi}{5}\right)n} + e^{-j\frac{\pi}{3}} e^{-j\left(8\frac{\pi}{5}\right)n}$$

$$= 2 + 4\cos\left[\left(4\pi\frac{n}{5}\right) + \frac{\pi}{6}\right] + 2\cos\left[\left(8\pi\frac{n}{5}\right) + \frac{\pi}{3}\right]$$

$$= 2 + 4\sin\left[\left(4\frac{\pi n}{5}\right) + 2\frac{\pi}{3}\right] + 2\sin\left[\left(8\frac{\pi n}{5}\right) + 5\frac{\pi}{6}\right]$$

Problem 3.7: Determine the Fourier series coeffcients for each of the following discrete-time periodic signals. Plot the magnitude and phase of each set of coefficients a_k.

(*a*) Each $x[n]$ depicted in Figure P. 3.7 (*a*)–(*c*)

(*b*) $x[n] = \sin(2\pi n/3)\cos(\pi n/2)$

(*c*) $x[n]$ periodic with period 4 and

$$x[n] = 1 - \sin\frac{\pi n}{4} \text{ for } 0 \le n \le 3$$

(*d*) $x[n]$ periodic with period 12 and

$$x[n] = 1 - \sin\frac{\pi n}{4} \text{ for } 0 \le n \le 11$$

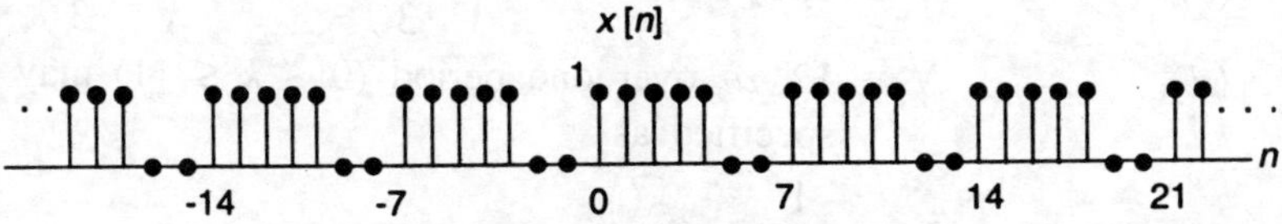

Fig. P. 3.7 (a)

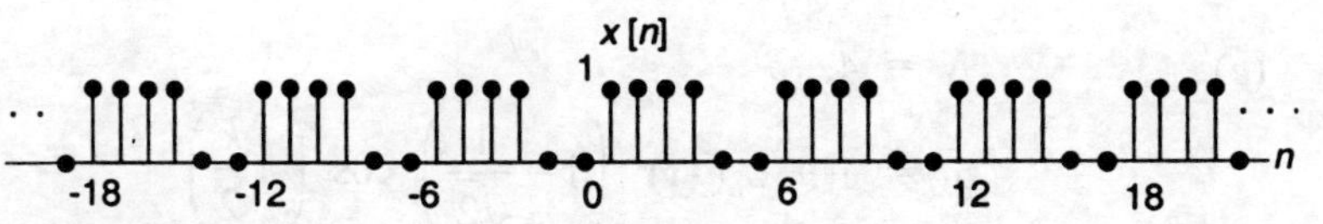

Fig. P. 3.7 (b)

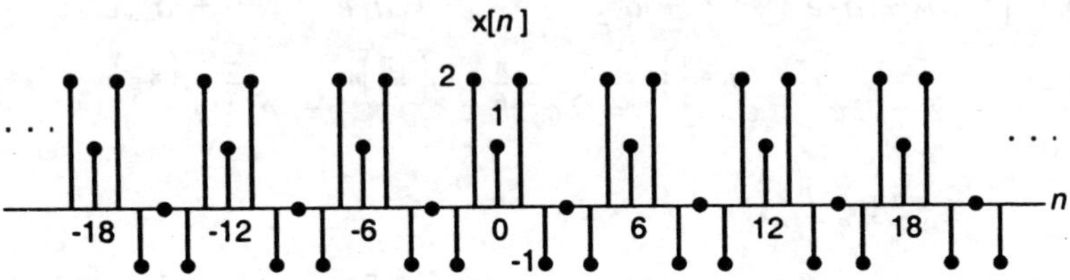

Fig. P. 3.7 (c)

Solution:

(*a*) $$N = 7,\ a_k = \frac{1}{7}\frac{e^{-j4\pi\frac{k}{7}}\sin\left(5\pi\frac{k}{7}\right)}{\sin\left(\pi\frac{k}{7}\right)}$$

(*b*) $N = 6$, a_k over one period ($0 \le k \le 5$) may be satisfied as

$$a_0 = \frac{4}{6}$$

$$a_k = \frac{1}{6}e^{-j\pi\frac{k}{2}}\frac{\sin\left(2\pi\frac{k}{3}\right)}{\sin\left(\pi\frac{k}{6}\right)} \quad 1 \le k \le 5$$

(*c*) $$N = 6,\ a_k = 1 + 4\cos\left(k\frac{\pi}{3}\right) - 2\cos\left(2\pi\frac{k}{3}\right)$$

(*d*) $N = 12$, a_k over one period ($0 \le k \le 11$) may be specified as

$$a_1 = \frac{1}{4j} = a_{11}^*$$

$$a_5 = -\frac{1}{4j} = a_7^*,\ a_k = 0 \text{ otherwise}$$

(*e*) $N = 4$,

$$a_k = 1 + 2(-1)^k\left(1 - \frac{1}{\sqrt{2}}\right)\cos\left(\pi\frac{k}{2}\right)$$

(*f*) $N = 12$,

$$a_k = 1 + \left(1 - \frac{1}{\sqrt{2}}\right) 2\cos\left(\pi\frac{k}{6}\right) + 2\left(1 - \frac{1}{\sqrt{2}}\right)$$

$$\cos\left(\frac{\pi k}{2}\right) + 2\left(1 + \frac{1}{\sqrt{2}}\right)\cos\left(\frac{5\pi k}{6}\right) + 2(-1)^k + 2\cos\left(2\pi\frac{k}{3}\right)$$

Problem 3.8: In each of the following, we specify the Fourier series coefficients of a signal that is periodic with period 8. Determine the signal $x[n]$ in each case.

(*a*) $a_k = \cos\left(\frac{k\pi}{4}\right) + \sin\left(\frac{3k\pi}{4}\right)$ (*b*) $a_k = \begin{cases} \sin\left(\frac{k\pi}{3}\right), & 0 \le k \le 6 \\ 0, & k = 7 \end{cases}$

(*c*) a_k as in Figure P. 3.7 (*a*)

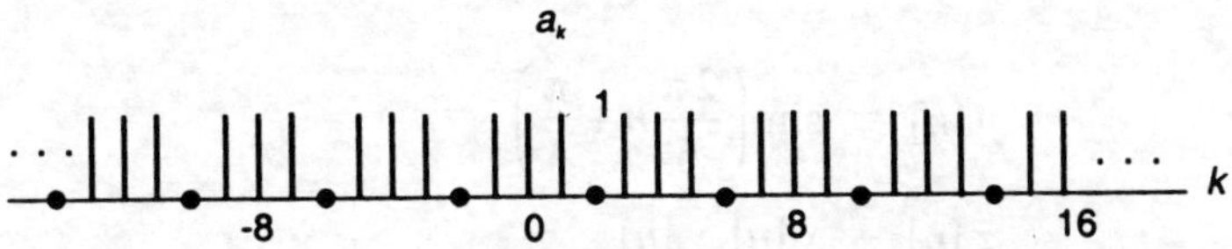

Fig. P. 3.8 (a)

(*d*) a_k as in Figure P. 3.8 (*b*)

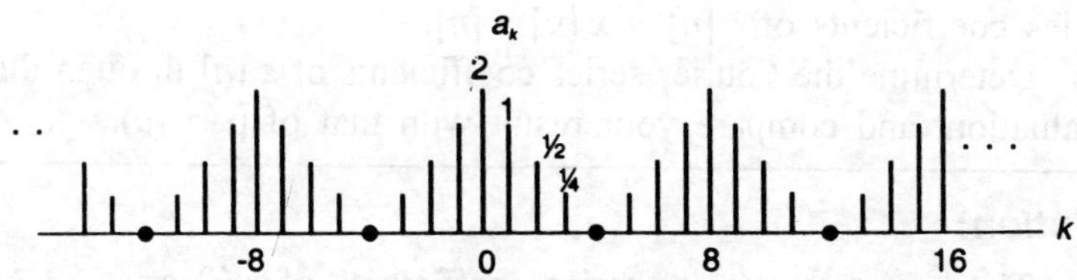

Fig. P. 3.8 (b)

Solution:

(*a*) $N = 8$,

over one period $(0 \le n \le 7)$

$$x[n] = \frac{1}{2j}\left[\frac{-e^{-j\frac{3\alpha n}{4}}\sin\left\{\frac{7}{2}\left(\pi\frac{n}{4} + \frac{\pi}{3}\right)\right\}}{\sin\left\{\frac{1}{2}\left(\pi\frac{n}{4} + \frac{\pi}{2}\right)\right\}} + e^{j\frac{3\pi n}{4}}\frac{\sin\left\{\frac{7}{2}\left[\frac{\pi n}{4} - \frac{\pi}{3}\right]\right\}}{\sin\left\{\frac{1}{2}\left(\frac{\pi n}{4} - \frac{\pi}{3}\right)\right\}}\right]$$

(*c*) $N = 8$, over one period $(0 \le n \le 7)$

$$x[n] = 1 + (-1)^n + 2\cos\left(\pi\frac{n}{4}\right) + 2\cos\left(\frac{3\pi n}{4}\right)$$

(*d*) $N = 8$, over one period $(0 \le n \le 7)$

$$x[n] = 2 + 2\cos\left(\frac{\pi n}{4}\right) + \cos\left(\frac{\pi n}{2}\right) + \frac{1}{2}\cos\left(3\frac{\pi n}{4}\right)$$

Problem 3.9: Consider the following three discrete-time signals with a fundamental period of 6:

$$x[n] = 1 + \cos\left(\frac{2\pi}{6}n\right)$$

$$y[n] = \sin\left(\frac{2\pi}{6}n + \frac{\pi}{4}\right)$$

$$z[n] = x[n]\,y[n]$$

(*a*) Determine the Fourier series coefficients of $x[n]$.
(*b*) Determine the Fourier series coefficients of $y[n]$.
(*c*) Use the results of parts (*a*) and (*b*), along with the multiplication property of the discrete-time Fourier series, to determine the Fourier series coefficients of $z[n] = x[n]\,y[n]$.
(*d*) Determine the Fourier series coefficients of $z[n]$ through direct evaluation, and compare your result with that of part (*c*).

Solution:

(*a*) The non zero Fourier series coefficients of $x(t)$ are

$$a_0 = 1,$$

$$a_1 = a_{-1} = \frac{1}{2}$$

(*b*) The non zero Fourier series coefficients of $x(t)$ are

$$b_1 = b_{-1}^{*} = \frac{e^{-J\frac{\pi}{4}}}{2}$$

(*c*) Using the multiplication property,

We know that

$$z[n] = x[n]\, y[n] \overset{Fs}{\longleftrightarrow}$$

$$c_k = \sum_{l=-2}^{2} a_L b_{k-L}$$

This implies that the non zero, Fourier series coefficients of $z[n]$ are

$$c_0 = \cos\frac{\left(\frac{\pi}{4}\right)}{2} \quad c_1 = c_{-1}^{*} = \frac{e^{-j\frac{\pi}{4}}}{2}$$

$$c_2 = c_{-2}^{*} = \frac{e^{-j\frac{\pi}{4}}}{4}$$

(*d*) We have

$$z[n] = \sin\left(2\frac{\pi}{6}n + \frac{\pi}{4}\right) + \sin\left(2\frac{\pi}{6}n + \frac{\pi}{4}\right)\cos\left(2\frac{\pi}{6}n\right)$$

$$= \sin\left(2\frac{\pi}{6}n + \frac{\pi}{4}\right) + \frac{1}{2}\left[\sin\left(4\frac{\pi}{6}n + \frac{\pi}{4}\right) + \sin\left(\frac{\pi}{4}\right)\right]$$

This implies that,

The non zero Fourier series coefficients of $z[n]$ are

$$c_0 = c_0 p\frac{\left(\frac{\pi}{4}\right)}{2}$$

$$c_1 = c_{-1}^{*} = \frac{e^{-j\frac{\pi}{4}}}{2} = c_{-2}^{*} = \frac{e^{-j\frac{\pi}{4}}}{4}$$

Problem 3.10: Let

$$x[n] = \begin{cases} 1, & 0 \le n \le 7 \\ 0, & 8 \le n \le 9 \end{cases}$$

be a periodic signal with fundamental period $N = 10$ and Fourier series coefficients a_k. Also, let

$$g[n] = x[n] - x[n-1]$$

(*a*) Show that $g[n]$ has a fundamental period of 10.

(*b*) Determine the Fourier series coefficients of $g[n]$.

(*c*) Using the Fourier series coefficients of $g[n]$ and the First-Difference property determine a_k for $k \neq 0$.

Solution:

(a) $g[n]$ is as shown in figure below.

Clearly $g[n]$ has a fundamental period of 10.

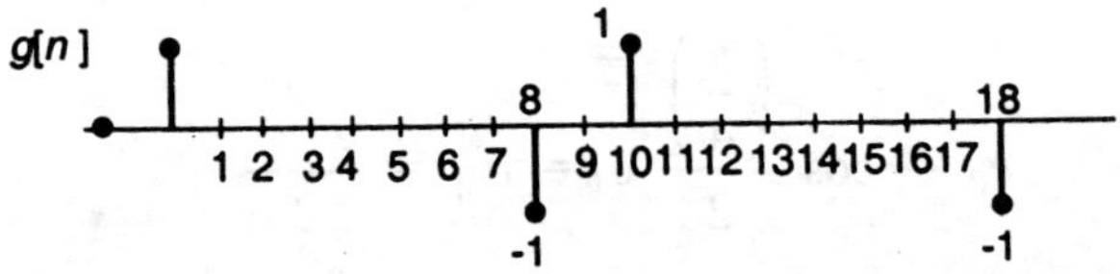

Fig. P. 3.10 (a)

(b) The Fourier series coefficients of $g[n]$ are

$$b_k = \left(\frac{1}{10}\right)\left[1 - e^{-j\left(\frac{2\pi}{10}\right)8k}\right]$$

(c) Since $\quad g[n] = x[n] - x[n-1]$,

The Fourier series coefficients a_k and b_k must be related as

$$b_k = a_k - e^{-j\left(\frac{2\pi}{10}\right)k} a_k$$

Therefore,

$$a_k = \left(\frac{1}{10}\right)\frac{\left[1 - e^{-j\left(\frac{2\pi}{10}\right)8}{}_k\right]}{1 - e^{-j\left(\frac{2\pi}{10}\right)}{}_k}$$

Problem 3.11: Consider the signal $x[n]$ depicted in Fig. P. 3.11. This signal is periodic with period $N = 4$. The signal can be expressed in terms of a discrete-time Fourier series as

$$x[n] = \sum_{k=0}^{3} a_k e^{jk(2\pi/4)n}$$

To determine the Fourier series coefficients as a set of four linear equations (for $n = 0, 1, 2, 3$) in four unknowns (a_0, a_1, a_2 and a_3).

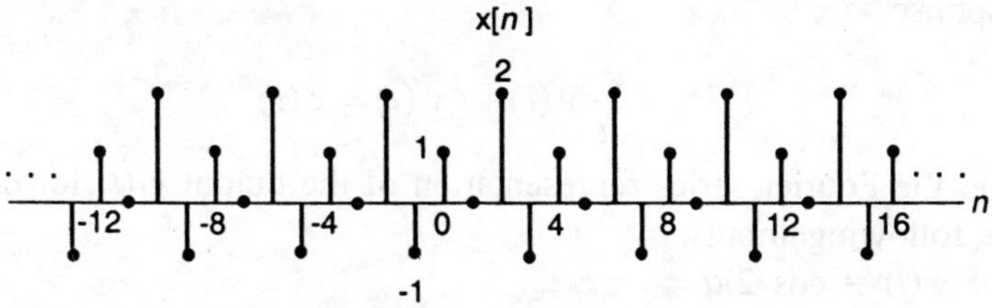

Fig. P. 3.11

(*a*) Write out these four equations explicitly, and solve them directly using any standard technique for solving four equations in four unknowns.

(*b*) check your answer by claculating the a_k directly, using the discrete-time Fourier series analysis equation

$$a_k = \frac{1}{4}\sum_{n=0}^{3} x[n]e^{-jk(2\pi/4)n}$$

Solution:

(*a*) The four equation are

$$a_0 + a_1 + a_2 + a_3 = 1$$
$$a_0 + ja_1 - a_2 - ja_3 = 0$$
$$a_0 - a_1 + a_2 - a_3 = 2$$
$$a_0 - ja_1 - a_2 + ja_3 = -1$$

Solving, these four equations,

The we get

$$a_0 = 2, \qquad a_1 = -\frac{1+j}{4}$$

$$a_2 = -1 \quad \text{and} \quad a_3 = -\frac{1-j}{4}$$

(*b*) By direct calculations

$$a_x = \frac{1}{4}\left[1 + 2\,e^{-Jk\pi} - e^{-Jk3\frac{\pi}{2}}\right]$$

This is the same as the answer we obtained in part (a),
For $0 \le k \le 3$

Problem 3.12: Consider a continuous-time LTI system with impulse response

$$\frac{d}{dt} y(t) + 4y(t) = x(t)$$

Find the Fourier series representation of the output $y(t)$ for each of the following inputs:

(*a*) $x(t) = \cos 2\pi t$

(*b*) $x(t) = \sin 4\pi t + \cos(6\pi t + \pi/4)$

Solution:

We will first evaluate the frequency response of the system:
Consider an input $x(t)$ of the form $e^{j\omega t}$
We know that,
The response of this input, will be

$$y(t) = H(j\omega)\, e^{j\omega t},$$

Therefore, substituting these in the given differential equation we get

$$H(j\omega)\, j\omega\, e^{j\omega t} + 4e^{5\omega t} = e^{j\omega t}$$

Therefore, $$h(j\omega) = \frac{1}{j\omega + 4}$$

Also

We know that, $$y(t) = \sum_{K=-\infty}^{\infty} a_k\, H(jk\omega_o)\, e^{jk\omega_o t}$$

When the input is $x(t)$,
$x(t)$ has the Fourier series coefficients a_k and fundamental frequency ω_o.

Therefore, the Fourier series coefficients of $y(t)$ and $H(jk\omega_o)$

(*a*) Here,

$\omega_o = 2\pi$ and the non zero Fourier series coefficients of $x(t)$ are

$$a_1 = a_{-1} = \frac{1}{2}$$

Therefore
The non zero Fourier series coefficients of $y(t)$ are

$$b_1 = a_1\, H(j2\pi) = \frac{1}{2(4 + j2\pi)}$$

$$b_{-1} = a_{-1}\, H(-j2\pi) = \frac{1}{2\,(4 - j2\pi)}$$

(*b*) Here, $\omega_o = 2\pi$

and the non zero Fourier series coefficieints of $x\,(t)$ are

$$a_2 = a_{-2}^{*} = \frac{1}{2j}$$

and

$$a_3 = a_{-3} = \frac{e^{j\frac{\pi}{4}}}{2}$$

Therefore, The non zero Fourier series coefficients of $y\,(t)$ are

$$b_2 = a_2\, H\,(j4\pi) = \frac{1}{2j(4 + j4\pi)}$$

$$b_{-2} = a_{-2}\, H\,(-j\pi) = -\frac{1}{2j(4 - j4\pi)}$$

$$b_3 = a_3\, H\,(j6\pi) = \frac{e^{j\frac{\pi}{4}}}{2(4 + j6\pi)}$$

$$b_{-3} = a_{-3}\, H\,(-j6\pi) = -\frac{e^{-j\frac{\pi}{4}}}{2(4 - j6\pi)}$$

Problem 3.13: Consider a continuous-time LTI system with impulse response

$$h\,(t) = e^{-4|t|}$$

Find the Fourier series representation of the output $y\,(t)$ for each of the following inputs:

(*a*) $x\,(t) = \sum_{n=-\infty}^{+\infty} \delta(t - n)$

(*b*) $x\,(t) = \sum_{n=-\infty}^{+\infty} \delta(-1)^n \delta(t - n)$

(*c*) $x\,(t)$ is the periodic wave depicted in Figure P. 3.13.

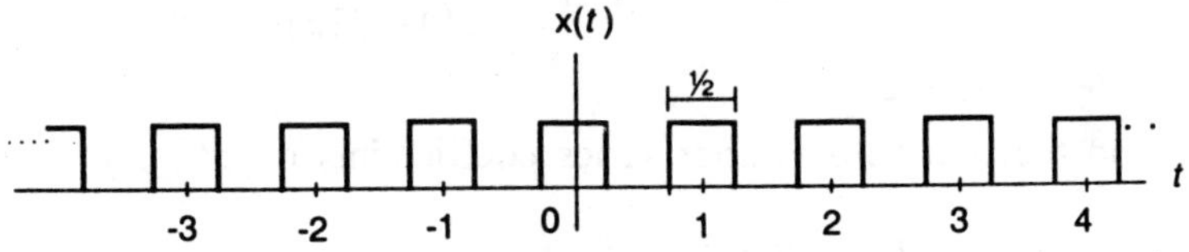

Fig. P. 3.13

Solution:

The frequency response of the system is given by

$$H(j\omega) = \int_{-\infty}^{\infty} e^{-4|t|}\, e^{-j\omega t}\, dt = \frac{1}{4+j\omega} + \frac{1}{4-j\omega}$$

(*a*) Here,

$$T = 1 \quad \text{and} \quad \omega_o = 2\pi$$

and $a_k = 1$ for all k.

The Fourier series coefficients of the output are

$$b_k = a_k\, H(jk\omega_o) = \frac{1}{4+j2\pi k} + \frac{1}{4-j2\pi k}$$

(*b*) Here, $T = 2$ and $\omega_o = \pi$

and

$$a_k = \begin{cases} 0 & k\text{-even} \\ 1 & k\text{-odd} \end{cases}$$

Therefore, the *Fs* coefficients of the output are

$$b_k = a_k\, H(jk\omega_o) = \begin{cases} 0 & k\text{-even} \\ \dfrac{1}{4+jk\pi} + \dfrac{1}{4-jk\pi} & k\text{-odd} \end{cases}$$

(*c*) $T = 1, \quad \omega_o = 2\pi$

and

$$a_k = \begin{cases} \dfrac{1}{2} & k = 0 \\ 0 & k\text{-even } k \neq 0 \\ \dfrac{\sin\left(\pi \dfrac{k}{2}\right)}{\pi k} & k\text{-odd} \end{cases}$$

Therefore, the Fourier series coefficients of the output are

$$b_k = a_k H(jk\omega_o)$$

$$= \begin{cases} \dfrac{1}{4} & k = 0 \\ 0 & k \text{ even} \quad k \neq 0 \\ \dfrac{\sin\left(\pi \dfrac{k}{2}\right)}{\pi k}\left[\dfrac{1}{4 + j2\pi k} + \dfrac{1}{4 - j2\pi k}\right] & k\text{ - odd} \end{cases}$$

Problem 3.14: Consider a continuous-time LTI system S whose frequency response is

$$H(j\omega) = \begin{cases} 1, & |\omega| \geq 250 \\ 0, & \text{otherwise} \end{cases}$$

When the input to this system is a signal $x(t)$ with fundamental period $T = \pi/7$ and Fourier series coefficients a_k, it is found that the output $y(t)$ is identical to $x(t)$.

For what values of k is it guaranteed that $a_k = 0$?

Solution:

We know that, the Fourier series coefficients of $y(t)$ are

$$b_k = H(jk\omega_n)\, c_k$$

Where ω_n is the fundamental frequency of $x(t)$ are the Fourier series coefficients of $x(t)$.

If $y(t)$ is identical to $x(t)$, then $b_k = a_k$ for all k.

Note that

$$H(j\omega) = 0 \text{ for } |\omega| \geq 250$$

We know that

$$H(jk\omega_o) = 0 \text{ for } |k| \geq 18$$

Because $\omega_o = 14$,

Therefore a_k must be zero for $|k| \geq 18$ **Ans.**

Problem 3.15: Consider a causal discrete-time LTI system whose input $x[n]$ and output $y[n]$ are related by the following difference equation:

$$y[n] - \frac{1}{4}y[n-1] = x[n]$$

Find the Fourier series representation of the output $y[n]$ for each of the following inputs:

(*a*) $x[n] = \sin\left(\frac{3\pi}{4}n\right)$

(*b*) $x[n] = \cos\left(\frac{\pi}{4}n\right) + 2\cos\left(\frac{\pi}{2}n\right)$

Solution:

We will first evaluate the frequency response of the system. Consider an input $x[n]$ of the form $e^{j\omega n}$.

We know that,

The response of this input will be

$$y[n] = H(e^{j\omega})\, e^{j\omega n}$$

Therefore,

Substituting in the given differential equation, we get

$$H(e^{j\omega})\, e^{j\omega n} - \frac{1}{4}e^{-j\omega}\, e^{j\omega n}\, H(e^{j\omega}) = e^{j\omega n}$$

Therefore,

$$H(j\omega) = \frac{1}{1-\frac{1}{4}e^{-j\omega}}$$

We know that

$$y[n] = \sum_{K=\langle n\rangle} a_k H\left(e^{j2\pi\frac{k}{N}}\right) e^{jk\left(2\frac{\pi}{N}\right)n}$$

When the input is $x[n]$,

$x[n]$ has the Fourier series coefficients a_k and fundamental frequency $2\frac{\pi}{N}$.

Therefore, the Fourier series coefficients of $y[n]$ are $a_k\, H\left(e^{j2\pi\frac{k}{N}}\right)$

(*a*) Here, $N = 4$,

and the nonzero, Fourier series coefficients of $x[n]$ are

$$a_3 = a_{-3}^* = \frac{1}{2j}$$

Therefore, the non zero Fourier series coefficients of $y[n]$ are

$$b_3 = a_{-1}\, H\left(e^{j3\frac{\pi}{4}}\right) = \frac{1}{2j\left(1-\left(\frac{1}{4}\right)e^{-j3\frac{\pi}{4}}\right)}$$

$$b_{-3} = a_{-1}\, H\left(e^{-j\frac{\pi}{4}}\right) = \frac{-1}{2j\left[1-\left(\frac{1}{4}\right)e^{j3\frac{\pi}{4}}\right]}$$

(*b*) Here, $N = 8$,

and the non zero Fourier series coefficients of $x[n]$ are

$$a_1 = a_{-1} = \frac{1}{2} \text{ and}$$

$$a_2 = a_{-2} = 1,$$

Therefore, the non zero Fourier series coefficients of $y(t)$ are

$$b_1 = a_1\, H\left(e^{j\frac{\pi}{4}}\right) = \frac{1}{2\left(1-\left(\frac{1}{4}\right)e^{j\frac{\pi}{4}}\right)}$$

$$b_{-1} = a_{-1}\, H\left(e^{-j\frac{\pi}{4}}\right) = \frac{1}{2\left(1-\left(\frac{1}{4}\right)e^{j\frac{\pi}{4}}\right)}$$

$$b_2 = a_2\, H\left(e^{j\frac{\pi}{2}}\right) = \frac{1}{\left[1-\left(\frac{1}{4}\right)e^{-j\frac{\pi}{2}}\right]}$$

$$b_{-2} = a_{-2} H\left(e^{-j\frac{\pi}{2}}\right) = \frac{1}{\left[1-\left(\frac{1}{4}\right)e^{j\frac{\pi}{2}}\right]}$$

Problem 3.16: Consider a discrete-time LTI system with impulse response

$$h[n] = \left(\frac{1}{2}\right)^{|n|}$$

Find the Fourier series representation of the output $y[n]$ for each of the following inputs:

(*a*) $x[n] = \sum_{k=-\infty}^{\infty} \delta[n-4k]$

(*b*) $x[n]$ is periodic with period 6 and

$$x[n] = \begin{cases} 1, & n = 0, \pm 1 \\ 0, & n = \pm 2, \pm 3 \end{cases}$$

Solution:

The frequency response of the system may be easily shown to be

$$H(e^{j\omega}) = \frac{1}{1-\frac{1}{2}e^{-j\omega}} - \frac{1}{1-2e^{-j\omega}}$$

(*a*) The Fourier series coefficients of $x[n]$ are $a_k = \frac{1}{4}$ for all k.

Also, $N = 4$, therefore the Fourier series coefficients of $y[n]$ are

$$b_k = a_k H\left(e^{j2k\frac{\pi}{N}}\right) = \frac{1}{4}\left[\frac{1}{1-\frac{1}{2}e^{-jk\frac{\pi}{2}}} - \frac{1}{1-2e^{-jk\frac{\pi}{2}}}\right]$$

(*b*) In this case,

The Fourier series coefficients of $x[n]$ are

$$a_k = \frac{1}{6}\left[1+2\cos\left(k\frac{\pi}{3}\right)\right] \text{ for all } k.$$

Also, $N = 6,$

Therefore, the Fourier series coefficients of $y[n]$ are

$$b_k = H\left(e^{j2k\frac{\pi}{N}}\right) = \frac{1}{6}\left[1+2\cos\left(k\frac{\pi}{3}\right)\right]\left[\frac{1}{1-\frac{1}{2}e^{-jk\frac{\pi}{3}}} - \frac{1}{1-2e^{-jk\frac{\pi}{3}}}\right]$$

Problem 3.17: Consider a discrete-time LTI system with impulse response

$$h[n] = \begin{cases} 1, & 0 \le n \le 2 \\ -1, & -2 \le n \le -1 \\ 0, & \text{otherwise} \end{cases}$$

Given that the input to this system is

$$x[n] = \sum_{k=-\infty}^{+\infty} \delta[n-4k],$$

Determine the Fourier series coefficients of the output $y[n]$

Solution:

The frequency response of the system may be evaluated as

$$H(e^{j\omega}) = -e^{2j\omega} - e^{j\omega} + 1 + e^{-j\omega} + e^{-2j\omega}$$

For,

$$x[n],\ N = 4, \text{ and } \omega_o = \frac{\pi}{2}$$

The Fourier series coefficients of the input $x[n]$ are

$$a_k = \frac{1}{4} \text{ for all } n.$$

Therefore, the Fourier series coefficients of the output are

$$b_k = a_k\, H\left(e^{jk\omega_o}\right) = -\left[1 - e^{jk\frac{\pi}{2}} + e^{-jk\frac{\pi}{2}}\right]$$

Problem 3.18: Consider a discrete-time LTI system S whose freuqncy response is

$$H(e^{j\omega}) = \begin{cases} 1, & |\omega| \le \frac{\pi}{8} \\ 0, & \frac{\pi}{8} < |\omega| < \pi \end{cases}$$

Show that if the input $x[n]$ to this system has a period $N = 3$, the output $y[n]$ has only one nonzero Fourier series coefficient per period.

Solution:

Let Fourier series coefficients of the input be a_k
The Fourier series coefficients of the output are of the form

$$b_k = a_k H(e^{jk\omega_o})$$

Where $\omega_o = 2\frac{\pi}{3}$

Note that, in the range $0 \le -k \le 2$,

$$H(e^{jk\omega_o}) = 0 \text{ for } k = 1, 2$$

Therefore,
Only b_0 has a nonzero value among b_k in the range $0 \le k \le 2$

Problem 3.19: Let $x(t)$ be a periodic signal with fundamental period T and Fourier series coefficients a_k. Derive the Fourier series coefficients of each of the following signals in terms of a_k:

(a) $x(t - t_0) + x(t + t_0)$

(b) $\varepsilon v\ \{x(t)\}$

(c) $(\Re e\ \{x(t)\}$

(d) $\frac{d^2x(t)}{dt^2}$

(e) $x(3t - 1)$ [for this part, first determine the period of $x(3t - 1)$]

Solution:

Let the Fourier series coefficients of $x(t)$ be a_k.

(a) $x(t - t_o)$ is also periodic, with period T. The Fourier series coefficients b_k of $x(t - t_o)$ are

$$b_k = \frac{1}{T}\int_T x(t - t_0)\, e^{jk\left(2\frac{\pi}{T}\right)t} \cdot dt = \frac{e^{-jk\left(2\frac{\pi}{T}\right)t_0}}{T}\int_T x(t)\, e^{-jk\left(2\frac{\pi}{T}\right)t} \cdot dt$$

Similarly, the Fourier series coefficients of $x(t + t_0)$ are

$$c_k = e^{jk\left(2\frac{\pi}{T}\right)t_0} \cdot a_k$$

Finally, the Fourier series coefficients of
$x(t - t_0) + x(t + t_0)$ are

$$d_k = b_k + c_k = e^{-jk\left(2\frac{\pi}{T}\right)t_0} a_k + e^{jk\left(2\frac{\pi}{T}\right)t_0} a_k = 2\cos\left(k\, 2\pi \frac{t_0}{T}\right) a_k$$

(*b*) Note that

$$\varepsilon\, u(x(t)) = \frac{[x(t) + x(-t)]}{2}$$

The Fourier series coefficients of $x(-t)$ are

$$b_k = \frac{1}{T}\int_T x(-t)\, e^{-jk\left(2\frac{\pi}{T}\right)t} \cdot dt = \frac{1}{T}\int_T x(t)\, e^{jk\left(2\frac{\pi}{T}\right)t} \cdot dt = a_{-k}$$

Therefore, the Fourier series coefficients of $\varepsilon\, v\{x(t)\}$ are

$$c_k = \frac{a_k + b_k}{2} = \frac{a_k + a_{-k}}{2}$$

(*c*) Note that $Re\,\{x(t)\} = \dfrac{[x(t) + x^*(t)]}{2}$

The Fourier series coefficients of $x^*(t)$ are

$$b_k = \frac{1}{T}\int_T x^*(t)\, e^{-jk\left(2\frac{\pi}{T}\right)t} \cdot dt$$

Conjugating both sides we get

$$b_k^* = \frac{1}{T}\int_T x(t)\, e^{jk\left(\frac{2\pi}{T}\right)t} dt = a_{-k}$$

Therefore, the Fourier series coefficients of $Re\,\{x(t)\}$ are

$$c_k = \frac{a_k + b_k}{2} = \frac{a_k + a^*_{-k}}{2}$$

(*d*) The Fourier series synthesis equation gives

$$x(t) = \sum_{K=-\infty}^{\infty} a_k e^{j\left(2\frac{\pi}{T}\right)kt}$$

Differentiating both sides with respect to '*t*', twice,
Then we get

$$\frac{d^2 x(t)}{dt^2} = \sum_{K=-\infty}^{\infty} -k^2 \frac{4\pi^2}{T^2} a_k e^{j\left(2\frac{\pi}{T}\right)kt}$$

By inspection, we know that the Fourier series coefficients of

$$\frac{d^2 x(t)}{dt^2} \text{ are } -k\frac{4\pi^2}{T^2} a_k$$

(*e*) The period of $x(3t)$ is a third of the period of $x(t)$. Therefore, the signal $x(3t - 1)$ is period, with period $\frac{T}{3}$.

The Fourier series coefficients of $x(3t)$ are still a_k.
Using the analysis in part (*a*),
we know that, the Fourier series coefficients of $x(3t - 1)$ is

$$e^{-jk\left(6\frac{\pi}{T}\right)} a_k$$

Problem 3.20: Let $x(t)$ be a real-valued signal with fundamental period T and Fourier series coefficients a_k.

(*a*) Show that $a_k = a^*_{-k}$ and a_0 must be real.

(*b*) Show that if $x(t)$ is even, then its Fourier series coefficients must be real and even.

(*c*) Show that if $x(t)$ is odd, then its Fourier series coefficients are imaginary and odd and $a_0 = 0$.

(*d*) Show that the Fourier coefficients of the even part of $x(t)$ are equal to $Re\{a_k\}$.

(*e*) Show that the Fourier coefficients of the odd part of $x(t)$ are equal to $jIm\{a_k\}$.

Solution:

We know that

The Fourier series coefficients of $x^*(t)$ are a^*_{-k}.

Now, we know that $x(t)$ is real,

Then, $x(t) = x^*(t)$

Therefore, $a_k = a^*_{-k}$

This implies $a_0 = a^*_0$

Therefore

a_0 must be real.

(*b*) We know that

The Fourier series coefficients of $x(-t)$ are a_{-k}. If $x(t)$ is even

Then $x(t) = x(-t)$

This implies that $a_k = a_{-k}$

This implies that the Fourier series coefficients are even, from the previous part,

we know that

If $x(t)$ is real then $a_k = a^*_{-k}$

We know that $a_k = a^*_k$,

Therefore a_k is real for all k. We may conclude that

a_k is real and even.

(*c*) We know that Fourier series coefficients of $x(-t)$ are a_{-k}. If $x(t)$ is odd, then

$$x(t) = -x(-t)$$

This implies that $a_k = -a_{-k}$

This implies that the Fourier series coefficients are odd, from the previous part.

We know that

If $x(t)$ is real then $a_k = a^*_{-k}$

Therefore a_k is imaginary for all k. We may conclude that

a_k is real and even.

Note that $a_0 = -a_0$

We may conclude that $a_0 = 0$

(*d*) Note that

$$\varepsilon v\{x(t)\} = \frac{[x(t) + x(-t)]}{2}$$

From previous part, we know that the Fourier series coefficients of $\varepsilon v\{x(t)\}$ will be

$$\frac{[a_k + a_{-k}]}{2}$$

We may write the Fourier series coefficients of

$\varepsilon v\ \{x(t)\}$ as $\dfrac{[a_k + a_k^*]}{2} = Re\ \{a_k\}$

(*e*) Note that

$$\text{odd}\ \{x(t)\} = \frac{[x(t) - x(-t)]}{2}$$

We may also know that the Fourier series coefficients of odd $\{x(t)\}$ will be $\dfrac{[a_k - a_{-k}]}{2}$

We may write the Fourier series coefficients of odd $\{x(t)\}$ as $\dfrac{[a_k - a_k^*]}{2} = jIm\ \{a_k\}$

> **Problem 3.21:** Suppose we are given the following information about a continuous-time periodic signal with period 3 and Fourier coefficients a_k:
>
> 1. $a_k = a_{k+2}$
> 2. $a_k = a_{-k}$
> 3. $\int_{-0.5}^{0.5} x(t)\,dt = 1$
> 4. $\int_{-0.5}^{1.5} x(t)\,dt = 2$
>
> Determine $x(t)$.

Solution:

Since $a_k = a_{-k}$ we require that $x(t) = x(-t)$

Also note that since $a_k = a_{k+2}$

We require that

$$x(t) = x(t)\,e^{-J\left(4\frac{\pi}{3}\right)t}$$

This is turn implies that $x(t)$ may have non zero value only for $t = 0, \pm 1.5, \pm 3, \pm 4.5, \ldots$

Since $\int_{-0.5}^{0.5} x(t) = 1,$

we may conclude that $x(t) = \delta(t)$ for $-0.5 \le t \le 0.5$

Also, since

$$\int_{-0.5}^{1.5} x(t)\,dt = 2$$

We may conclude that $x(t) = \delta(t)$ for $-0.5 \le t \le 0.5$

Also, since

$$\int_{0.5}^{1.5} x(t)\,dt = 2$$

We may conclude that

$$x(t) = 2\delta\left(t - \frac{3}{2}\right)$$

in the range $0.5 \le t \le \frac{3}{2}$

Therefore, $x(t)$ may be written as

$$x(t) = \sum_{K=-\infty}^{\infty} \delta(t - 3k) + 2\sum_{K=-\infty}^{\infty} \delta\left(t - 3k - \frac{3}{2}\right)$$

Problem 3.22: (*a*) A continuous-time periodic signal $x(t)$ with period T is said to be *odd harmonic* if, in its Fourier series representation

$$x(t) = \sum_{k=-\infty}^{+\infty} a_k e^{jk(2\pi/T)t} \tag{1}$$

$a_k = 0$ for every non-zero even integer k.

(*i*) Show that if $x(t)$ is odd harmonic, then

$$x(t) = -x\left(t + \frac{T}{2}\right) \tag{2}$$

(*ii*) Show that if $x(t)$ satisfies eq. 2, then it is odd harmonic.

(*b*) Suppose that $x(t)$ is an odd-harmonic periodic signal with period 2 such that

$$x(t) = t \text{ for } 0 < t < 1$$

Sketch $x(t)$ and find its Fourier series coefficients.

(*c*) Analogously, to an odd-harmonic signal, we could define an even-harmonic signal as a signal for which $a_k = 0$ for k odd in the representation in eq. 1. Could T be the fundamental period for such a signal? Explain your answer.

(*d*) More generally, show that T is the fundamental period of $x(t)$ in eq. 1 if one of two things happens:

(1) Either a_1 or a_{-1} is nonzero;

or

(2) There are two integers k and l that have no common factors and are such that both a_k and a_l are nonzero.

Solution:

(*a*) (*i*) We know that

$$x(t) = \sum_{\text{odd } k} a_k e^{jk\frac{2\pi}{T}t}$$

Therefore

$$x\left(t+\frac{T}{2}\right) = \sum_{\text{odd } k} a_k e^{jk\frac{2\pi}{T}t} \cdot e^{jk\pi}$$

Since $\quad e^{jk\pi} = -1$ for k odd

$$x\left(t+\frac{T}{2}\right) = -x(t)$$

(*ii*) The Fourier series coefficients of $x(t)$ are

$$a_k = \frac{1}{T}\int_0^{T/2} x(t)\, e^{-jk\omega_o t} \cdot dt + \frac{1}{t}\int_{T/2}^{T} x(t)\, e^{-jk\omega_o t} \cdot dt$$

$$= \frac{1}{T}\int_0^{T/2} \left[x(t) + x\left(t+\frac{T}{2}\right)e^{-jk\pi}\right] e^{-jk\omega_o t} \cdot dt$$

Note that,

The right hand side of the above equation evaluate to zero, for even values of k

If $\quad x(t) = -x\left(t+\frac{T}{2}\right)$

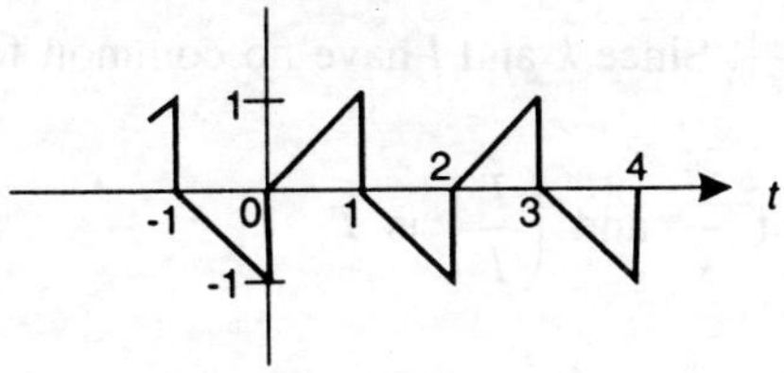

Fig. P. 3.22 (a)

(*b*) Note that,

$$T = 2, \text{ and } \omega_0 = \pi, \text{ therefore}$$

$$a_k = \begin{cases} 0 & k\text{ - even} \\ \dfrac{1}{jk\pi} + \dfrac{2}{k^2\pi^2} & k\text{ - odd} \end{cases}$$

(*c*) No, for the even harmonies signal we may follow the reasoning of part (*a*) (*i*) to show that

$$x(t) = -x\left(t + \frac{T}{2}\right)$$

In this case,

The funadmental period is $\dfrac{T}{2}$.

(*d*) (*i*) If a_1 or a_{-1} is non zero, then

$$x(t) = a_{\pm 1}\, e^{\pm j2\pi\frac{t}{T}} + \ldots$$

and

$$x(t + t_0) = a_{\pm 1}\, e^{\pm j\frac{2\pi}{T}(t+t_0)} + \ldots$$

The smallest value of $|t_0|$ other than $|t_0| = 0$ for which $e^{\pm j\frac{2\pi}{T}t_0}$ = 1 is the fundamental period only then is

$$x(t + t_0) = a_{\pm 1} e^{\pm j2\pi\frac{t}{T}} + \ldots = x(t)$$

Therefore,

t_0 has to be the fundamental period.

(*ii*) The period of $x(t)$ is the least common multiple of the periods of $e^{jk\left(\frac{2\pi}{T}\right)t}$ and $e^{jl\left(\frac{2\pi}{T}\right)t}$. The period of $e^{jk\left(\frac{2\pi}{T}\right)t}$ is $\dfrac{T}{k}$ and the period of $e^{jl\left(\frac{2\pi}{T}\right)t}$ and $\left(\frac{T}{l}\right)$. Since k and l have no common features the least common multiple of $\frac{T}{k}$ and $\left(\frac{T}{l}\right)$ is T.

Problem 3.23: Suppose we are given the following information about a signal $x(t)$:

1. $x(t)$ is a real signal.
2. $x(t)$ is periodic with period $T = 6$ and has Fourier coefficients a_k.
3. $a_k = 0$ for $k = 0$ and $k > 2$.
4. $x(t) = -x(t-3)$.
5. $\frac{1}{6}\int_{-3}^{3} |x(t)|^2 \, dt = \frac{1}{2}$.
6. a_1 is a positive real number.

Show that $x(t) = A\cos(Bt + C)$, and determine the values of the constants A, B and C.

Solution:

The only unknown Fourier series coefficients are a_1, a_{-1} a_2 and a_{-2}.

Since $x(t)$ is real, $a_1 = a_{-1}^*$ and $a_2 = a_{-2}$, since a_1 is real $a_1 = a_{-1}$

Now, $x(t)$ is of the form

$$x(t) = A_1 \cos(\omega_0 t) + A_2 \cos(2\omega_0 t + \theta)$$

Where $\omega_0 = 2\frac{\pi}{6}$, from this we get

$$x(t-3) = A_1 \cos(\omega_0 t - 3\omega_0) + A_2 \cos(2\,\omega_0 t + \theta - 6\omega_0)$$

Now,

$x(t) = -x(t-3)$, then $3\omega_0$ and $6\omega_0$ should both be odd multiple of π clearly, this is impossible,

Therefore $a_2 = a_{-2} = 0$ and

$$x(t) = A_1 \cos(\omega_0 t)$$

Then we get

$$\sum_{K=-\infty}^{\infty} |a_k|^2 = |a_1|^2 + |a_{-1}|^2 = \frac{1}{2}$$

Therefore,

$$|a_1| = \frac{1}{2}, \text{ since } a_1 \text{ is positive, we have}$$

$$a_1 = a_{-1} = \frac{1}{2}$$

$$\text{Therefore } x(t) = \cos\left(\frac{\pi t}{3}\right)$$

Here, $A = 1$, $B = \frac{\pi}{3}$ and $C = 0$.

Problem 3.24: Let $x(t)$ be a real periodic signal with Fourier series representation given in the sine-cosine form of eq.; i.e.,

$$x(t) = a_0 + 2\sum_{k=1}^{\infty}\left[B_k \cos k\omega_0 t - C_k \sin k\omega_0 t\right]$$

(*a*) Find the exponeneial Fourier series representation of the even and odd parts of $x(t)$; that is, find the coefficients α_k and β_k so that

$$Ev\{x(t)\} = \sum_{k=-\infty}^{+\infty} \alpha_k e^{jk\omega_0 t},$$

$$Od\{x(t)\} = \sum_{k=-\infty}^{+\infty} \beta_k e^{jk\omega_0 t},$$

(*b*) What is the relationship between α_k and α_{-k} in part (*a*)? What is the relationship between β_k and β_{-k}?

(*c*) Suppose that the signals $x(t)$ and $z(t)$ shown in Figure P. 3.24 (*a*) and (*b*) have the sine-cosine series representations

$$x(t) = a_0 + 2\sum_{k=1}^{\infty}\left[B_k \cos\left(\frac{2\pi kt}{3}\right) - C_k \sin\left(\frac{2\pi kt}{3}\right)\right]$$

$$z(t) = d_0 + 2\sum_{k=1}^{\infty}\left[E_k \cos\left(\frac{2\pi kt}{3}\right) - F_k \sin\left(\frac{2\pi kt}{3}\right)\right]$$

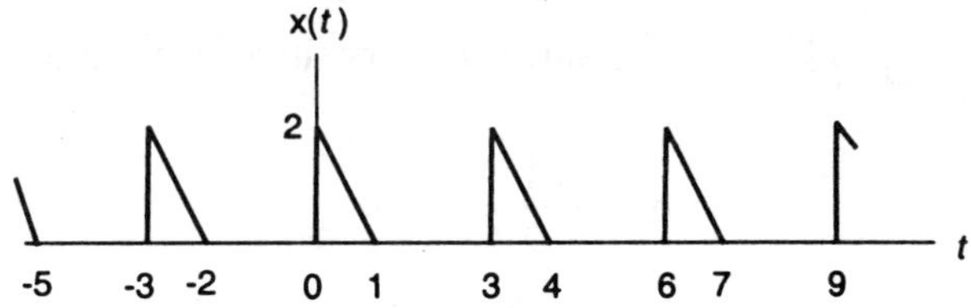

Fig. P. 3.24 (a)

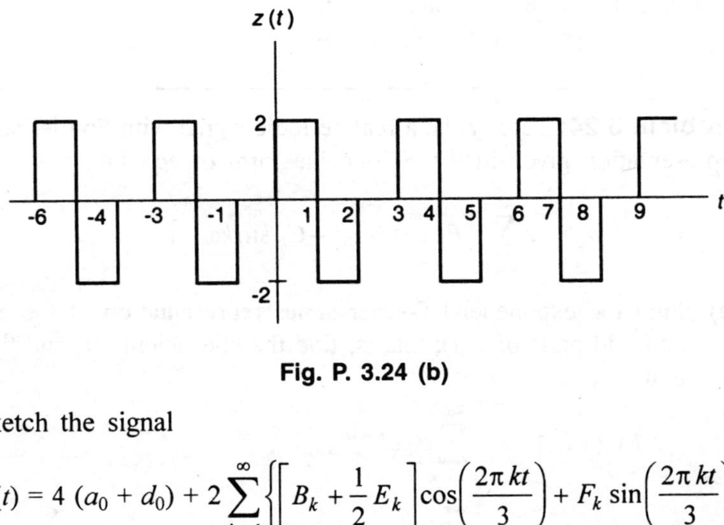

Fig. P. 3.24 (b)

Sketch the signal

$$y(t) = 4(a_0 + d_0) + 2\sum_{k=1}^{\infty}\left\{\left[B_k + \frac{1}{2}E_k\right]\cos\left(\frac{2\pi kt}{3}\right) + F_k \sin\left(\frac{2\pi kt}{3}\right)\right\}$$

Solution:

By inspection,

we may conclude that the Fourier series coefficients of $x(t)$ are

$$\gamma_k = \begin{cases} a_o & k = 0 \\ B_k + jc_k & k > 0 \\ B_k - jc_k & k < 0 \end{cases}$$

(*a*) We know that,

If $x(t)$ is real, the Fourier series coefficients of $\text{E}(x(t))\}$ are real $Re\ \{\gamma_k\}$, therefore

$$a_0 = a_0$$
$$a_k = B_k$$

We also know that

If $x(t)$ is real, the Fourier series coefficients of Od $\{x(t)\}$ are *Im* $\{\gamma_k\}$, therefore

$$B_0 = 0, \; B_k = \begin{cases} jc_k & k > 0 \\ -jc_k & k < 0 \end{cases}$$

(*b*) $a_k = a_{-k}$ and $B_k = -B_{-k}$

(*c*) The signal is

$$y(t) = 1 + \text{E}v\{x(t)\} + \frac{1}{2}\text{E}v\{\tau(t) - \text{odd}[z(t)]\}$$

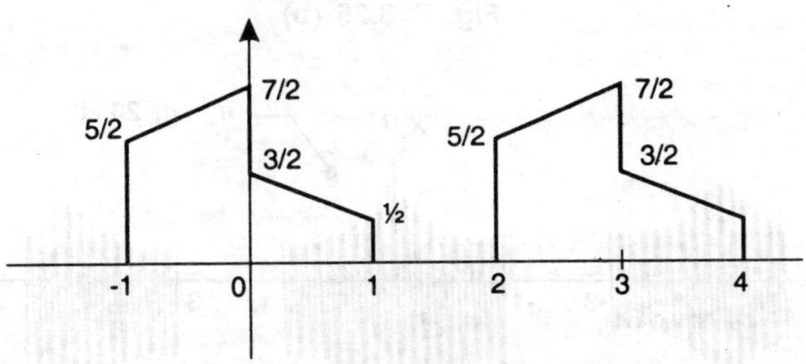

Fig. P. 3.24 (c)

Problem 3.25: In this problem, we derive two important properties of the continuous-time Fourier series; the multiplication property and Parseval's relation. Let $x(t)$ and $y(t)$ both be continuous-time periodic signals having period T_0 and with Fourier series representations given by

$$x(t) = \sum_{k=-\infty}^{+\infty} a_k e^{jk\omega_0 t}, \quad y(t) = \sum_{k=-\infty}^{+\infty} b_k e^{jk\omega_0 t}, \qquad (1)$$

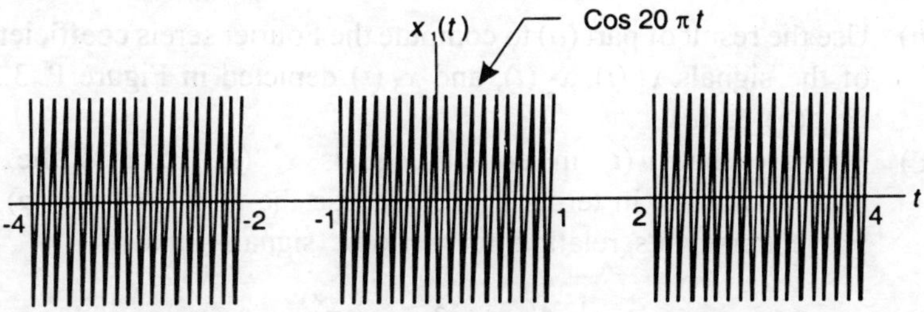

Fig. P. 3.25 (a)

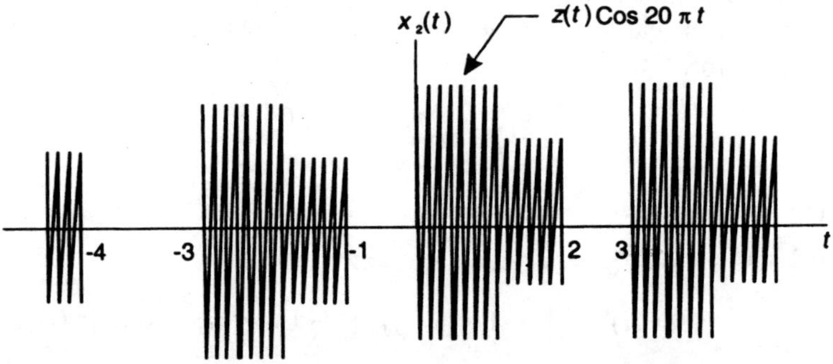

Fig. P. 3.25 (b)

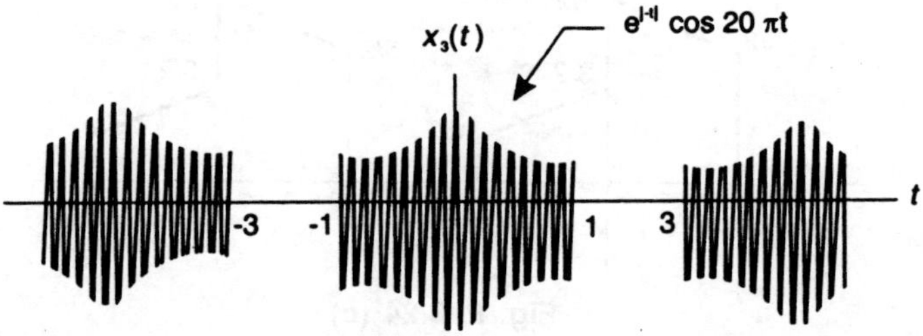

Fig. P. 3.25 (c)

(*a*) Show that the Fourier series coefficients of the signal

$$z(t) = x(t)\ y(t) = \sum_{k=-\infty}^{+\infty} c_k e^{jk\omega_0 t}$$

are given by the discrete convolution

$$c_k = \sum_{n=-\infty}^{+\infty} a_n b_{k-n}$$

(*b*) Use the result of part (*a*) to compute the Fourier sereis coefficients of the signals $x_1(t)$, $x_2(t)$, and $x_3(t)$ depicted in Figure P. 3.25 (*a*), (*b*) & (*c*).

(*c*) Suppose that $y(t)$ in eq. (1) equals $x^*(t)$. Express the b_k in the equation in terms of a_k, and use the result of part (*a*) to prove Parseval's relation for periodic signals—that is,

$$\frac{1}{T_0}\int_0^{T_0} |x(t)|^2\, dt = \sum_{k=-\infty}^{+\infty} |a_k|^2$$

Solution:

(*a*) The Fourier series coefficients of $z(t)$ are

$$c_k = \frac{1}{T}\int_T \sum_n \sum_l a_n\, b_n\, e^{j(n+l)^{-at}}\; e^{-jk\omega_o t}\cdot dt$$

$$c_k = \frac{1}{T}\sum_n \sum_l \; a_n\, b_l\, \delta\,[k-(n+l)] = \frac{1}{T}\sum_n a_n\, b_{k-n}$$

(*b*) (*i*) Here $T_0 = 3$ and $\omega_0 = 2\frac{\pi}{3}$

Therefore

$$c_k = \left[\frac{1}{2}\delta(k-30)+\frac{1}{2}\delta(k+30)\right]\cdot\frac{2\sin\left(k\frac{2\pi}{3}\right)}{3k\left(\frac{2\pi}{3}\right)}$$

Simplifying

$$c_k = \frac{\sin\left\{(k-30)\,2\frac{\pi}{3}\right\}}{3(k-30)\,2\frac{\pi}{3}}+\frac{\sin\left\{(k+30)2\frac{\pi}{3}\right\}}{3(k+30)\frac{2\pi}{3}}$$

and $\quad c \pm 30 = \dfrac{1}{3}$

(*ii*) We may express $x_2(t)$ as

$x_2(t)$ = sum of two shifted square wave $\cos(20\pi t)$

Here,

$$T_o = 3,\ \omega_o = \frac{2\pi}{3},\ \text{Therefore}$$

$$c_k = \frac{1}{3}\; e^{-j(k-30)\frac{2\pi}{3}}\;\frac{\sin\left\{(k-30)\frac{2\pi}{3}\right\}}{(k-30)\frac{2\pi}{3}} +$$

$$\frac{1}{3}\; e^{-j(k+30)\frac{2\pi}{3}}\cdot\frac{\sin\left\{(k+30)\frac{2\pi}{3}\right\}}{(k+30)\frac{2\pi}{3}} +$$

$$\frac{1}{3}\, e^{-j(k-30)\frac{\pi}{3}} \cdot \frac{\sin\left\{(k-30)\frac{\pi}{3}\right\}}{(k-30)\frac{2\pi}{3}} +$$

$$\frac{1}{3}\, e^{-j(k+30)\left(\frac{\pi}{3}\right)} \frac{\sin\left\{(k+30)\frac{\pi}{3}\right\}}{(k+30)\frac{2\pi}{3}}$$

(*iii*) Here, $T_0 = 4 \qquad \omega_0 = \frac{\pi}{2}$

Therefore

$$c_k = \left[\frac{1}{2}\delta(k-40) + \frac{1}{2}\delta(k+40)\right] * \frac{j\left(k\omega_0 + e^{-1}\left(\sin(k\omega_0) - \cos(k\omega_0)\right)\right)}{2\left(1+(k\omega_0)^2\right)}$$

Simplyfing,

$$c_k = \frac{j\left[(k-40)\,\omega_0 + e^{-1}\left\{\sin(k-40)\,\omega_0 - \cos(k-40)\,\omega_0\right\}\right]}{2\left(1+(k\omega_0)^2\right)} +$$

$$\frac{j\left[(k+40)\,\omega_0 + e^{-1}\left\{\sin(k+40)\,\omega_0 - \cos(k+40)\,\omega_0\right\}\right]}{4\,[1+\{(k+40)\,\omega_0\}^2]}$$

(*c*) We know that

$$b_k = a^*_{-k}$$

From part (*a*), we know that the Fourier series coefficients of $z(t) = x(t)\, y(t)$

$$= x(t) \cdot x^*(t) = [x(t)]^2 \text{ will be}$$

$$c_k = \sum_{k=-\infty}^{\infty} a_k b_{k-n} = \sum_{n=-\infty}^{\infty} a_k a^*_{n+k}$$

putting $n = 0$, on this equation, we get

$$\frac{1}{T_0}\int_0^{T_0} |x(t)|^2\, dt = \sum_{k=-\infty}^{\infty} |a_k|^2$$

Problem 3.26: Consider the signal

$$x(t) = \cos 2\pi t.$$

Since $x(t)$ is periodic with a fundamental period of 1, it is also periodic with a period of N, where N is any positive integer. What are the Fourier series coefficients of $x(t)$ if we regard it as a periodic signal with period 3?

Solution:

Considering $x(t)$ to be periodic, with period 1 the non zero Fourier series of $x(t)$ are

$$a_1 = a_{-1} = \frac{1}{2}$$

If we now consider $x(t)$ to be periodic, with period 3, then the non zero Fourier series of $x(t)$ are $b_3 = b_{-3} = \frac{1}{2}$.

Problem 3.27: Let $x[n]$ be a periodic sequence with period N and Fourier series representation

$$x[n] = \sum_{k=\langle N\rangle} a_k e^{jk(2\pi/N)n} \qquad (1)$$

The Fourier series coefficients for each of the following signals can be expressed in terms of a_k in eq. (1). Derive the expressions.

(*a*) $x[n - n_0]$

(*b*) $x[n] - x[n-1]$

(*c*) $x[n] - x\left[n - \frac{N}{2}\right]$ (assume that N is even)

(*d*) $x[n] + x\left[n + \frac{N}{2}\right]$ (assume that N is even; note that this signal is periodic with period $N/2$)

(*e*) $x^*[-n]$

(*f*) $(-1)^n x[n]$ (assume that N is odd; note that this signal is periodic with period $2N$)

(*g*) $y[n] = \begin{cases} x[n], & n \text{ even} \\ 0, & n \text{ odd} \end{cases}$

Solution:

(a) The Fourier series coefficients of $x[n - n_0]$ are

$$a_k = \frac{1}{N}\sum_{n=0}^{N-1} x[n - n_0]\, e^{-j2\pi n\frac{k}{N}}$$

$$= \frac{1}{N} e^{-j\frac{2\pi n_0 k}{N}} \sum_{n=0}^{N-1} x[n] e^{-j2\pi n\frac{k}{N}} = e^{-j2\pi k\frac{n_0}{N}} a_k$$

(b) Using the result of part (a),

The Fourier series coefficients of $x[n] - x[n - 1]$ are given by

$$a_k = a_k - e^{-j2\pi\frac{k}{n}} a_k = \left[1 - e^{-j2\pi\frac{k}{n}}\right] a_k$$

(c) Using the result of part (a),

The Fourier series coefficient of $x[n] - x\left[n - \frac{N}{2}\right]$ are given by

$$a_k = a_k\left[1 - e^{-jk\pi}\right] = \begin{cases} 0 & k\text{ - even} \\ 2a_k & k\text{ - odd} \end{cases}$$

(d) Note that

$x[n] + x\left[n + \frac{N}{2}\right]$ has a period of $\frac{N}{2}$. The Fourier series coefficients of $x[n] + x\left[n - \frac{N}{2}\right]$ are given by

$$a_k = \frac{2}{N}\sum_{n=0}^{\frac{N}{2}-1}\left[x[n] + x\left[n + \frac{N}{2}\right]\right] e^{-j4\pi n\frac{K}{N}} = 2a_{2k} \text{ for } 0 \le k \le \left(\frac{N}{2} - 1\right)$$

(e) The Fourier series coefficient of $x^*[-n]$ are

$$a_k = \frac{2}{N}\sum_{n=0}^{n-1} x[n] e^{-j\left(2\frac{\pi n}{N}\right)\left(k-\frac{N}{2}\right)} = a_{k-\frac{N}{2}}$$

(f) With N odd the period of $(-1)^{-n} x[n]$ is $2N$. Therefore the Fourier series coefficients are

$$a_k = \frac{1}{2N}\left[\sum_{n=0}^{N-1} x[n]\, e^{-j\frac{2\pi n}{N}\left(\frac{k-N}{2}\right)} + \sum_{n=0}^{N-1} x[n]\, e^{-j\frac{2\pi n}{N}\left(\frac{k-N}{2}\right)} e^{-j\pi(k-N)}\right]$$

Note that

For k odd, $\frac{k-N}{2}$ is an integer and $k - N$ is an even integer. Also for k even $k - N$ is an odd integer and $e^{-j\pi(k-N)} = -1$

Therefore,

$$a_k = \begin{cases} a_{\frac{k-N}{2}} & k\text{-odd} \\ 0 & k\text{-even} \end{cases}$$

(*g*) Here

$$y[n] = \frac{1}{2}\left[x[n] + (-1)^n x[n]\right]$$

$$\text{for } N \text{ even } a_k = \frac{1}{2}\left[a_k + a_{k-\frac{N}{2}}\right]$$

$$\text{For } N \text{ odd } a_{(k)} = \begin{cases} \frac{1}{2}\left[a_k + a_{k-\frac{N}{2}}\right] & k\text{-even} \\ \frac{1}{2} a_k & k\text{-odd} \end{cases}$$

Problem 3.28: Let $x[n]$ be a periodic sequence with period N and Fourier series representation

$$x[n] = \sum_{k=\langle N\rangle} a_k e^{jk(2\pi/N)n} \qquad (1)$$

(*a*) Suppose that N is even and that $x[n]$ in eq. (1) satisfies

$$x[n] = -x\left[n + \frac{N}{2}\right] \text{ for all } n.$$

Show that $a_k = 0$ for all even integers k.

(*b*) Suppose that N is divisible by 4. Show that if

$$x[n] = -x\left[n+\frac{N}{4}\right] \text{ for all } n.$$

then $a_k = 0$ for evern value of k that is a multiple of 4.

(*c*) *More generally, suppose that* N is divisible by an integer M. Show that if

$$\sum_{r=0}^{(N/M)-1} x\left[n+r\frac{N}{M}\right] = 0 \text{ for all } n.$$

then $a_k = 0$ for every value of k that is a multiple of M.

Solution:

(*a*) The Fourier series coefficients are given by

$$a_k = \frac{1}{N}\sum_{n=0}^{N-1} x[n]\, e^{-j\frac{2\pi nk}{N}}$$

$$= \frac{1}{N}\sum_{n=0}^{\left(\frac{N}{2}-1\right)} x[n]\, e^{-j\frac{2\pi nk}{N}} + \frac{1}{N}\sum_{n=\frac{N}{2}}^{N-1} x[n]\, e^{-j\frac{2\pi nk}{N}}$$

$$= \frac{1}{N}\sum_{n=0}^{\left(\frac{N}{2}-1\right)} x[n]\, e^{-\frac{j2\pi nk}{N}} + \frac{e^{-j\pi k}}{N}\sum_{n=0}^{\left(\frac{N}{2}-1\right)} x\left[n+\frac{N}{2}\right] e^{-\frac{j2\pi nk}{N}} = 0$$

$$= \frac{1}{N}\sum_{n=0}^{\left(\frac{N}{2}-1\right)} x[n]\, e^{-\frac{j2\pi nk}{N}} - \frac{e^{-j\pi k}}{N}\sum_{n=0}^{\left(\frac{N}{2}-1\right)} x[n]\, e^{-\frac{j2\pi nk}{N}} = 0 \text{ for } k \text{ even}$$

(*b*) By adopting an approach similar to part (*a*), we may show that

$$a_k = \frac{1}{N}\left[\sum_{n=0}^{\left(\frac{N}{4}-1\right)}\left\{1 - e^{-jk\frac{\pi}{2}} + e^{-j\pi k} - e^{-\frac{j3k\pi}{2}}\right\} x[n]\, e^{-\frac{j2\pi nk}{N}}\right]$$

$$= 0 \text{ for } k = 4r,\ r\ \varepsilon\ 1$$

(c) If $\frac{N}{m}$ is an integer, we may generalize the approach of part (a), to show that

$$a_k = \frac{1}{N}\left[\sum_{k=0}^{B-1}\left\{1 - e^{-j2\pi r} + e^{-j4\pi r} \ldots + e^{-j2\pi(m-1)r}\right\} x[n]\, e^{-\frac{j2\pi nk}{N}}\right]$$

Where,

$$B = \frac{N}{m}, \text{ and } r = \frac{k}{m},$$

From, the above equation,
It is clear that

$$a_k = 0, \quad \text{if } k = rm \qquad r \,\varepsilon\, 1$$

> **Problem 3.29:** Suppose we are given the following information about a periodic signal $x[n]$ with period 8 and Fourier coefficients a_k;
> 1. $a_k = -a_{k-4}$
> 2. $x[2n+1] = (-1)^n$.
>
> Sketch one period of $x[n]$.

Solution:

We know that if $x[n] \xleftrightarrow{Fs} a_k$
Then,

$$(-1)^n x[n] = e^{j\left(\frac{2\pi}{N}\right)\left(\frac{N}{2}\right)n} x[n] \xleftrightarrow{Fs} a_{k-\frac{N}{2}}$$

In this case, $N = 8$, therefore

$$(-1)^n x[n] \xleftrightarrow{Fs} a_{k-4}$$

since it is given that $a_k = -a_{k-4}$
We have

$$x[n] = -(-1)^n x[n]$$

This implies that

$$x[0] = x[\pm 2] = x[\pm 4] \ldots = 0$$

We are also given that $x[1] = x[5] = \ldots = 1$ and

$$x[3] = x[7] = -1$$

Therefore one period of $x[n]$ is as shown in figure below.

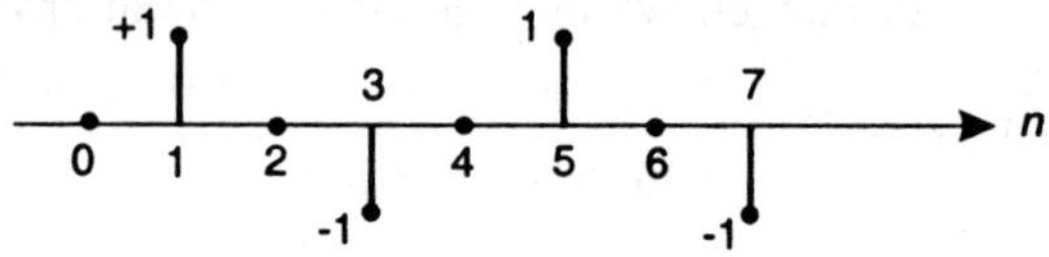

Fig. P. 3.29 (a)

Problem 3.30: Let $x[n]$ be a periodic signal with period $N = 8$ and Fourier series coefficients $a_k = -a_{k-4}$. A signal

$$y[n] = \left(\frac{1+(-1)^n}{2}\right)x[n-1]$$

with period $N = 8$ is generated. Denoting the Fourier series coefficient of $y[n]$ by b_k, find a function $f[k]$ such that

$$b_k = f[k]\, a_k.$$

Solution:

We have

$$e^{j4\left(\frac{2\pi}{8}\right)n} x[n] = e^{j\pi n} x[n] = (-1)^n x[n] \xleftrightarrow{Fs} a_{k-4}$$

and therefore

$$(-1)^{n+1} x[n] \xleftrightarrow{Fs} -a_{k-4}$$

If $a_k = -a_{k-4}$, then $x[0] = x[+2] = x[\pm 4] = \ldots = 0$

Now

Note that in the signal $p[n] = x[n-1]$

$$p(\pm 1) = p(\pm 3) = \ldots = 0$$

Now let us plot,

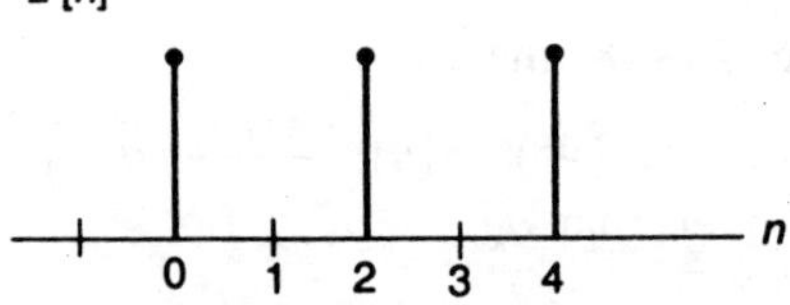

Fig. P. 3.30 (a)

Clearly, the signal $y[n] = z[n]\,p[n] = p[n]$ because $p[n]$ is zero, whenever $z[n]$ is zero. Therefore, $y[n] = x[n-1]$, The Fourier series coefficients of $y[n]$ are $a_k e^{-j\left(\frac{2\pi}{6}\right)}$

Problem 3.31: $x[n]$ is real periodic signal with period N and complex Fourier series coefficients a_k. Let the Cartesian form for a_k be denoted by

$$a_k = b_k + jc_k,$$

where b_k and c_k are both real

(*a*) Show that $a_{-k} = a_k^*$. What is the relation between b_k and b_{-k}? What is the relation between c_k and c_{-k}?

(*b*) Suppose that N is even. Show that $a_{N/2}$ is real.

(*c*) Show that $x[n]$ can also be expressed as a trigonometric Fourier series of the form

$$x[n] = a_0 + 2\sum_{k=1}^{(N-1)/2} b_k \cos\left(\frac{2\pi kn}{N}\right) - c_k \sin\left(\frac{2\pi kn}{N}\right)$$

if N is odd or as

$$x[n] = (a_0 + a_{N/2}(-1)^n) + 2\sum_{k=1}^{(N-2)/2} b_k \cos\left(\frac{2\pi kn}{N}\right) - c_k \sin\left(\frac{2\pi kn}{N}\right)$$

if N is even.

(*d*) Show that if the polar form of a_k is $A_k e^{j\theta k}$, then the Fourier series representation for $x[n]$ can also be written as

$$x[n] = a_0 + 2\sum_{k=1}^{(N-1)/2} A_k \cos\left(\frac{2\pi kn}{N} + \theta_k\right)$$

if N is odd or as

$$x[n] = (a_0 + a_{N/2}(-1)^n) + 2\sum_{k=1}^{(N/2)-1} A_k \cos\left(\frac{2\pi ka}{N} + \theta_k\right)$$

if N is even

(*e*) Suppose that $x[n]$ and $z[n]$, as depicted in Figure P. 3.31, have the sine-cosine series representations

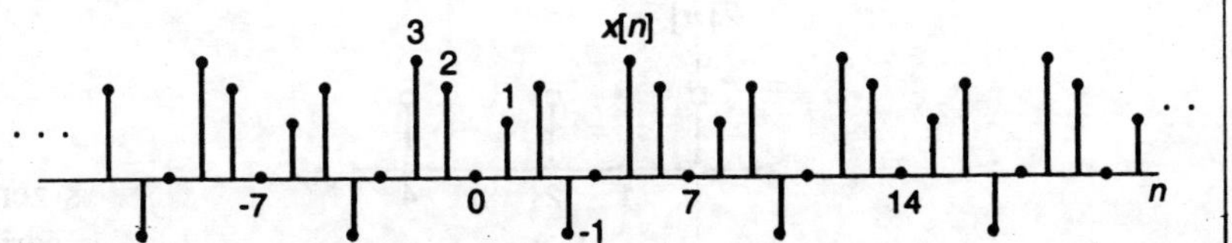

Fig. P. 3.31 (a)

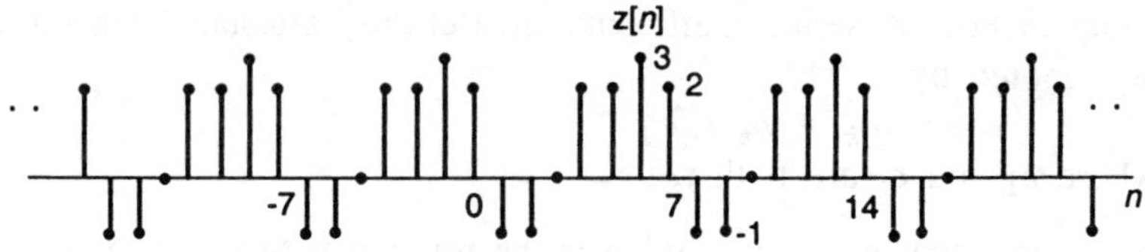

Fig. P. 3.31 (b)

$$x[n] = a_0 + 2\sum_{k=1}^{3}\left\{b_k \cos\left(\frac{2\pi kn}{7}\right) - c_k \sin\left(\frac{2\pi kn}{7}\right)\right\}$$

$$x[n] = a_0 + 2\sum_{k=1}^{3}\left\{d_k \cos\left(\frac{2\pi kn}{7}\right) - f_k \sin\left(\frac{2\pi kn}{7}\right)\right\}$$

Sketch the signal

$$y[n] = a_0 - d_0 + 2\sum_{k=1}^{3}\left\{d_k \cos\left(\frac{2\pi kn}{7}\right) + (f_k - c_k)\sin\left(\frac{2\pi kn}{7}\right)\right\}$$

Solution:

(*a*) If $x[n]$ is real,

$$x[n] = x^*[n], \text{ therefore}$$

$$a_{-k} = \sum_n x[n]\, e^{j2\pi n\frac{k}{N}} = a_k^*$$

From this result,
We get

$$b_{-k} = b_k \text{ and } c_{-k} = -c_k$$

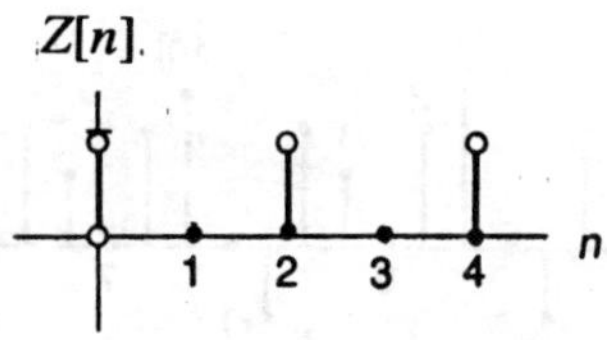

Fig. P. 3.31 (c)

(*b*) If N is given then

$$a_{\frac{N}{2}} = \frac{1}{N}\sum_{n} x[n]\ e^{-j\pi n} = \frac{1}{N}\sum_{n}(-1)^n\ x[n] = \text{real}$$

(*c*) If N is odd, then,

$$x[n] = \sum_{k=-\frac{(N-1)}{2}}^{\frac{(N-1)}{2}} a_k\ e^{j\left(\frac{2\pi}{N}\right)kn} = \sum_{k=0}^{\frac{(N-1)}{2}} a_k\ e^{j\left(\frac{2\pi}{N}\right)kn} + \sum_{k=1}^{\frac{(N-1)}{2}} a_k^*\ e^{-j\left(\frac{2\pi}{N}\right)kn}$$

$$= a_0 + 2\sum_{k=1}^{\frac{(N-1)}{2}} b_k \cos\left(2\pi\frac{kn}{N}\right) - c_k \sin\left(2\pi\frac{kn}{N}\right)$$

If N is even, then

$$x[n] = \sum_{k=0}^{N-1} a_k\ e^{j\left(\frac{2\pi}{N}\right)kn}$$

$$= a_0 + (-1)^n\ a_{\frac{N}{2}} + 2\sum_{k=1}^{\frac{(N-1)}{2}} a_k\ e^{j\left(2\frac{\pi}{N}\right)kn} + a_{N-k}\ e^{J\left(2\frac{\pi}{N}\right)(N-k)n}$$

$$= a_0 + (-1)^n\ a_{\frac{N}{2}} + 2\sum_{k=1}^{\frac{(N-2)}{2}} a_k\ e^{j\left(2\frac{\pi}{N}\right)kn} - a_k^* e^{-j\left(\frac{2\pi}{N}\right)kn}$$

$$= a_0 + (-1)^n\ a_{\frac{N}{2}} + 2\sum_{k=1}^{\frac{(N-2)}{2}} b_k \cos\left[2\pi\frac{kn}{N}\right] - c_k \sin\left(2\pi\frac{kn}{N}\right)$$

(*d*) If $\quad a_k = A_k\ e^{j\theta_k}$ then

$$b_k = A\cos(\theta_k)$$

and $\quad c_k = A\sin(\theta_k)$, substituting in the result of the previous part, we get for N odd,

$$x[n] = a_o + 2\sum_{k=1}^{\frac{(N-1)}{2}} a\cos(\theta_k)\cos\left(2\pi\frac{kn}{N}\right) - c_k\sin(\theta_k)\cdot\sin\left(2\pi\frac{kn}{N}\right)$$

$$= a_o + 2\sum_{k=1}^{\frac{(N-1)}{2}} A_k \cos\left(\frac{2\pi nk}{N} + \theta_k\right)$$

Similarly for N even,

$$x[n] = a_o + (-1)^n a_{\frac{N}{2}} + 2\sum_{k=1}^{\left(\frac{N-1}{2}\right)} A\cos(\theta_k) \times$$

$$\cos\left(2\pi\frac{kn}{N}\right) - c_k \sin(\theta_k)\cdot\sin\left(2\pi\frac{kn}{N}\right)$$

$$= a_o + (-1)^n a_{\frac{N}{2}} + 2\sum_{k=1}^{\left(\frac{N}{2}-1\right)} A_k\cdot\cos\left(\frac{2\pi nk}{N}+\theta_k\right)$$

(*e*) The signal is

$$y[n] = d\cdot c\,\{x[n]\} - d\cdot c\,\{\tau[n]\} + \varepsilon v\,\{\tau\} +$$
$$\text{Odd}\cdot\{x\} - z\ \text{Od}\,\{\tau\}$$

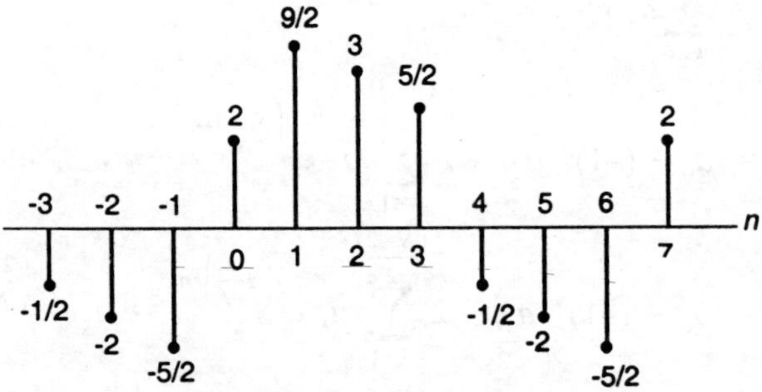

Fig. P. 3.31 (c)

> **Problem 3.32:** Let $x[n]$ be a real periodic signal with period N and Fourier coefficients a_k.
>
> (*a*) Show that if N is even, at least two of the Fourier coefficients within one period of a_k are real.
>
> (*b*) Show that if N is odd, at least one of the Fourier coefficients within one period of a_k is real.

Solution:

We know that

$$a_k = \frac{1}{N}\sum_{\langle N\rangle} x[n]\,e^{-j\left(\frac{2\pi}{N}\right)kn}$$

Note that

$$a_0 = \frac{1}{N}\sum_{\langle N\rangle} x[n]$$

Which is real if $x[n]$ is real

(*a*) If N is even, then

$$a_{\frac{N}{2}} = \frac{1}{N}\sum_{\langle N\rangle} x[n]\, e^{-j\pi n} = \frac{1}{N}\sum_{\langle N\rangle} x[n]\,(-1)^n$$

Clearly $a_{\frac{N}{2}}$ is also real if $x[n]$ is real.

(*b*) Let N is odd, only a_0 is guaranteed to be real.

Problem 3.33: Consider the function

$$a[k] = \sum_{n=0}^{N-1} e^{j(2\pi/N)kn}$$

(*a*) Show that $a[k] = N$ for $k = 0, \pm N, \pm 2N, \pm 3N, \ldots$

(*b*) Show that $a[k] = 0$ whenever k is not an integer multiple of N.

(*c*) Repeat parts (*a*) and (*b*) if

$$a[k] = \sum_{n=\langle N\rangle} e^{j(2\pi/N)kn}$$

Solution:

(*a*) Let $k = pN,\ p \in 1,$

Then, $a(pN) = \sum_{n=0}^{N-1} e^{j\left(\frac{2\pi}{N}\right)pNn} = \sum_{n=0}^{N-1} e^{j(2\pi pn)} = \sum_{n=0}^{N-1} 1 = N$

(*b*) Using the finite sum, formula, We get

$$a[k] = \frac{1-e^{j2\pi k}}{1-e^{j\left(\frac{2\pi}{N}\right)k}} = 0 \text{ if } k \neq pN,\ P \in 1$$

(c) Let $a[k] = \sum_{n=q}^{q+N-1} e^{j\left(\frac{2\pi}{N}\right)kn}$

Where q is some arbitrary integer. By putting $k = pN$, we may again easily show that

$$a(pN) = \sum_{n=q}^{q+N-1} e^{j\left(\frac{2\pi}{N}\right)pN_n} = \sum_{n=q}^{q+N-1} e^{j(2\pi pn)} = \sum_{n=q}^{q+N-1} 1 = N$$

Now,

$$a[k] = e^{j\left(\frac{2\pi}{N}\right)kq} \sum_{n=0}^{N-1} e^{j\left(\frac{2\pi}{N}\right)kN}$$

Using part (b), we may argue that

$$a[k] = 0 \text{ for } k \neq pN, \; p \in 1$$

Problem 3.34: Let $x[n]$ be a periodic signal with fundamental period N and Fourier series coefficients a_k. In this problem, we derive the time-scaling property

$$x_{(m)}[n] = \begin{cases} x\left[\dfrac{n}{m}\right], & n = 0, \pm m, \pm 2m, \ldots \\ 0, & \text{elsewhere} \end{cases}$$

(a) Show that $x_{(m)}[n]$ has period of mN.

(b) Show that if

$$x[n] = v[n] + w[n],$$

then

$$x_{(m)}[n] = v_{(m)}[n] + w_{(m)}[n].$$

(c) Assuming that $x[n] = e^{j2\pi k_0 n/N}$ for some integer k_0, verify that

$$x_{(m)}[n] = \frac{1}{m}\sum_{l=0}^{m-1} e^{j2\pi(k_0+lN)n/(mN)}$$

That is, one complex exponential in $x[n]$ becomes a linear combination of m complex exponentials in $x_{(m)}[n]$.

(d) Using the results of parts (a), (b), and (c) show that if $x[n]$ has the Fourier coefficients a_k, then $x_{(m)}[n]$ must have the Fourier coefficients $\dfrac{1}{m}a_k$.

Solution:

(a) Note that

$$a_m\ (n + mN) = \begin{cases} x\left(\dfrac{n}{m} + N\right) & m = 0, \pm m \\ 0 & \text{otherwise} \end{cases}$$

$$= \begin{cases} x\left(\dfrac{n}{m}\right) & n = 0 \pm m \\ 0 & \text{otherwise} \end{cases} = x_m\,[n]$$

Therefore $x_m\,[n]$ is periodic with period mN.

(*b*) The time scaling operation discussed in this problem is a linear operation. Therefore if

$$x\,[n] = x\,[n] + w\,[n] \text{ then,}$$
$$x_m\,[n] = v_m\,[n] + w_m\,[n]$$

(*c*) Let us consider

$$y\,[n] = \frac{1}{m}\sum_{0}^{m-1} e^{j\left(\frac{2\pi}{mN}\right)(k_0+1.N)n} = \frac{1}{m}\sum_{0}^{m-1} e^{j\left(\frac{2\pi}{mN}\right)k_0 n}\sum_{l=0}^{m-1} e^{j\left(\frac{2\pi}{m}\right)N}$$

This may be written as,

$$y\,[n] = \begin{cases} e^{j\left(\frac{2\pi}{mN}\right)k_0 n} & n = 0, \pm N, \pm 2N \\ 0 & \text{otherwise} \end{cases}$$

Now,

Also, note that by applying time scaling on $x\,[n]$ we get

$$x_m\,[n] = \begin{cases} e^{j\left(\frac{2\pi}{mN}\right)k_0 n} & n = 0, \pm N, \pm 2N \\ 0 & \text{otherwise} \end{cases}$$

(*d*) We know that,

$$b_k = \frac{1}{mN}\sum_{n=0}^{mN-1} x_{(m)}[n]\, e^{-j\left(\frac{2\pi}{mN}\right)kn}$$

We know that every *m*th value in the above summation is non zero, therefore

$$b_k = \frac{1}{mN}\sum_{n=0}^{N-1} x_{(m)}[nm]\, e^{-j\left(\frac{2\pi}{mN}\right)kmn} = \frac{1}{mN}\sum_{n=0}^{N-1} x_{(m)}[nm]\, e^{-j\left(\frac{2\pi}{N}\right)kn}$$

Note that

$$x_m[Nm] = x[n],$$

Therefore

$$b_k = \sum_{n=0}^{N-1} x_{[n]}\, e^{-j\left(\frac{2\pi}{N}\right)kn} = \frac{a_k}{m}$$

Problem 3.35: (*a*) Let

$$x[n] = \sum_{k=0}^{N-1} a_k e^{jk(2\pi/N)n} \qquad (1)$$

and

$$y[n] = \sum_{k=0}^{N-1} b_k e^{jk(2\pi/N)n}$$

be periodic signals. Show that

$$x[n]\, y[n] = \sum_{k=0}^{N-1} c_k e^{jk(2\pi/N)n}$$

where

$$c_k = \sum_{l=0}^{N-1} a_l b_{k-l} = \sum_{l=0}^{N-1} a_{k-l} b_l .$$

(*b*) Generalize the result of part (*a*) by showing that

$$c_k = \sum_{l=\langle N\rangle} a_l b_{k-1} = \sum_{l=\langle N\rangle} a_{k-l} b_l .$$

(*c*) Use the result of part (*b*) to find the Fourier series representation of the following signals, where $x[n]$ is given by eq. (1).

(*i*) $x[n]\cos\left(\dfrac{6\pi n}{N}\right)$

(*ii*) $x[n]\sum_{r=-\infty}^{+\infty} \delta[n - rN]$

(*iii*) $x[n]\left(\sum_{r=-\infty}^{+\infty} \delta\left[n - \dfrac{rN}{3}\right]\right)$

(assume that N is divisible by 3)

(*d*) Find the Fourier series representation for the signal $x[n]\,y[n]$, where

$$x[n] = \cos(\pi n/3)$$

and

$$y[n] = \begin{cases} 1, & |n| \le 3 \\ 0, & 4 \le |n| \le 6 \end{cases} \text{ is periodic with period 12.}$$

(*e*) Use the result of part (*b*) to show that

$$\sum_{n=\langle N\rangle} x[n]\,y[n] = N \sum_{l=\langle N\rangle} a_l b_{-l}\,,$$

and from this expression, derive Parseval's relation for discrete–time periodic signals.

Solution:

(*a*) We know that

$$x[n]\,y[n] = \sum_{k=0}^{N-1} \sum_{l'=0}^{(k+N-1)} a_k\, b_{l'-k}\, e^{j\left(\frac{2\pi}{N}\right)l'}$$

But since both $b_{l'-k}$ and $e^{j\left(\frac{2\pi}{N}\right)l'n}$ are periodic with period N. We may rewrite this

$$x[n]\,y[n] = \sum_{k=0}^{N-1} a_k b_{l'-k} e^{j\left(\frac{2\pi}{N}\right)l'n} = \sum_{l=0}^{N-1}\left[\sum_{k=0}^{N-1} a_k\, b_{l-k}\right] e^{j\left(\frac{2\pi}{N}\right)ln}$$

Therefore,

$$c_k = \sum_{k=0}^{N-1} a_k b_{l-k}$$

By interchanging a_k and b_k, we may show that

$$c_k = \sum_{k=0}^{N-1} b_k a_{l-k}$$

(*b*) Note that since both a_k and b_k are periodic with period N, we may reqrite the above summation as,

$$c_k = \sum_{\langle N \rangle} a_k b_{l-k} = \sum_{\langle N \rangle} b_k a_{l-k}$$

(*c*)(*i*) Here,

$$c_k = \sum_{l=0}^{N-1} \frac{1}{2} [\delta(l-3) + \delta(l-N+3)]\, a_{k-l}$$

Therefore

$$c_k = \frac{1}{2} a_{k-3} + \frac{1}{2} a_{k-3-N}$$

(*ii*) Period $= N$, Also $b_k = \dfrac{1}{N}$ for all k

Therefore

$$c_k = \frac{1}{N} \sum_{l=0}^{N-1} a_l$$

(*iii*) Here,

$$b_k = \frac{1}{N}\left[1 + e^{-j2\pi\frac{k}{3}} + e^{-j4\pi\frac{k}{3}}\right]$$

Therefore,

$$c_k = \frac{1}{N}\left[1 + e^{-j2\pi\frac{l}{3}} + e^{-j4\pi\frac{l}{3}}\right] a_{k-l}$$

(*d*) period = 12 also,

$$x[n] \xleftrightarrow{Fs} a_2 = a_2 = a_{10} = \frac{1}{2}$$

All other,

$$a_k = 0,\ 0 \le k \le 11$$

and

$$y[n] \xleftrightarrow{Fs} b_k = \left(\frac{1}{12}\right) \frac{\sin\left(7\pi \dfrac{k}{12}\right)}{\sin\left(\pi \dfrac{k}{12}\right)} \quad 0 \le k \le 11$$

Therefore one period of c_k is

$$c_k = \frac{1}{24}\left[\frac{\dfrac{\sin\{7\pi(k-2)\}}{12}}{\dfrac{\sin\{\pi(k-2)\}}{12}} + \frac{\dfrac{\sin\{7\pi(k-10)\}}{12}}{\dfrac{\sin\{\pi(k-10)\}}{12}}\right], 0 \le k \le 11$$

(*e*) Using the Fourier series coefficient equation we have

$$N \sum_{L=\langle N\rangle} a_l b_{-l} = \sum_{\langle N\rangle} x[n]\, y[n]$$

Now let

$$y[n] = x^*_{[n]} \text{ then}$$

$$b_l = a^*_{-l} \text{ therefore}$$

$$N \sum_{l=\langle N\rangle} a_l a_l^* = \sum_{\langle N\rangle} x[n]\, x^*[n]$$

Therefore

$$N \sum_{l=\langle N\rangle} |a_L|^2 = \sum_{\langle N\rangle} [x(n)]^2$$

Problem 3.36: Let $x[n]$ and $y[n]$ be periodic signals with common period N, and let

$$z[n] = \sum_{r=\langle N\rangle} x[r]\, y[n-r]$$

be their periodic convolution.

(*a*) Show that $z[n]$ is also periodic with period N.

(*b*) Verify that if a_k, b_k, and c_k are the Fourier coefficients of $x[n]$, $y[n]$, and $z[n]$, respectively, then

$$c_k = N a_k b_k.$$

(*c*) Let $\quad x[n] = \sin\left(\dfrac{3\pi n}{4}\right)$

and $\quad y[n] = \begin{cases} 1, & 0 \le n \le 3 \\ 0, & 4 \le n \le 7 \end{cases}$

be two signals that are periodic with period 8. Find the Fourier series representation for the periodic convolution of these signals.

(*d*) Repeat part (*c*) for the following two periodic signals that also have period 8.

$$x[n] = \begin{cases} \sin\left(\frac{3\pi n}{4}\right), & 0 \le n \le 3 \\ 0, & 4 \le n \le 7 \end{cases}$$

$$y[n] = \left(\frac{1}{2}\right)^n;\ 0 \le n \le 7.$$

Solution:

(*a*) We know that

$$z[n + N] = \sum_{\langle l \rangle} x[r]\, y[n + N - r]$$

Since $y[n]$ is periodic with period N,
$y[n + N - r] = y[n - r]$
therefore

$$z[n + N] = \sum_{\langle l \rangle} x[r]\, y[n - r] = z[n]$$

Therefore,
$z[n]$ is also periodic with period N

(*b*) The Fourier series coefficients of $z[n]$ are

$$c_1 = \frac{1}{N} \sum_{n=\langle N \rangle} \sum_{k=\langle N \rangle} a_k\, b_{n-k}\, e^{-j2\pi n \frac{l}{N}}$$

$$= \frac{1}{N} \sum_{k=\langle N \rangle} a_k e^{-j2\pi k \frac{l}{N}} \sum_{n=\langle N \rangle} b_{n-k} e^{-j2\pi\left(n - k\frac{L}{N}\right)} = \frac{1}{N}\, Na_l N_{bl} = N_{al} b_l$$

(*c*) Here $n = 8$, the non zero Fourier series coefficients in the range $0 \le k \le 7$

for $x[n]$ are

$$a_3 = a_5^* = \frac{1}{2j}$$

Note that for $y[n]$, we need only evaluate b_3 and b_5, we have

$$b_3 = b_5^* = \frac{1}{4\left(1-e^{-j3\frac{\pi}{4}}\right)}$$

Therefore, the only non zero Fourier series coefficients in the range $0 \le k \le 7$ for the periodic correlation of these signal are

$$c_3 = 8a_3b_3$$

and $$c_5 = 8a_5b_5$$

(*d*) Here,

$$x[n] \overset{Fs}{\longleftrightarrow} \frac{1}{16j}\left[\frac{1-e^{0j\left(3\frac{\pi}{7}-\pi\frac{k}{4}\right)4}}{1-e^{-j\left(3\frac{\pi}{7}-\pi\frac{k}{4}\right)}} - \frac{1-e^{j\left(3\frac{\pi}{7}+\pi\frac{k}{4}\right)4}}{1-e^{-j\left(3\frac{\pi}{7}+\pi\frac{k}{4}\right)4}}\right]$$

and $$y[n] \overset{Fs}{\longleftrightarrow} b_k = \frac{1}{8}\left[\frac{1-\left(\frac{1}{2}\right)^8}{1-\frac{1}{2}e^{-jk\frac{\pi}{4}}}\right]$$

therefore

$$z[n] = x[n]\,y[n] \overset{Fs}{\longleftrightarrow} 8a_kb_k$$

Problem 3.37: (*a*) Suppose $x[n]$ is a periodic signal with period N. Show that the Fourier series coefficients of the periodic signal

$$g(t) = \sum_{k=-\infty}^{\infty} x[k]\,\delta(t-kT)$$

are periodic with period N.

(*b*) Suppose that $x(t)$ is a periodic signal with period T and Fourier series coefficients a_k with period N. Show that there must exist a periodic squence $g[n]$ such that

$$x(t) = \sum_{k=-\infty}^{\infty} g[k]\,\delta(t-kT/N)$$

(*c*) Can a continuous periodic signal have periodic Fourier coefficients?

Solution:

(*a*) Note that the signal $x(t)$ is periodic with period NT. The Fourier series coefficients of $x(t)$ are

$$a_k = \frac{1}{NT}\int_0^{NT}\left[\sum_{p=-\infty}^{\infty} x[p]\,\delta(t-pT)\right]e^{-j\left(\frac{2\pi}{NT}\right)kt}\cdot dt$$

Note that,

The limits of the summation may be changed in accordance wth the limits of the integration, so that we get

$$a_k = \frac{1}{NT}\int_0^{NT}\left[\sum_{p=0}^{N-1} x[p]\,\delta(t-pT)\right]e^{-j\left(\frac{2\pi}{NT}\right)kt}\cdot dt$$

Interchanging the summation and integration and simplifying

$$a_k = \left(\frac{1}{NT}\right)\sum_{p=0}^{N-1} x[p]\int_0^{NT}\delta(t-pT)\;e^{-j\left(\frac{2\pi}{NT}\right)^{kt}}\cdot dt$$

$$= \left(\frac{1}{NT}\right)\sum_{p=0}^{N-1} x[p]\;e^{-j\left(\frac{2\pi}{N}\right)pk} = \left(\frac{1}{T}\right)\left[\left(\frac{1}{N}\right)\sum_{p=0}^{N-1} x[p]\,e^{-j\left(\frac{2\pi}{N}\right)pk}\right]$$

Note that, the term within bracket on the RHS of the above equation constitutes the Fourier series coefficients of the signal $x[n]$. Since this is periodic with period N, a_k must also be periodic with period N.

(*b*) If the Fourier series coefficients of $x(t)$ are periodic with period N, then $a_k = a_{k-N}$

This implies that $x(t) = x(t)\,e^{j\left(\frac{2\pi}{T}\right)Nt}$

This is possible only if $x(t)$ is zero, for all t, other than when

$$\left(\frac{2\pi}{T}\right)Nt = 2\pi k \text{ where } k\ \varepsilon\ 1$$

Therefore $x(t)$ is of the form

$$x(t) = \sum_{k=-\infty}^{\infty} g[k]\;\delta\left(t-\frac{kT}{N}\right)$$

(*c*) A simple example would be

$$x(t) = \sum_{k=-\infty}^{\infty}\delta(t-kT)$$

Problem 3.38: Consider the following pairs of signals $x[n]$ and $y[n]$. For each pair, determine whether there is a discrete-time LTI system for which $y[n]$ is the output when the corresponding $x[n]$ is the input. If such a system exists, determine whether the system is unique (i.e., whether there is more than one LTI system with the given input-output pair). Also, determine the frequency response of an LTI system with the desired behaviuor. If no such LTI system exists for a given $x[n]$, $y[n]$ pair, explain why?

(*a*) $x[n] = \left(\frac{1}{2}^{n}\right), y[n] = \left(\frac{1}{4}^{n}\right)$

(*b*) $x[n] = \left(\frac{1}{2}^{n}\right)u[n], y[n] = \left(\frac{1}{4}^{n}\right)u[n]$

(*c*) $x[n] = \left(\frac{1}{2}^{n}\right)u[n], y[n] = 4^{n}u[-n]$

(*d*) $x[n] = e^{jn/8}, y[n] = 2e^{jn/8}$

(*e*) $x[n] = e^{jn/8}\, u[n], y[n] = 2e^{jn/8}\, u[n]$

(*f*) $x[n] = j^{n}, y[n] = 2\, j^{n}(1-j)$

(*g*) $x[n] = \cos(\pi n/3), y[n] = \cos(\pi n/3) + \sqrt{3}\sin(\pi n/3)$

(*h*) $x[n]$ and $y_1[n]$ as in Figure P. 3.38

(*i*) $x[n]$ and $y_2[n]$ as in Figure P. 3.38

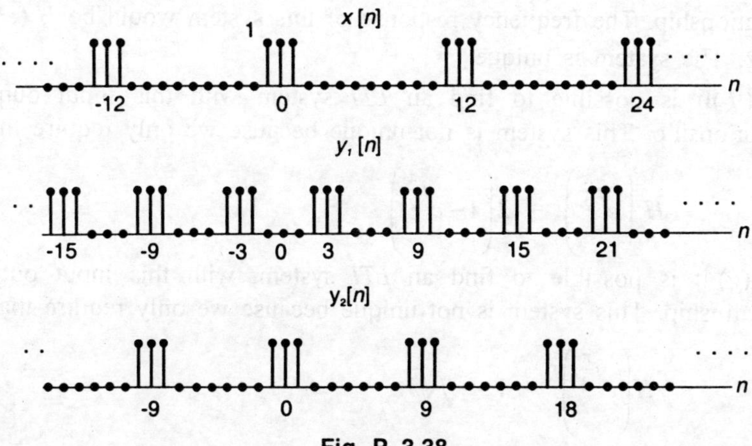

Fig. P. 3.38

Solution:

(*a*) The system is *LTI.* $\left(\frac{1}{2}\right)^n$ is an eigen function of *LTI* system, therefore, the output have been of the form $k\left(\frac{1}{2}\right)^n$, where *k* is a complex constant.

(*b*) It is possible to find an *LTI* system, with this input-output relationship. The frequency response of this system would be

$$H(e^{j\omega}) = \frac{\left[1-\left(\frac{1}{2}\right)e^{-j\omega}\right]}{\left(1-\left(\frac{1}{4}\right)e^{j\omega}\right)}$$ The system is unique.

(*c*) it is possible to find an *LTI* system with this input output relationship. The trequency response of this system would be

$$H(e^{j\omega}) = \frac{\left[1-\left(\frac{1}{2}\right)e^{-j\omega}\right]}{\left(1-\left(\frac{1}{4}\right)e^{j\omega}\right)}$$

(*d*) It is possible to find an *LTI* system with this input output relationship. The system is not unique because we only require that

$$H(e^{j8}) = 2$$

(*e*) It is possible to find an *LTI* system with this input output relationship. The frequency response of this system would be $H(e^{j\omega}) = 2$, The system is unique.

(*f*) It is possible to find an *LTI* system with this input output relationship. This system is not unique because we only require that

$$H\left(e^{j\frac{\pi}{2}}\right) = 2\left(1-e^{j\frac{\pi}{2}}\right)$$

(*g*) It is possible to find an *LTI* system, with this input output relatinship. This system is not unique because we only require that

$$H\left(e^{j\frac{\pi}{3}}\right) = 1 - j\sqrt{3}$$

(*h*) Note that,

$x[n]$ and $y[n]$ are periodic with the same frequency. Therefore it is possible to find an *LTI* system with this input output relationship without violating the eigen function property. This system is not unique because $H(e^{j\omega})$ needs to be have specific value only for $H\left(e^{j\left(\frac{2\pi}{12}\right)k}\right)$

The rest of $H(e^{j\omega})$ may be chosen arbitrarily.

(*i*) Note that, $x[n]$ and $y[n]$ are not periodic, with the same frequency. Further more note that $y_2[n]$ has $\frac{2}{3}$, the period of $x[n]$. Therefore $y[n]$ will be made up of complex exponsntials which are not present in $x[n]$. This violates the eigen function property of *LTI* system. Therefore the system be *LTI*.

Problem 3.39: As we have seen, the techniques of Fourier analysis are of value in examining continuous-time LTI systems because periodic complex exponentials are eigenfunctions for LTI systems. In this problem, we wish to substantiate the following statement: Although some LTI systems may have additional eigenfunctions, the complex exponentials are the *only* signals that are eigenfunctions of *every* LTI system.

(*a*) What are the eigenfunctions of the LTI system with unit impulse response $h(t) = \delta(t)$? What are the associated eigenvalues?

(*b*) Consider the LTI system with unit impulse response $h(t) = \delta(t - T)$. FInd a signal that is not of the form e^{st}, but that is an eigenfunction of the system with eigenvalue 1. Similarly, find the eigenfunctions with eigenvalues 1/2 and 2 that are not complex exponentials.

(*c*) Consider a stable LTI system with impulse response $h(t)$ that is real and even. Show that $\cos \omega t$ are eigenfunctions of this system.

(*d*) Consider the LTI system with impulse response $h(t) = u(t)$. Suppose that $\phi(t)$ must satisfy, and solve the equation. This result, together with those of parts (*a*) through (*c*), should prove the validity of the statement made at the beginning of the problem.

Solution:

(*a*) For this system $x(t) \to \delta(t) \to x(t)$

Therefore, all functions are eigen fucntion siwth an eigen value of one.

(*b*) The following is an eigen function with an eignen value of 1.

$$x(t) = \sum_k \delta(t - kT)$$

The following is an eigen function with an eigen value of $\frac{1}{2}$

$$x(t) = \sum_k \left(\frac{1}{2}\right)^k \delta(t - kT)$$

The following is an eigen function with an eigen value of 2.

$$x(t) = \sum_k (2)^k \delta(t - kT)$$

(*c*) If $h(t)$ is real and even then $H(j\omega)$ is real and even

$$e^{j\omega t} \to \boxed{H(j\omega)} \to H(j\omega)\, e^{j\omega t}$$

and

$$e^{-j\omega t} \to \boxed{H(j\omega)} \to H(-j\omega)\, e^{j\omega t} = H(j\omega)\, e^{j\omega t}$$

From these two statements,

we may argue that

$$\cos wt = \frac{1}{2}\left(e^{j\omega t} + e^{-j\omega t}\right) \to H(j\omega) \to H(j\omega)\cos(\omega t)$$

Therefore, $\cos(\omega t)$ is an eigen function. We may similarly show that $\sin(\omega t)$ is an eigen function.

(*d*) We have

$$\phi(t) \to u(t) \to \lambda\phi(t)$$

Therefore,

$$\lambda\phi(t) = \int_{-\infty}^{t} \phi(T)\, dT$$

Differentiating both sides w.r. to *t*,

We get

$$\lambda\phi'(t) = \phi(t)$$

Let

$$\phi(0) = \phi 0$$

Then

$$\phi(t) = \phi_0 \frac{t}{\lambda}$$

Problem 3.40: Exponential Sequence. FInd the DTFT of the sequence $x[n] = \alpha^n u[n]$. ·

Solution:

We know

$$X(e^{j\Omega}) = \sum_{n=-\infty}^{\infty} \alpha^n u[n]\, e^{-j\Omega n} = \sum_{n=0}^{\infty} \alpha^n\, e^{-j\Omega n}$$

This sum diverges for $|\alpha| \geq 1$. For $|\alpha| < 1$ we have the convergent geometric series:

$$X(e^{j\Omega}) = \sum_{n=0}^{\infty} (\alpha e^{-j\Omega})^n$$

$$= \frac{1}{1-\alpha e^{-j\Omega}} \quad |\alpha| < 1$$

If α is real valued, we may write the magnitude and phase of $X(e^{j\Omega})$ as

$$|X(e^{j\Omega})| = \frac{1}{\left((1-\alpha\cos\Omega)^2 + \alpha^2\sin^2\Omega\right)^{1/2}}$$

$$= \frac{1}{\left(\alpha^2 + 1 - 2\alpha\cos\Omega\right)^{1/2}}$$

$$\text{art}\,\{X(e^{j\Omega})\} = -\arctan\left(\frac{\alpha\sin\Omega}{1-\alpha\cos\Omega}\right)$$

The magnitude and phase of $X(e^{j\Omega})$ are depicted graphically in Fig. P. 3.40 (a) for $\alpha = 0.5$. Note that both are 2π periodic.

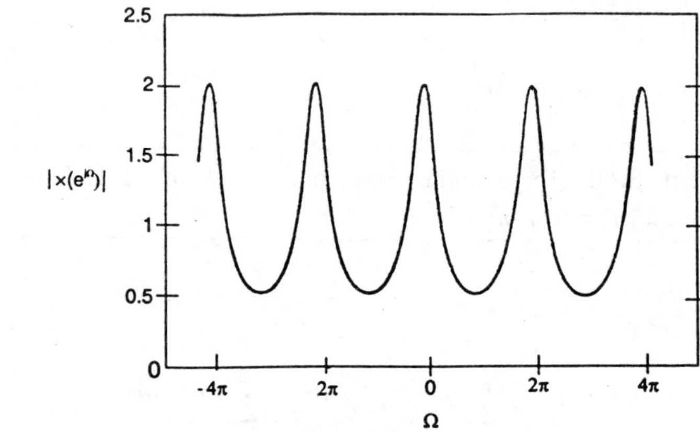

Fig. P. 3.40 (a)

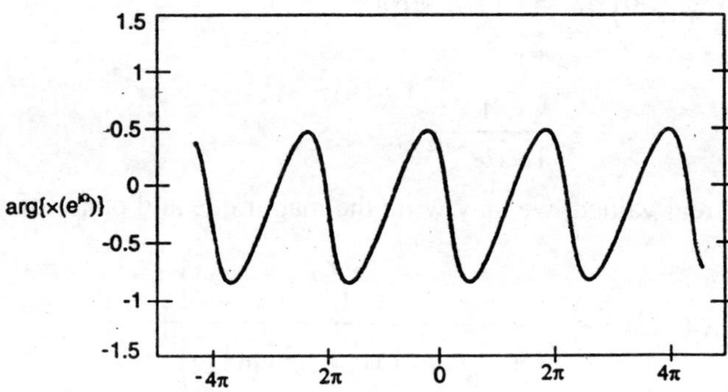

Fig. P. 3.40 (b)

Problem 3.41: Rectangular Pulse. Let

$$x[n] = \begin{cases} 1, & |n| \le M \\ 0, & |n| > M \end{cases}$$

as depicted in Fig. 3.41 (*a*). Find the DTFT of $x[n]$.

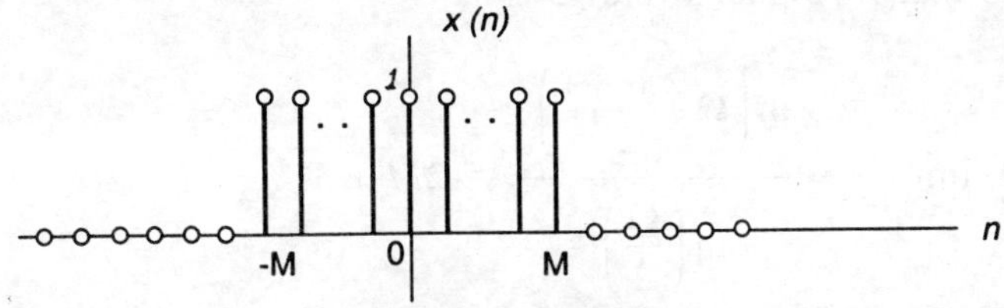

Fig. P. 3.41 (a)

Solution:

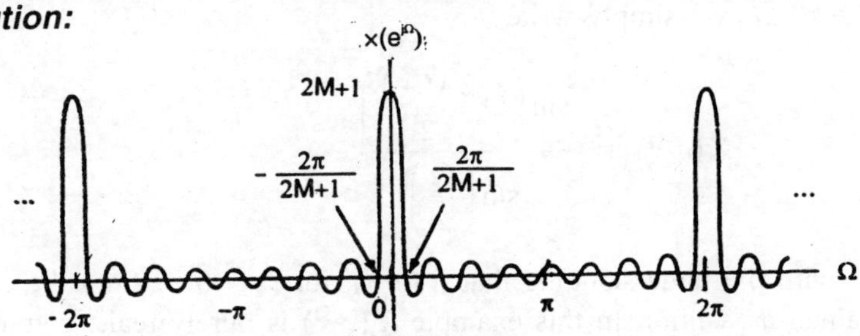

Fig. P. 3.41 (b)

$$X(e^{j\Omega}) = \sum_{n=-M}^{M} 1 e^{-j\Omega n}$$

Now perform the change of vairable, $m = n + M$, obtaining

$$X(e^{j\Omega}) = \sum_{m=0}^{2M} e^{-j\Omega(m - M)} = e^{j\Omega M} \sum_{m=0}^{2M} e^{-j\Omega M}$$

$$= \begin{cases} e^{j\Omega M} \dfrac{1-e^{-j\Omega(2M+i)}}{1-e^{-j\Omega}}, & \Omega \neq 0, \pm 2\pi, \pm 4\pi, \ldots \\ 2M+1, & \Omega = 0, \pm 2\pi, \pm 4\pi, \ldots \end{cases}$$

The expression for $X(e^{j\Omega})$, when $\Omega \neq 0, \pm 2\pi, \pm 4\pi, \ldots$, may be simplified by symmetrizing the powers of the exponential in the numerator and denominator, as shown by

$$X(e^{j\Omega}) = e^{j\Omega M} \frac{e^{-j\Omega(2M+1)/2} e^{j\Omega(2M+1)/2} - e^{-j\Omega(2M+1)/2}}{e^{-j\Omega/2}\left(e^{j\Omega/2} - e^{-j\Omega/2}\right)}$$

$$= \frac{\sin\left(\dfrac{\Omega}{2}(2M+1)\right)}{\sin\left(\dfrac{\Omega}{2}\right)}$$

Note that *L*'Hopital's rule gives

$$\lim_{\Omega \to 0, \pm 2\pi, \pm 4\pi, \ldots} \frac{\sin\left(\Omega \frac{2M+1}{2}\right)}{\sin\left(\frac{\Omega}{2}\right)} = 2M + 1$$

Hence rather than writing $X(e^{j\Omega})$ as two forms dependent on the value of Ω, we simply write

$$X(e^{j\Omega}) = \frac{\sin\left(\Omega \frac{2M+1}{2}\right)}{\sin\left(\frac{\Omega}{2}\right)}$$

with the understanding that $X(e^{j\Omega})$, for $\Omega = 0, \pm 2\pi, \pm 4\pi, \ldots$, is obtained as a limit. In this example $X(e^{j\Omega})$ is purely real. A graph of $X(e^{j\Omega})$ as a function of Ω is given in Fig. 3.41 (b). We see that as M increases, the time extent of $x[n]$ increases while the energy in $X(e^{j\Omega})$ becomes more concentrated near $\Omega = 0$.

Problem 3.42: Discrete-Time sinc Function. FInd the inverse DTFT

$$X(e^{j\Omega}) = \begin{cases} 1, & |\Omega| \le W \\ 0, & W < |\Omega| < \pi \end{cases}$$

as depicted in Fig. 3.42 (*a*).

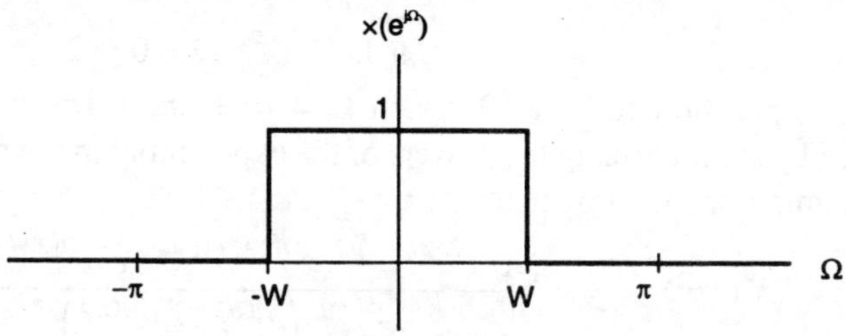

Fig. P. 3.42 (a)

Solution:

First note that $X(e^{j\Omega})$ is specified only for $-\pi < \Omega < \pi$. This is all that is needed, since $X(e^{j\Omega})$ is always 2π periodic and the inverse DTFT depends only on the values in the interval $-\pi < \Omega < \pi$.

Here, $$x[n] = \frac{1}{2\pi}\int_{-W}^{W} e^{j\Omega n}\, d\Omega = \frac{1}{2\pi n j} e^{j\Omega n}\Big|_{-W}^{W}, \quad n \neq 0$$

$$= \frac{1}{\pi n} \sin(Wn), \; n \neq 0$$

For $n = 0$, the integrand is unity and we have $x[0] = W/\pi$. It is easy to show using L'Hopital's rule that

$$\lim_{n\to 0} \frac{1}{\pi n} \sin(Wn) = \frac{W}{\pi}$$

and thus we usually write

$$x[n] = \frac{1}{\pi n} \sin(Wn)$$

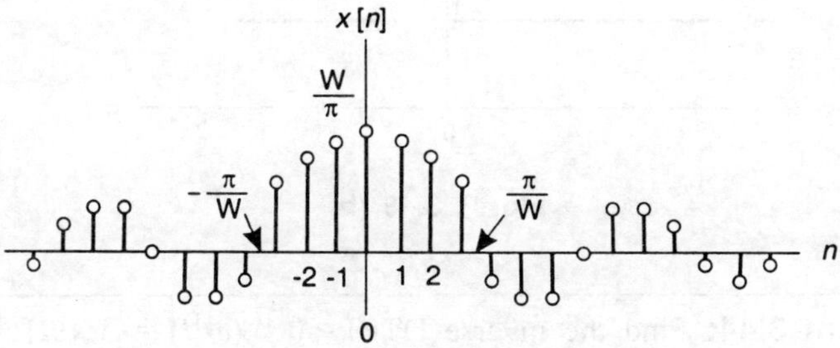

Fig. P. 3.42 (b)

as the inverse DTFT of $X(e^{j\Omega})$ with the understanding that the value at $n = 0$ is obtained as the limit. We may also write

$$x[n] = \frac{W}{\pi} \operatorname{sinc}\left(\frac{Wn}{\pi}\right)$$

using the sinc function notation. A graph depicting $x[n]$ is given in Fig. 3.42 (b).

Problem 3.43: The Impulse. Find the DTFT of $x[n] = \delta[n]$.

Solution:

For $x[n] = \delta[n]$, we have

$$X(e^{j\Omega}) = \sum_{n=-\infty}^{\infty} \delta[n]\, e^{-j\Omega n} = 1$$

Hence

$$\delta[n] \xleftrightarrow{DTFT} 1$$

This DTFT pair is depicted in Fig. 3.43.

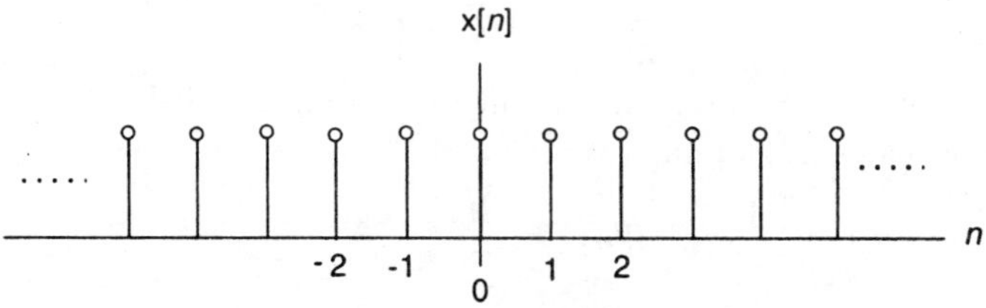

Fig. P. 3.43 (a)

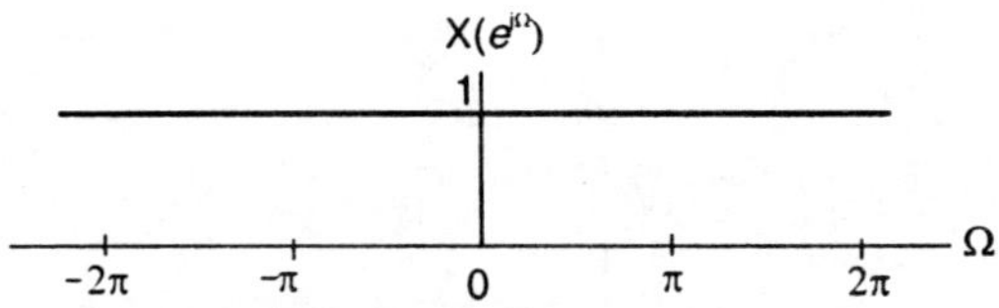

Fig. P. 3.43 (b)

> **Problem 3.44:** Find the inverse DTFT of $X(e^{j\Omega}) = \delta\{\Omega\}$, $-\pi < \Omega \leq \pi$.

Solution: By definition, $x[n] = \dfrac{1}{2\pi}\displaystyle\int_{-\pi}^{\pi} \delta(\Omega)\, e^{j\Omega n}\, d\Omega$

Use the sifting property of the impulse function to obtain

$x[n] = \dfrac{1}{2}\pi$, and thus write

$$\frac{1}{2\pi} \xleftrightarrow{DTFT} \delta(\Omega),\ -\pi < \Omega \leq \pi$$

In this example we have again defined only one period of $X(e^{j\Omega})$. We can alternatively define $X(e^{j\Omega})$ over all Ω by writing it as an infinite sum of delta functions shifted by integer multiples of 2π.

$$X(e^{j\Omega}) = \sum_{k=-\infty}^{\infty} \delta(\Omega - k2\pi)$$

Both definitions are common. This DTFT pair is depicted in Fig. P. 3.44.

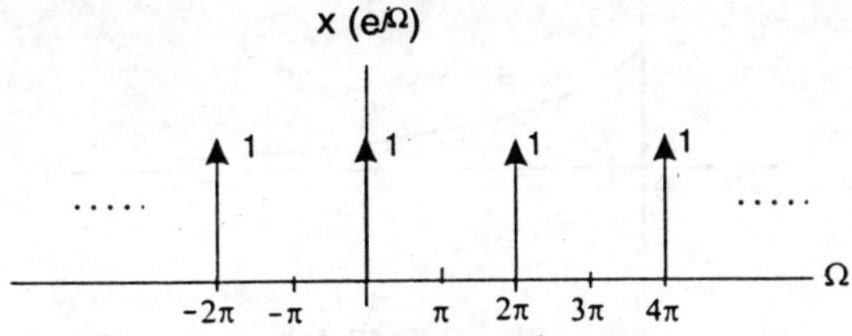

Fig. P. 3.44 (a)

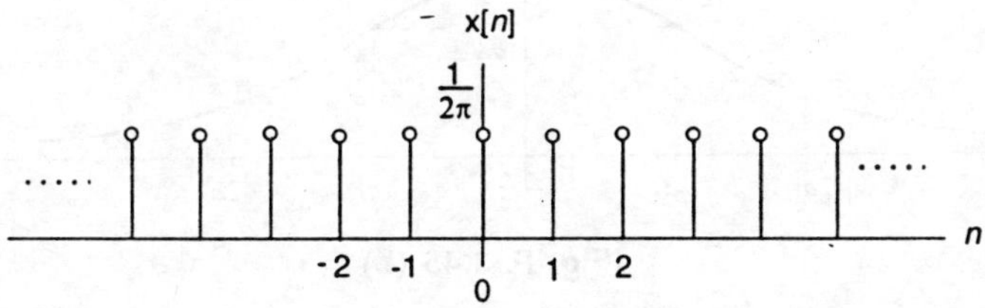

Fig. P. 3.44 (b)

Problem 3.45: Real Exponential. Find the FT of $x(t) = e^{-at}u(t)$.

Solution:

The *FT* does not converge for $a \leq 0$ since $x(t)$ is not absolutely integrable, as shown by

$$\int_0^\infty e^{-at}\, dt = \infty, \; a \leq 0$$

For $a > 0$, we have

$$X(j\omega) = \int_{-\infty}^{\infty} e^{-at}\, u(t)\, e^{-j\omega t}\, dt = \int_0^\infty e^{-(a+j\omega)t}\, dt$$

$$= -\frac{1}{a+j\omega} e^{-(a+j\omega)}\Big|_0^\infty = \frac{1}{a+j\omega}$$

Converting to polar form, the magnitude and phase of $X(j\omega)$

$$|X(j\omega)| = \frac{1}{(a^2+\omega^2)^{1/2}} \text{ arg }\{X(j\omega)\} = -\text{ arg tan}\left(\frac{\omega}{a}\right)$$

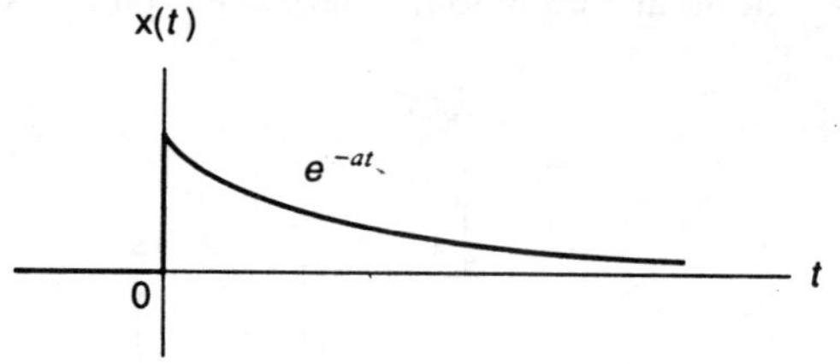

Fig. P. 3.45 (a)

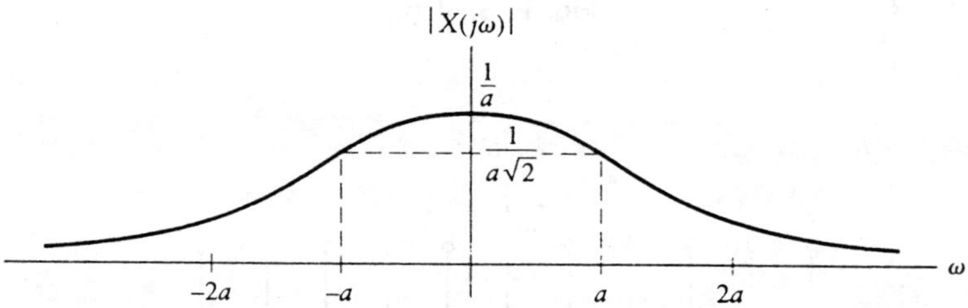

Fig. P. 3.45 (b)

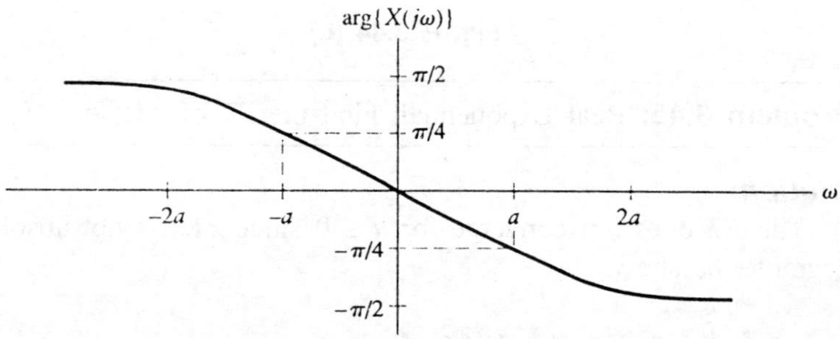

Fig. P. 3.45 (c)

Problem 3.46: Rectangular Pulse. Consider the rectangular pulse depicted in Fig. 3.46 (*a*) and defined as

$$x(t) = \begin{cases} 1, & -T \leq t \leq T \\ 0, & |t| > T \end{cases}$$

Find the FT of $x(t)$.

Solution: The rectangular pulse $x(t)$ is absolutely integrable provided $T < \infty$. For $\omega \neq 0$ we have

$$X(j\omega) = \int_{-\infty}^{\infty} x(t)\, e^{-j\omega t}\, dt = \int_{-T}^{T} e^{-j\omega t}\, dt = -\frac{1}{j\omega} e^{-j\omega t}\Big|_{-T}^{T} = \frac{2}{\omega} \sin(\omega T),\ \omega \neq 0$$

For $\omega = 0$, the integral simplifies to $2T$. It is straightforward to show using L'Hopital's rule that

$$\lim_{\omega \to 0} \frac{2}{\omega} \sin(\omega T) = 2T$$

Thus we write for all ω, $X(j\omega) = \dfrac{2}{\omega} \sin(\omega T)$

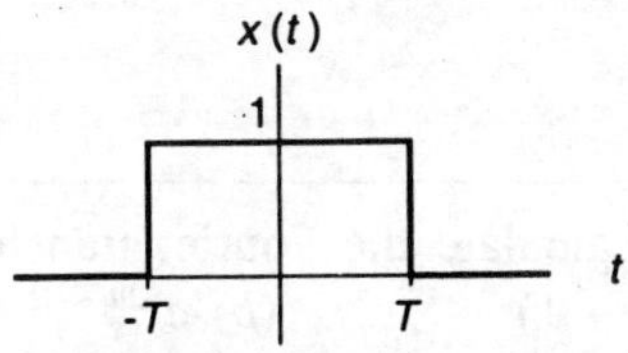

Fig. P. 3.46 (a)

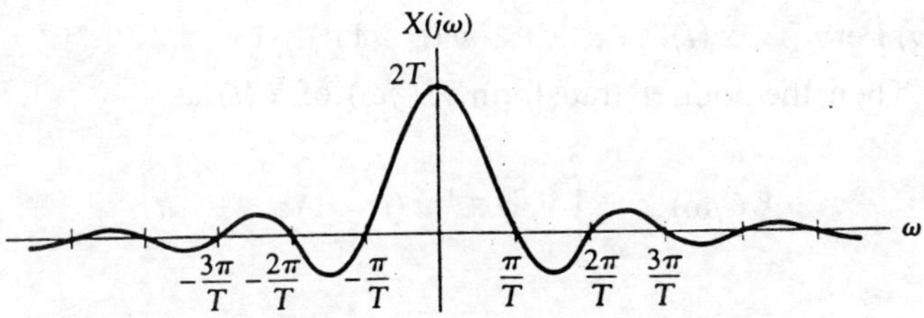

Fig. P. 3.46 (b)

with the understanding that the value at $\omega = 0$ is obtained by evaluating a limit. In this case $X(j\omega)$ is real. It is depicted in Fig. 3.46 (b). The magnitude spectrum is

$$|X(j\omega)| = 2\left|\frac{\sin(\omega T)}{\omega}\right|$$

and the phase spectrum is

$$\arg\{X(j\omega)\} = \begin{cases} 0, & \dfrac{\sin(\omega T)}{\omega} > 0 \\ \pi, & \dfrac{\sin(\omega T)}{\omega} < 0 \end{cases}$$

We may write $X(j\omega)$ using the sin c function notation as

$$X(j\omega) = 2T \sin c\left(\frac{\omega T}{\pi}\right)$$

Chapter 4

Fourier Transform

Problem 4.1: Calculate the Fourier transforms of:
(*a*) $e^{-2(t-1)}u(t-1)$ (*b*) $e^{-2|t-1|}$
Sketch and label the magnitude of each Fourier transform.

Solution :

(*a*) Let $x(t) = e^{-2(t-1)}u(t-1)$

Then the fourier transform $X(j\omega)$ of $x(t)$ as

$$X(j\omega) = \int_{-\infty}^{\infty} e^{-2(t-1)}u(t-1)e^{-j\omega t} \cdot dt$$

$$= \int_{1}^{\infty} e^{-2(t-1)}e^{-j\omega t} \cdot dt = \frac{e^{-j\omega}}{2+j\omega}$$

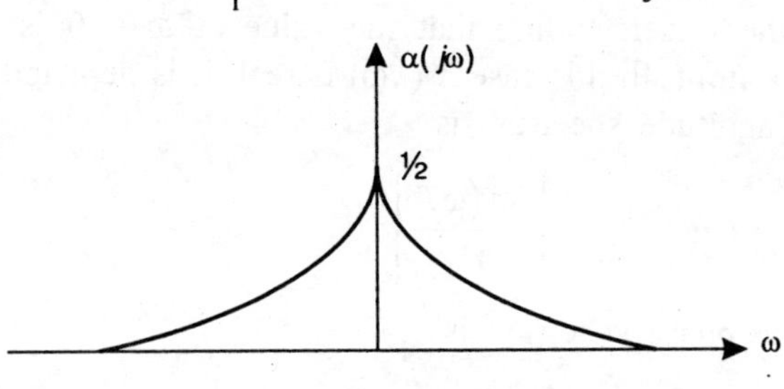

Fig. P. 4.1 (a)

(*b*) let $x(t) = e^{-2|t-1|}$

Then the fourier transform $X(j\omega)$ of $x(t)$ is

$$X(j\omega) = \int_{-\infty}^{\infty} e^{-2|t-1|}e^{-j\omega t} \cdot dt$$

$$= \int_{1}^{\infty} e^{-2|t-1|} e^{-j\omega t} \cdot dt + \int_{-\infty}^{1} e^{2|t-1|} e^{-j\omega t} \cdot dt$$

$$= \frac{e^{-j\omega}}{(2+j\omega)} - \frac{e^{-j\omega}}{(2-j\omega)} = \frac{4\,e^{-j\omega}}{(4+\omega^2)}$$

$|X(j\omega)|$ is as shown in figure below

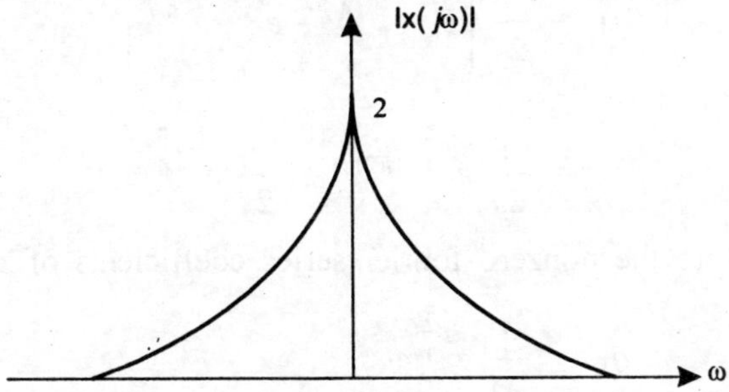

Fig. P. 4.1 (b)

> **Problem 4.2:** Use the Fourier transform analysis equation to calculate the Fourier transforms of:
>
> $$\delta(t+1) + \delta(t-1)$$

Solution:

Let $x(t) = \delta(t+1) + \delta(t-1)$

The fourier transform $X(j\omega)$ of $x(t)$ is

$$\alpha_1(j\omega) = \int_{-\infty}^{\infty} [\delta(t+1) + \delta(t-1)]\; e^{-j\omega t} \cdot dt = 2 \cos w$$

> **Problem 4.3:** Determine the Fourier transform of each of the following periodic signals:
>
> (*a*) $\sin\left(2\pi t + \frac{\pi}{4}\right)$ (*b*) $1 + \cos\left(6\pi t + \frac{\pi}{8}\right)$

Solution:

(a) The signal $x_1(t) = \sin(2\pi t + \frac{\pi}{4})$ is periodic with a fundamental period of $T = 1$, This translates to a fundamental frequency of $\omega_0 = 2\pi$.

The non zero fourier series coefficients of this signal may be found by writing it in the form

$$x_1(t) = \frac{1}{2j}\left(e^{j\left(2\pi t - \frac{\pi}{4}\right)} - e^{j\left(2\pi t + \frac{\pi}{4}\right)}\right)$$

$$= \frac{1}{2j} e^{j\frac{\pi}{4}} e^{j(2\pi t)} - \frac{1}{2j} e^{-j\frac{\pi}{4}} e^{-j2\pi t}$$

Therefore the nonzero fourier series coefficients of $x_1(t)$ are

$$a_1 = \frac{1}{2j} e^{j\frac{\pi}{4}} e^{j2\pi t}$$

$$a_{-1} = -\frac{1}{2j} e^{-j\frac{\pi}{4}} e^{-j2\pi t}$$

We know that for periodic signals the fourier transform consists of a train of impulse occurring at $k\omega_0$. Furthermore the area under each impulse of 2π times the fourier series coefficients a_k.

Therefore for $x_1(t)$ the corresponding fourier transform $X_1(j\omega)$ is given by

$$X_1(j\omega) = 2\pi a_1 \delta(\omega - \omega_0) - 2\pi a_{-1}\,\delta(\omega - \omega_0)$$

$$= \left(\frac{\pi}{j}\right) e^{j\frac{\pi}{4}} \delta(\omega - 2\pi) - \left(\frac{\pi}{j}\right) e^{-j\frac{\pi}{4}} \delta(\omega - 2\pi)$$

(b) The signal

$x_2(t) = 1 - \cos\left(6\pi t + \frac{\pi}{8}\right)$ is periodic with a fundamental period

of $T = \frac{1}{3}$. This translates to a fundamental frequency of $\omega_o = 6\pi$.

The non zero fourier series coefficients of this signal may be found in writing it in the form

$$x_2(t) = 1 + \frac{1}{2}\left(e^{j\left(6\pi t+\frac{\pi}{8}\right)} + e^{-j\left(6\pi t+\frac{\pi}{8}\right)}\right)$$

$$= 1 + \frac{1}{2}e^{j\frac{\pi}{8}}e^{j6\pi t} + \frac{1}{2}e^{-j\frac{\pi}{8}}e^{-j6\pi t}$$

Therefore, the non zero fourier series coefficients of $x_2(t)$ are

$$a_o = 1 \qquad a_1 = \frac{1}{2}e^{j\frac{\pi}{8}}e^{j6\pi t} \qquad a_2 = \frac{1}{2}e^{-j\frac{\pi}{8}}e^{-j6\pi t}$$

We know that for periodic signal, the fourier transform consists of a train of impulses occuring at $k\omega_0$.

Furthermore the area under each impulse is 2π times the fourier series coefficients a_k.

Therefore for $x_2(t)$, the corresponding fourier transform $X_2(j\omega)$ is given by

$$X_2(j\omega) = 2\pi a_o\delta(\omega) + 2\pi a_1\delta(\omega - \omega_0) + 2\pi a_{-1}\,\delta(\omega + \omega_0)$$

$$= 2\pi\delta(\omega) + \pi e^{j\frac{\pi}{8}}\delta(\omega - 6\pi) + \pi e^{-j\frac{\pi}{8}}\delta(\omega - 6\pi)$$

Problem 4.4: Use the Fourier transform synthesis equation to determine the inverse Fourier transforms of:

(*a*) $X_1(j\omega) = 2\pi\delta(\omega) + \pi\delta(\omega - 4\pi) + \pi\delta(\omega + 4\pi)$

(*b*) $X_2(j\omega) = \begin{cases} 2, & 0 \le \omega \le 2 \\ -2, & -2 \le \omega < 0 \\ 0, & |\omega| > 2 \end{cases}$

Solution:

(*a*) The inverse fourier transform is

$$x_1(t) = \left(\frac{1}{2\pi}\right)\int_{-\infty}^{\infty}[2\pi\delta(\omega) + \pi\delta(\omega - 4\pi) + \pi\delta(\omega + 4\pi)]\ e^{-j\omega t}\cdot d\omega$$

$$= \left(\frac{1}{2\pi}\right)[2\pi e^{j0t} + \pi e^{j4\pi t} + \pi e^{-j4\pi t}]$$

$$= 1 + \left(\frac{1}{2}\right) e^{j4\pi t} + \left(\frac{1}{2}\right) e^{-j4\pi t} = 1 + \cos(4\pi t)$$

(*b*) The inverse fourier transform is

$$x_2(t) = \left(\frac{1}{2\pi}\right) \int_{-\infty}^{\infty} X_2(j\omega)\, e^{jwt} \cdot d\omega$$

$$= \left(\frac{1}{2\pi}\right) \int_{0}^{2} 2\, e^{jwt}\, d\omega + \left(\frac{1}{2\pi}\right) \int_{-2}^{0} (-2)\, e^{jwt} \cdot d\omega$$

$$= \frac{(e^{j2t} - 1)}{\pi jt} - \frac{(1 - e^{-j2t})}{\pi jt} = -\frac{(4j \sin^2 t)}{\pi t}$$

Problem 4.5: Use the Fourier transform synthesis equation to determine the inverse Fourier transform of $X(j\omega) = |X(j\omega)e^{j\sphericalangle X(j\omega)}$, where

$$|X(j\omega)| = 2\{u(\omega + 3) - u(\omega - 3)\},$$

$$\sphericalangle X(j\omega) = -\frac{3}{2}\omega + \pi.$$

Use your answer to determine the values of t for which $x(t) = 0$.

Solution:

From the given information

$$x(t) = \left(\frac{1}{2\pi}\right) \int_{-\infty}^{\infty} X(j\omega)\, e^{j\omega t} \cdot dw$$

$$= \left(\frac{1}{2\pi}\right) \int_{-\infty}^{\infty} |X(j\omega)|\, e^{j\sphericalangle X(j\omega)}\, e^{j\omega t} \cdot d\omega$$

$$= \left(\frac{1}{2\pi}\right) \int_{-3}^{3} 2\, e^{-\frac{3}{2}\omega t}\, e^{-j\omega t} \cdot dt = \frac{-2}{\pi\left(t - \frac{3}{2}\right)} \sin\left[3\left(t - \frac{3}{2}\right)\right]$$

The signal $x(t)$ is zero, when $3\left(t - \frac{3}{2}\right)$ is a non zero integer multiple of π. This gives

$$= k\frac{\pi}{2} + \frac{3}{2} \text{ for } k \in I, \text{ and } k \neq 0$$

Problem 4.6: Given that $x(t)$ has the Fourier transform $X(j\omega)$, express the Fourier transforms of the signals listed below in terms of $X(j\omega)$.

(*a*) $x_1(t) = x(1-t) + x(-1-t)$

(*b*) $x_2(t) = x(3t-6)$

(*c*) $x_3(t) = \dfrac{d^2}{dt^2} x(t-1)$

Solution:

Throughout this problem we assume that

$$x(t) \xleftrightarrow{FT} X(j\omega)$$

(*a*) Using the time reversal property, we have

$$x(-t) \xleftrightarrow{FT} X(-j\omega)$$

Using the time shifting property on this, we have

$$x(-t+1) \xleftrightarrow{FT} e^{-j\omega t} \cdot X(-j\omega)$$

and

$$x(-t-1) \xleftrightarrow{FT} e^{Jwt}\, X(-j\omega)$$

Therefore

$$x_1(t) = x(-t+1) + x(-t-1) \xleftrightarrow{FT}$$

$$e^{-j\omega t}\, X(-j\omega) + e^{j\omega t}\, X(-j\omega)$$

$$\xleftrightarrow{FT} 2\, X(-j\omega) \cos \omega.$$

(*b*) Using the time scaling property,

we have

$$x(3t) \xleftrightarrow{FT} \frac{1}{3} X\left(j\frac{\omega}{3}\right)$$

Using the time shifting property on this we have

$$x_2(t) = x[3(t-2)] \xleftrightarrow{FT} e^{-2j\omega}\, \frac{1}{3} X\left(j\frac{\omega}{3}\right)$$

(*c*) Using the differentiation in time property we have

$$\frac{dx(t)}{dt} \xleftrightarrow{FT} j\omega X(j\omega)$$

Applying the property again,
we have

$$\frac{d^2x(t)}{dt^2} \xleftrightarrow{FT} -\omega^2 X(j\omega)$$

Using the time shifting property,
we have

$$x_3(t) = \frac{d^2x(t-1)}{dt^2} \xleftrightarrow{FT} -\omega^2 X(j\omega)\, e^{-j\omega t}$$

Problem 4.7: For each of the following Fourier transforms, use Fourier transform properties to determine whether the corresponding time-domain signal is (*i*) real, imaginary, or neither and (*ii*) even, odd, or neither. Do this without evaluating the inverse of any of the given transforms.

(*a*) $X_1(j\omega) = u(\omega) - u(\omega - 2)$

(*b*) $X_2(j\omega) = \cos(2\omega)\sin\left(\frac{\omega}{2}\right)$

(*c*) $X_3(j\omega) = A(\omega)e^{jB(\omega)}$, where $A(\omega) = (\sin 2\omega)/\omega$ and

$B(\omega) = 2\omega + \frac{\pi}{2}$

(*d*) $X(j\omega) = \sum_{k=-\infty}^{\infty}\left(\frac{1}{2}\right)^{|k|}\delta\left(\omega - \frac{k\pi}{4}\right)$

Solution:

(*a*) Since $X_1(j\omega)$ is not conjugate symmetric, the corresponding signal $x_1(t)$ is not real. Since $X_1(j\omega)$ is neither even nor odd, the corresponding signal $x_1(t)$ is neither even nor odd.

(*b*) The fourier transform of a real and odd signal is purely imaginary and odd. Therefore we may conclude that the fourier transform of a purely imaginary and odd signal is real and odd. Since $X_2(j\omega)$ is real and odd we may therefore conclude that the corresponding signal $x_2(t)$ is purely imaginary and odd.

(*c*) Consider a signal $y_3(t)$ whose magnitude of the Foureir transform is $|y_3(j\omega)| = A(\omega)$ and whose phase of the fourier transform is $\sphericalangle y_3(j\omega)\} = 2\omega$.

since $|y_3(j\omega)| = |y_3(-j\omega)|$ and

$$\measuredangle |y_3(j\omega)| = -\measuredangle |y_3(-j\omega)|$$

we may conclude that the signal $y_3(t)$ is real.

Now, consider the signal $x_3(t)$ with Fourier transform

$$X_3(j\omega) = y_3(j\omega)e^{j\frac{\omega}{2}} = j\,y_3(j\omega)$$

using the result from the previous paragraph and the lineaarity property of the fourier transform, we may conclude that $x_3(t)$ has to imaginary. Since the fourier transform $X_3(j\omega)$ is neither purely imaginery nor purely real. The signal $x_3(t)$ is neither even nor odd.

(*d*) Since $X_4(j\omega)$ is both real and even the corresponding signal $x_4(t)$ is real and even.

Problem 4.8: Consider the signal

$$x(t) = \begin{cases} 0, & t < -\frac{1}{2} \\ t + \frac{1}{2}, & -\frac{1}{2} \le t \le \frac{1}{2}. \\ 1, & t > \frac{1}{2} \end{cases}$$

(*a*) Use the differentiation and integration properties and the Fourier transform pair for the rectangular pulse to find a closed-form expression for $X(j\omega)$.

(*b*) What is the Fourier transform of $g(t) = x(t) - \frac{1}{2}$?

Solution:

(*a*) The signal $x(t)$ is as shown in figure

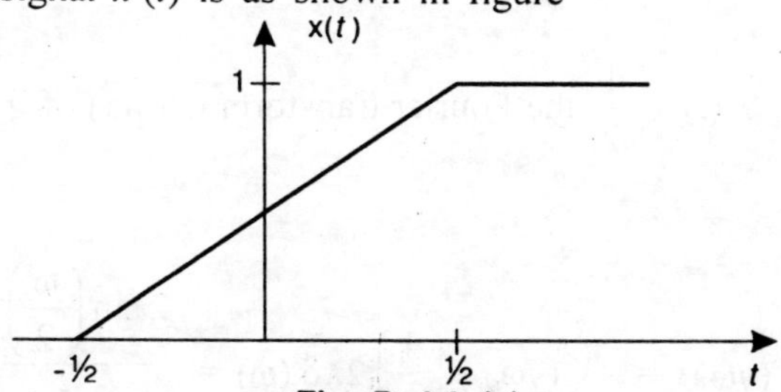

Fig. P. 4.8 (a)

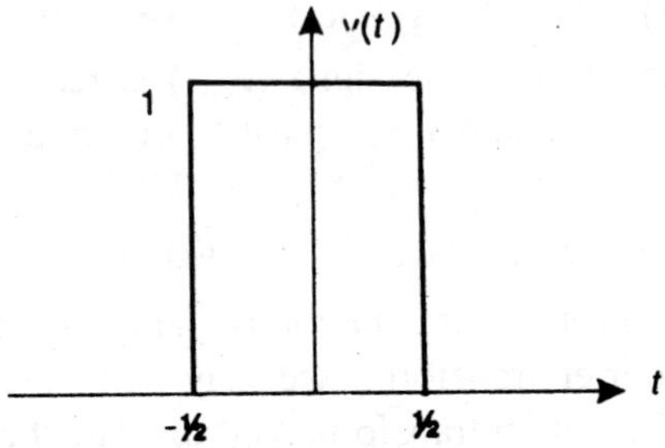

Fig. P. 4.8 (b)

We may express this signal as

$$x(t) = \int_{-\infty}^{t} y(t)\, dt$$

Where $y(t)$ is the rectangular pulse shown in fig. using the integration property of the fourier transform, we have

$$x(t) \xleftrightarrow{FT} X(j\omega) = \frac{1}{j\omega} y(j\omega) + \pi y(j0)\,\delta(\omega)$$

$$y(j\omega) = \frac{2\sin\left(\frac{\omega}{2}\right)}{\omega}$$

Therefore,

$$X(j\omega) = \frac{2\sin\left(\frac{\omega}{2}\right)}{j\omega^2} - \pi\delta(\omega)$$

(*b*) If $g(t) = x(t) + \frac{1}{2}$ the Fourier transform $G(j\omega)$ of $g(t)$ is given by

$$G(j\omega) = X(j\omega) - \left(\frac{1}{2}\right) 2\pi\delta(\omega) = \frac{2\sin\left(\frac{\omega}{2}\right)}{j\omega^2}$$

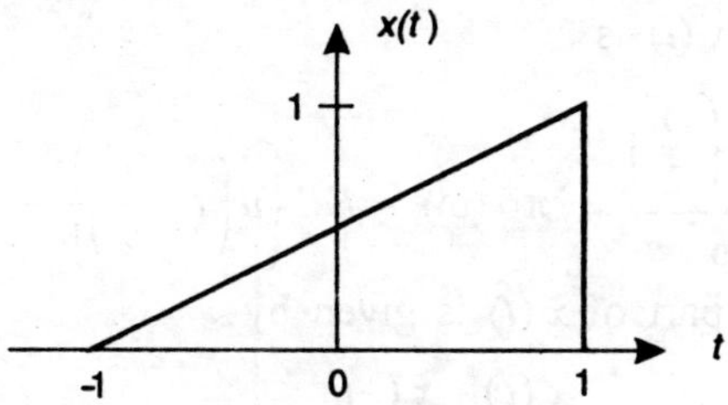

Fig. P. 4.8 (c)

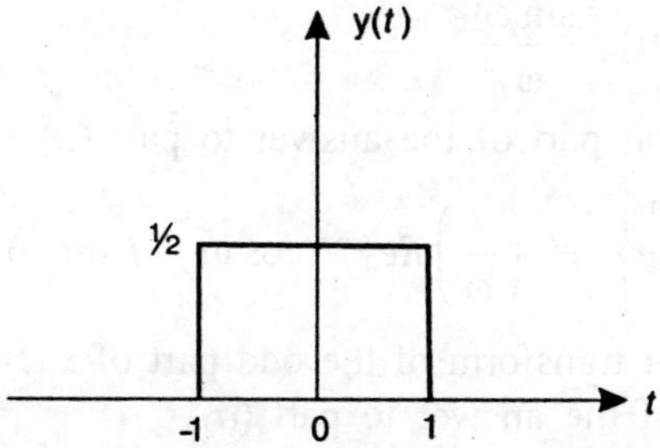

Fig. P. 4.8 (d)

Problem 4.9: Consider the signal

$$x(t) = \begin{cases} 0, & |t| > 1 \\ (t+1)/2, & -1 \le t \le 1 \end{cases}$$

(*a*) Determine the closed-form expression for $X(j\omega)$.
(*b*) Take the real part of your answer to part (*a*), and verify that it is the Fourier transform of the even part of $x(t)$.
(*c*) What is the Fourier transform of the odd part of $x(t)$?

Solution:

We see that this signal is very similar to the one considered in the previous problem.

In fact we may again express the signal $x(t)$ in term of the rectangular pulse $y(t)$ shown above as follows

$$x(t) = \int_{-\infty}^{t} y(t)\, dt - u\left(t - \frac{1}{2}\right) dt$$

Using the result obtained in part (a) of the pervious problem. The Fourier theorem

$X(j\omega)$ of $x(t)$ is

$$X(j\omega) = \frac{2\sin\left(\frac{\omega}{2}\right)}{j\omega^2} + \pi\delta(\omega) - FT\left\{u\left(t-\frac{1}{2}\right)\right\} = \frac{\sin\omega}{+j\omega^2} - \frac{e^{-j\omega}}{j\omega^2}$$

(*b*) The even part of $x(t)$ is given by

$$\varepsilon v\ \{x(t)\} = \frac{x(t) + x(-t)}{2}$$

Therefore

$$FT\ \{\varepsilon v\ [x(t)]\} = \frac{\sin\omega}{\omega}$$

Now, the real part of the answer to part (*a*), is

$$Re\left\{\frac{-e^{-j\omega}}{j\omega}\right\} = \left(\frac{1}{\omega}\right) Re\{j\ (\cos\omega - j\sin\omega)\} = \frac{\sin\omega}{\omega}$$

(*c*) The fourier transform of the odd part of $x(t)$ is same is j times. Imaginary part of the answer to part (*a*),

We have

$$Im\left\{\frac{\sin\omega}{j\omega^2} - \frac{e^{(-j\omega)}}{j\omega}\right\} = -\frac{\sin\omega}{\omega^2} + \frac{\cos\omega}{\omega}$$

Therefore,

The desired result is

$$FT\ \{\text{odd part of } x(t)\} = \frac{\sin\omega}{j\omega^2} - \frac{\cos\omega}{j\omega}$$

Problem 4.10: (*a*) Determine the Fourier transform of the following signal:

$$x(t) = t\left(\frac{\sin t}{\pi t}\right)^2$$

(*b*) Use Parseval's relation and the result of the previous part to determine the numerical value of

$$A = \int_{-x}^{+x} t^2 \left(\frac{\sin t}{\pi t}\right)^4 dt$$

Solution: (*a*) We know from $\frac{\sin t}{\pi t} \xleftrightarrow{FT}$ Rectangular function $y\,(j\omega)$

Therefore, $\left(\frac{\sin t}{\pi t}\right)^2 \xleftrightarrow{FT} \frac{1}{2\pi}$ [Rectangular function $y\,(j\omega)$.]

This is a triangular function $y_1\,(j\omega)$ as shown in the figure.

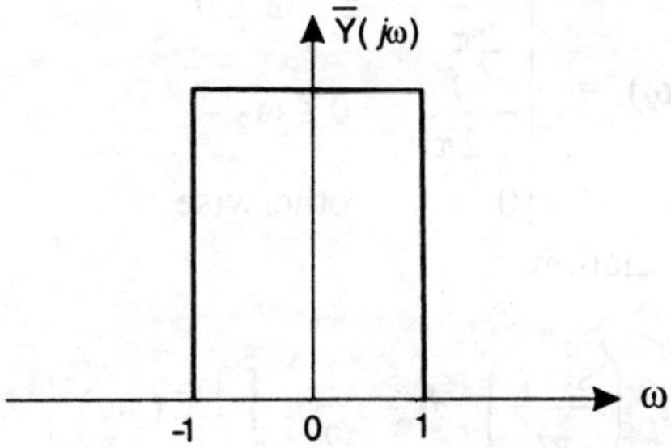

Fig. P. 4.10 (a)

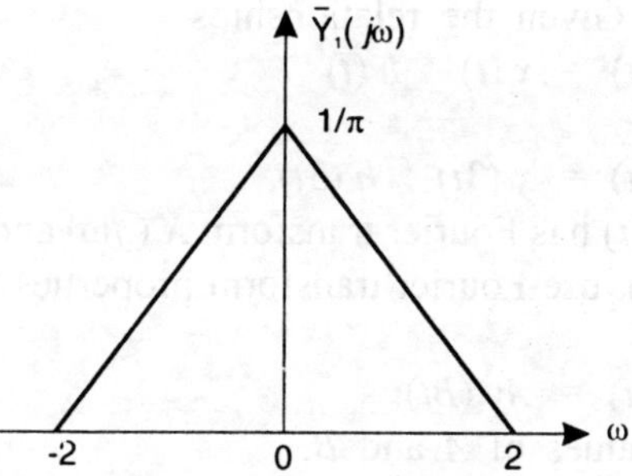

Fig. P. 4.10 (b)

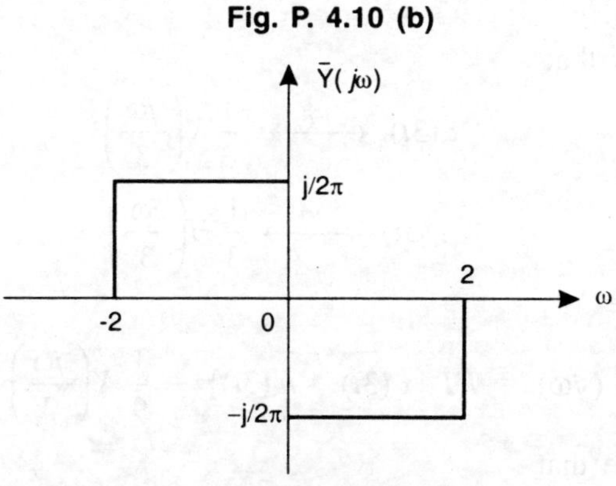

Fig. P. 4.10 (c)

We may write

$$t\left(\frac{\sin t}{\pi t}\right)^2 \xleftrightarrow{FT} X(j\omega) = J\frac{d}{d\omega} y_1(j\omega)$$

This is shown in figure,

$X(j\omega)$ may be expressed mathematically as

$$X(j\omega) = \begin{cases} \dfrac{j}{2\pi} & -2 \leq \omega \leq 0 \\ -\dfrac{j}{2\pi} & 0 \leq \omega_2 \\ 0 & \text{otherwise} \end{cases}$$

(*b*) Using the relation

$$A = \int_{-\infty}^{+\infty} t^2 \left(\frac{\sin t}{\pi t}\right)^4 dt = \frac{1}{2}\pi \int_{-\infty}^{\infty} \left|X(j\omega)^2\right| d\omega = \frac{1}{2\pi}$$

Problem 4.11: Given the relationships

$$y(t) = x(t) * h(t)$$

and

$$g(t) = x(3t) * h(3t),$$

and given that $x(t)$ has Fourier transform $X(j\omega)$ and $h(t)$ has Fourier transform $H(j\omega)$, use Fourier transform properties to show that $g(t)$ has the form

$$g(t) = Ay(Bt).$$

Determine the values of A and B.

Solution:

We know that

$$x(3t) \xleftrightarrow{FT} \frac{1}{3} X\left(\frac{j\omega}{3}\right)$$

$$h(3t) \xleftrightarrow{FT} \frac{1}{3} H\left(\frac{j\omega}{3}\right)$$

Therefore

$$G(j\omega) = FT\{x(3t) * h(3t)\} = \frac{1}{9} X\left(\frac{j\omega}{3}\right) H\left(\frac{j\omega}{3}\right)$$

Now note that

$$y(j\omega) = FT\{X(t) * h(t)\} = X(j\omega)\, H(j\omega)$$

From this

We may write

$$y\left(j\frac{\omega}{3}\right) = X\left(j\frac{\omega}{3}\right) \cdot H\left(j\frac{\omega}{3}\right)$$

Using this in equation (**) we have

$$G(j\omega) = \frac{1}{9}\, y\left(j\frac{\omega}{3}\right)$$

and $$g(t) = \frac{1}{3}\, y(3t)$$

Therefore,

$$A = \frac{1}{3} \text{ and } B = 3$$

Problem 4.12: Consider the Fourier transform pair

$$e^{-|t|} \xleftrightarrow{FT} \frac{1}{1+\omega^2}$$

(*a*) Use the appropriate Fourier transform properties to find the Fourier transform of $te^{-|t|}$

(*b*) Use the result from part (*a*), along with the duality property, to determine the Fourier transform of $\dfrac{4t}{\left(1+t^2\right)^2}$.

Solution:

We know that

$$e^{-|t|} = X(j\omega) \xleftrightarrow{FT} \frac{2}{1+\omega^2}$$

Using the differentiation in frequency property we have

$$t\, e^{-|t|} \xleftrightarrow{FT} j\frac{d}{d\omega}\left(\frac{2}{1+\omega^2}\right) = -\frac{4\, j\omega}{\left(1+\omega^2\right)^2}$$

(*b*) The duality property state that if

$$g(t) \xleftrightarrow{FT} G(j\omega)$$

Then

$$G(t) \xleftrightarrow{FT} 2\pi g(j\omega)$$

Now, since

$$t\, e^{-|t|} \xleftrightarrow{FT} -\frac{4\, j\omega}{\left(1+\omega^2\right)^2}$$

We may use duality to write

$$\frac{-4\, jt}{\left(1+t^2\right)^2} \xleftrightarrow{FT} 2\, \pi\omega e^{-|\omega|}$$

Multiplying both sides by (j)
Then we get

$$\frac{-4\, jt}{(1+t^2)^2} \xleftrightarrow{FT} j2\pi\omega\, e^{-|\omega|} \quad \textbf{Ans.}$$

Problem 4.13: Let $x(t)$ be a signal whose Fourier transform is

$$X(j\omega) = \delta(\omega) + \delta(\omega - \pi) + \delta(\omega - 5),$$

and let

$$h(t) = u(t) - u(t-2).$$

(*a*) Is $x(t)$ periodic?
(*b*) Is $x(t) * h(t)$ periodic?
(*c*) Can the convolution of two aperiodic signals be periodic?

Solution:

(*a*) Taking the inverse fourier transform of $X(j\omega)$, Then we get

$$x(t) = \frac{1}{2\pi} + \frac{1}{2\pi} e^{j\pi t} + \frac{1}{2\pi} e^{j5t}$$

The signal $x(t)$ is therefore a constant summed with two complex exponential whose fundamental frequencies are $2\frac{\pi}{5}$ rad/sec and 2 rad/sec. These two complex exponential are not harmonically related. That is the fundamental frequencies of these complex exponentials can never be integral multiplers of a common fundamental frequencies. Therefore the signal is not periodic.

(*b*) Consider the signal $y(t) = x(t) * h(t)$,

from the convolution property,

We know that

$$y(j\omega) = X(j\omega)\, H(j\omega)$$

Also, from $h(t)$

We know that

$$H(j\omega) = e^{-j\omega} \frac{2\sin\omega}{\omega}$$

The function $H(j\omega)$ is zero, when $w = k\pi$

where k is a non zero integer.

$$y(j\omega) = X(j\omega)\, H(j\omega) = \delta(\omega) + \delta(\omega - 5)$$

This gives,

$$y(t) = \frac{1}{2\pi} + \frac{1}{2\pi} e^{j5t}$$

Therefore,

$y(t)$ is a complex exponential summed with a constant. We know that a complex exponential is periodic. Adding a constant to a complex exponential does not affect its periodicity. Therefore, $y(t)$ will be a signal with a fundamental frequency of $2\pi/5$.

(*c*) From the result of part (*a*) and (*b*).

We see that the answer is yes.

Problem 4.14: Consider a signal $x(t)$ with Fourier transform $X(j\omega)$. Suppose we are given the following facts:

1. $x(t)$ is real and nonnegative.
2. $F^{-1}\{(1 + j\omega) X(j\omega)\} = Ae^{-2t} u(t)$, where A is independent of t.
3. $\int_{-\infty}^{\infty} |X(j\omega)|^2 \, d\omega = 2\pi$

Determine a closed-form expression for $x(t)$.

Solution:

Taking the fourier transform of both sides of the equation

$F^{-1}\{(1 + j\omega)\, X(j\omega)\} = A\, 2^{-2t} u(t)$

We obtain

$$X(j\omega) = \frac{A}{(1+j\omega)(2+j\omega)} = A\left\{\frac{1}{1+j\omega} - \frac{1}{2+j\omega}\right\}$$

Taking, the inverse fourier transform of the above equation

$$x(t) = Ae^{-t}u(t) - Ae^{-2t}u(t)$$

We know that

$$\int_{-\infty}^{\infty} |X(j\omega)|^2 \, d\omega = 2\pi \int_{-\infty}^{\infty} |x(t)|^2 \, dt$$

using the fact that

$$\int_{-\infty}^{\infty} |x(j\omega)|^2 \, dt = 2\pi, \text{ we have}$$

$$\int_{-\infty}^{\infty} |x(t)|^2 \, dt = 1$$

substituating the previously obtained expression for $x(t)$ in the above eqation,

We get

$$\int_{-\infty}^{\infty} |A^2 e^{-2t} + A^2 e^{-4t} - 2A^2 e^{-3t}| \, u(t) \, dt = 1$$

$$\int_{0}^{\infty} |A^2 e^{-2t} + A^2 e^{-4t} - 2A^2 e^{-3t}| \, dt = 1$$

$$A^2 \frac{1}{12} = 1$$

$$\Rightarrow \qquad A = \sqrt{12}$$

We choose A to be $\sqrt{12}$ instead of $-\sqrt{12}$ because we know that $x(t)$ is non negative.

Problem 4.15: Let $x(t)$ be a signal with Fourier transform $X(j\omega)$. Suppose we are given the following facts:

1. $x(t)$ is real.

2. $x(t) = 0$ for $t \leq 0$.
3. $\frac{1}{2\pi}\int_{-\infty}^{\infty} \text{Re}\{X(j\omega)\} e^{j\omega t} d\omega = |t| e^{-|t|}$.

Determine a closed-form expression for $x(t)$.

Solution:

Since $x(t)$ is real

$$\varepsilon v\{x(t)\} = \frac{x(t)+x(-t)}{2} \xleftrightarrow{FT} \text{Re}\{X(jw)\}$$

We are given that

$$FT\{Re[X(j\omega)]\} = |t|\ e^{-|t|}$$

Therefore

$$\varepsilon v\{x(t)\} = \frac{x(t)+x(-t)}{2} = |t|\ e^{-|t|}$$

We also know that $x(t) = 0$, for $t \leq 0$, This implies that $x(-t)$ is zero, for $t > 0$ we may conclude that

$$x(t) = 2(t)e^{-|t|} \text{ for } t \geq 0$$

Therefore

$$x(t) = 2te^{-t}u(t)$$

Problem 4.16: Consider the signal

$$x(t) = \sum_{k=-\infty}^{\infty} \frac{\sin\left(k\frac{\pi}{4}\right)}{\left(k\frac{\pi}{4}\right)} \delta\left(t - k\frac{\pi}{4}\right)$$

(*a*) Determine $g(t)$ such that

$$x(t) = \left(\frac{\sin t}{\pi t}\right) g(t).$$

(*b*) Use the multiplication property of the Fourier transform to argue that $X(j\omega)$ is periodic. Specify $X(j\omega)$ over one period.

Solution:

(*a*) We may write

$$x(t) = \sum_{k=-\infty}^{\infty} \frac{\sin\left(k\frac{\pi}{4}\right)}{k\frac{\pi}{4}} \delta\left(t - k\frac{\pi}{4}\right) = \frac{\sin t}{\pi t} \sum_{k=-\infty}^{\infty} \pi\delta\left(t - k\frac{\pi}{4}\right)$$

Therefore

$$g(t) = \sum_{k=-\infty}^{\infty} \pi\delta\left(t - k\frac{\pi}{4}\right)$$

(*b*) since $g(t)$ is an impulse that its fourier transform $G(j\omega)$ is also an impulse train

$$G(j\omega) = \pi\frac{2\pi}{\frac{\pi}{4}} \sum_{k=-\infty}^{\infty} \delta\left(\omega - \frac{2\pi k}{\frac{\pi}{4}}\right) = 8\pi \sum_{k=-\infty}^{\infty} \delta(\omega - 8k)$$

We see that $G(j\omega)$ is periodic, with a period of 8. using the multiplication property

We know that

$$X(j\omega) = \frac{1}{2\pi}\left[FT\left\{\frac{\sin t}{\pi t}\right\} \cdot G(j\omega)\right]$$

If we denote

$$FT\left(\frac{\sin t}{\pi t}\right) \text{ by } A(j\omega), \text{ then}$$

$$X(j\omega) = \left(\frac{1}{2\pi}\right)\left[A(jw) \cdot 8\pi \sum_{k=-\infty}^{\infty} A(\omega - 8k)\right] = 4\sum_{k=-\infty}^{\infty} A(j\omega - 8k)$$

$X(j\omega)$ may thus be viewed as a replication of $4A(j\omega)$ every 8 rad/sec.

This is obviously periodic.

$$A(j\omega) = \begin{cases} 1 & |\omega| \le 1 \\ 0 & \text{otherwise} \end{cases}$$

Therefore, we may specify $X(jw)$ over one period as

$$X(j\omega) = \begin{cases} 4 & |\omega| \le 1 \\ 0 & 1 < |\omega| \le 4 \end{cases}$$

> **Problem 4.17:** Determine whether each of the following statements is true or false. Justify your answers.
> (*a*) An odd and imaginary signal always has an odd and imaginary Fourier transform.
> (*b*) The convolution of an odd Fourier transform with an even Fourier transform is always odd.

Solution:

(*a*) We know that a real and odd signal $x(t)$ has a purely imaginary and odd fourier transform $X(j\omega)$. Let us now consider the purely imaginary and odd signal $jx(t)$.

Using linearity, we obtain the fourier transform of this signal to be $|X(j\omega)|$. The function $jX(j\omega)$ will clearly be real and odd. Therefore the given statement is false.

(*b*) An odd fourier transform corresponds to an odd signal, while an even fourier transform corresponds to an even signal. The convolution of an even fourier transform with an fourier may be viewed in the time domain as a multiplication of an even and the signal. Such a multiplication will always result in an odd the singla. The fourier transform of this odd signal will always be odd. Therefore the given statement *is true*.

> **Problem 4.18:** Consider a causal LTI system with frequency response
>
> $$H(j\omega) = \frac{1}{j\omega + 3}.$$
>
> For a particular input $x(t)$ this system is observed to produce the output
>
> $$y(t) = e^{-3t}\,u(t) - e^{-4t}\,u(t).$$
>
> Determine $x(t)$.

Solution:

We know that

$$H(j\omega) = \frac{Y(j\omega)}{X(j\omega)}$$

Since it is given that

$$y(t) = e^{-3t}\, u(t) - e^{-4t}\, u(t)$$

We can compute $y(j\omega)$

$$Y(j\omega) = \frac{1}{3+j\omega} - \frac{1}{4+j\omega} = \frac{1}{(3+j\omega)(4+j\omega)}$$

Since,

$$H(j\omega) = \frac{1}{(3+j\omega)}, \text{ we have}$$

$$X(j\omega) = \frac{Y(j\omega)}{H(j\omega)} = \frac{1}{(4+j\omega)}$$

Taking,
The inverse Fourier transform of $X(j\omega)$ we have,

$$x(t) = e^{-4t}\, u(t)$$

Problem 4.19: Compute the Fourier transform of each of the following signals:

(*a*) $[e^{-\alpha t} \cos \omega_0 t]\, u(t), \alpha > 0$

(*b*) $e^{-3|t|} \sin 2t$

(*c*) $x(t) = \begin{cases} 1 + \cos \pi t, & |t| \le 1 \\ 0, & |t| > 1 \end{cases}$

(*d*) $\sum_{k=0}^{\infty} \alpha^k\, \delta(t - kT), |\alpha| < 1$

(*e*) $[te^{-2t} \sin 4t]\, u(t)$

(*f*) $\left[\frac{\sin \pi t}{\pi t}\right]\left[\frac{\sin 2\pi(t-1)}{\pi(t-1)}\right]$

(*g*) $x(t)$ as shown in Figure P. 4.19 (*a*)

(*h*) $x(t)$ as shown in Figure P. 4.19 (*b*)

(*i*) $x(t) \begin{cases} 1 - t^2, & 0 < t < 1 \\ 0, & \text{otherwise} \end{cases}$

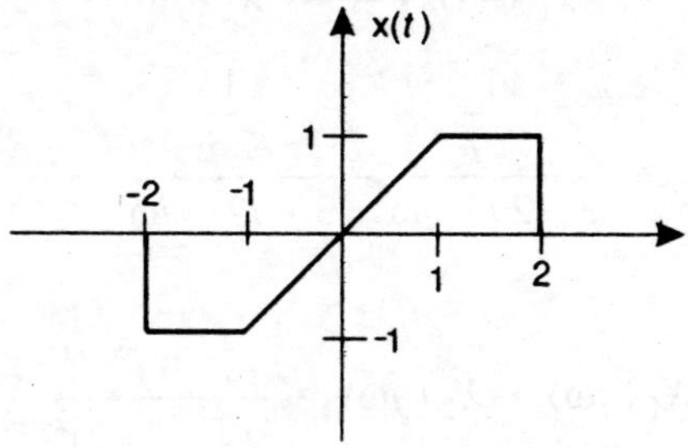

Fig. P. 4.19 (a)

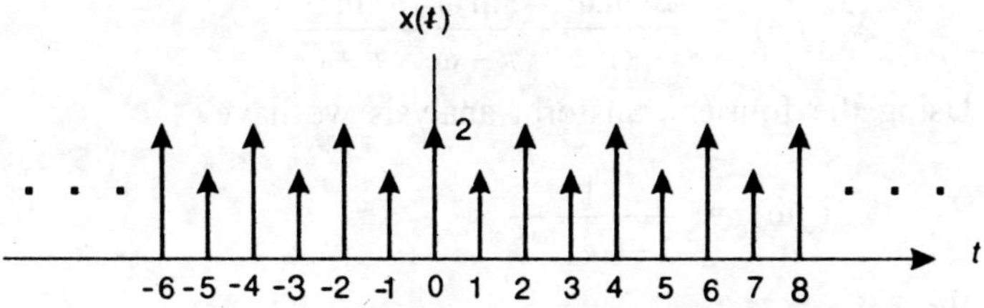

Fig. P. 4.19 (b)

Solution:

(*a*) The given signal is

$$e^{-\alpha t}\cos(\omega_o t)\,u(t) = \frac{1}{2}\,e^{-\alpha t}e^{j\omega_0 t}\,u(t) - \frac{1}{2}\,e^{-\alpha t}e^{j\omega_0 t}\,u(t)$$

Therefore

$$X(j\omega) = \frac{1}{2\left(\alpha - j\omega_0 + j\omega\right)} - \frac{1}{2\left(\alpha - j\omega_0 - j\omega\right)}$$

(*b*) The given signal is

$$x(t) = e^{-3t}\sin(2t)\,u(t) + e^{3t}\cdot\sin(2t)\,u(-t)$$

we have

$$x_1(t) = e^{-3t}\cdot\sin(2t)\cdot u(t) \overset{FT}{\longleftrightarrow} X_1(j\omega) = \frac{\frac{1}{2j}}{3-2j+j\omega} - \frac{\frac{1}{2j}}{3+j2-j\omega}$$

Also

$$x_2(t) = e^{3t}\sin(2t)\,u(-t)$$

$$= -x_1(-t) \xleftrightarrow{FT} X_2(j\omega) = -X_1(-j\omega)$$

$$= \frac{\frac{1}{2j}}{3-2j-j\omega} - \frac{\frac{1}{2j}}{3+j2-j\omega}$$

Therefore,

$$X(j\omega) = X_1(j\omega) + X_2(j\omega) = \frac{3j}{9+(\omega+2)^2} - \frac{3j}{9-(\omega+2)^2}$$

(c) Using the fourier transform analysis we have

$$X(j\omega) = \frac{2\sin\omega}{\omega} + \frac{\sin\omega}{\pi-\omega} - \frac{\sin\omega}{\pi+\omega}$$

(d) Using the fourier transform analysis we have

$$X(j\omega) = \frac{1}{1-\alpha e^{-j\omega T}}$$

(e) We have

$$x(t) = \left(\frac{1}{2j}\right) te^{-2t}e^{j4t} \cdot u(t) - \left(\frac{1}{2j}\right) te^{-2t} \cdot e^{-j4t} \cdot u(t)$$

Therefore,

$$X(j\omega) = \frac{\frac{1}{2j}}{(2-j4+j\omega)^2} - \frac{\frac{1}{2j}}{(2+j4-j\omega)^2}$$

(f) We have

$$x_1 t = \frac{\sin t}{\pi t} \xleftrightarrow{FT} X_1(j\omega) = \begin{cases} 1 & |\omega| < \pi \\ 0 & \text{otherwise} \end{cases}$$

Also,

$$x_2(t) = \frac{\sin 2\pi(t-1)}{\pi(t-1)} \xleftrightarrow{FT} X_2(j\omega) = \begin{cases} e^{-2\omega} & |\omega| < 2\pi \\ 0 & \text{otherwise} \end{cases}$$

$$x(t) = x_1(t) \cdot x_2(t) \xleftrightarrow{FT}$$

$$X(j\omega) = \frac{1}{2\pi}\{X_1(j\omega) * X_2(j\omega)\}$$

Therefore,

$$X(j\omega) = \begin{cases} e^{-j\omega} & |\omega| < \pi \\ \left(\frac{1}{2\pi}\right)(3\pi + \omega)e^{-j\omega} & -3\pi < \omega < -\pi \\ \left(\frac{1}{2\pi}\right)(3\pi - \omega)e^{-j\omega} & \pi < \omega < 3\pi \\ 0 & \text{otherwise} \end{cases}$$

(*g*) Using the Fourier transform analysis then. we get

$$X(j\omega) = \frac{2j}{\omega}\left[\cos 2\omega - \frac{\sin \omega}{\omega}\right]$$

(*h*) If

$$x_1(t) = \sum_{k=-\infty}^{\infty} \delta(t - 2k)$$

Then,

$$x(t) = 2\, x_1(t) + x_1(t - 1)$$

Therefore,

$$X(j\omega) = X_1(j\omega)\,[2 + e^{-\omega}] = \pi \sum_{k=-\infty}^{\infty} \delta(\omega - k\pi)\,[2 + (-1)^k]$$

(*i*) Using the Fourier transform analysis we get

$$X(j\omega) = \frac{1}{j\omega} + \frac{2e^{-j\omega}}{-\omega^2} - \frac{2e^{-j\omega} - 2}{j\omega^2}$$

where $X(j\omega)$ is the Fourier transform of one period of $x(t)$. That is

$$X(j\omega) = \frac{1}{1 - e^{-2}}\left[\frac{1 - e^{-2(1+j\omega)}}{1 + j\omega} - \frac{e^{-2}[1 - e^{-2(1-j\omega)}]}{1 - j\omega}\right]$$

Problem 4.20: Determine the continuous-time signal corresponding to each of the following transforms.

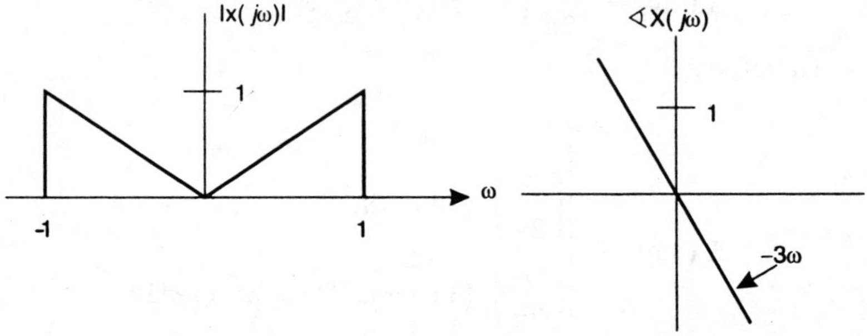

Fig. P. 4.20 (a)

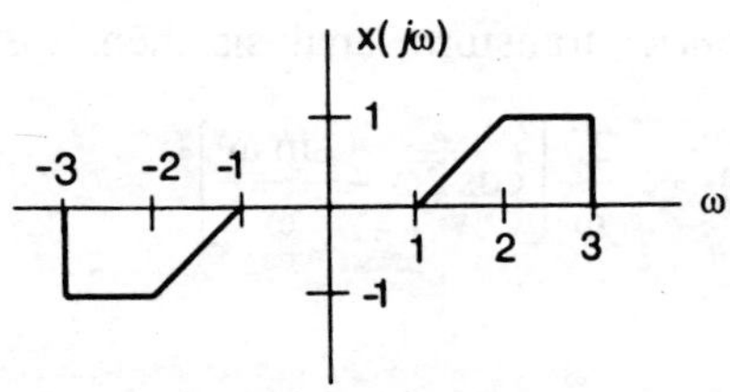

Fig. P. 4.20 (b)

(*a*) $X(j\omega) = \dfrac{2\sin[3(\omega - 2\pi)]}{(\omega - 2\pi)}$

(*b*) $X(j\omega) = \cos(4\omega + \pi/3)$

(*c*) $X(j\omega)$ as given by the magnitude and phase plots of Figure P. 4.20 (*a*)

(*d*) $X(j\omega) = 2[\delta(\omega - 1) - \delta(\omega - 2\pi)] + 3[\delta(\omega - 2\pi) + \delta(\omega + 2\pi)]$

(*e*) $X(j\omega)$ as in Figure P. 4.20 (*b*).

Solution:

(*a*)
$$x(t) = \begin{cases} e^{j2\pi t} & |t| < 3 \\ 0 & \text{otherwise} \end{cases}$$

(*b*)
$$x(t) = \frac{1}{2} e^{-j\frac{\pi}{3}} \delta(t - 4) + \frac{1}{2} e^{-j\frac{\pi}{3}} \delta(t + 4)$$

(*c*) The fourier transform synthesis may be written as

$$x(t) = \frac{1}{2\pi}\int_{-\infty}^{\infty} |X(j\omega)|\, e^{J\pi X(j\omega)}\, e^{j\omega t}\,.\, d\omega$$

From the given figure, we get

$$x(t) = \frac{1}{\pi}\left[\frac{\sin(t-3)}{t-3} + \frac{\cos(t-3)-1}{(t-3)^2}\right]$$

(*d*) $$x(t) = \frac{2j}{\pi}\sin t + \frac{3}{\pi}\cos(2\pi t)$$

(*e*) Using the transform synthesis equation

$$x(t) = \frac{\cos 3t}{j\pi t} + \frac{\sin t - \sin 2t}{j\pi t^2}$$

Problem 4.21: Consider the signal

$$x_0(t) = \begin{cases} e^{-t}, & 0 \le t \le 1 \\ 0, & \text{elsewhere} \end{cases}$$

Determine the Fourier transform of each of the signals shown in Figure P. 4.21. You should be able to do this by explicitly evaluating *only* the transform of $x_0(t)$ and then using properties of the Fourier transform.

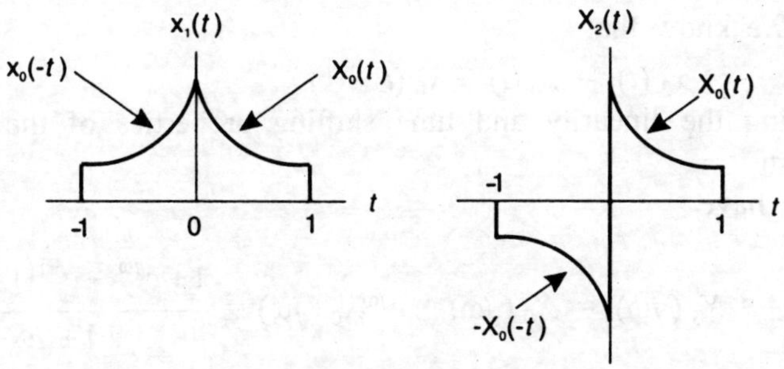

Fig. P. 4.21 (a)

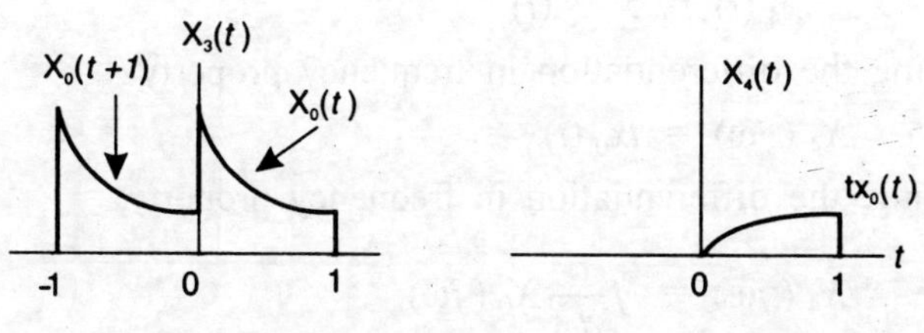

Fig. P. 4.21 (b)

Solution:

For the given signal $x_0(t)$, we use the fourier transform, to evaluate the corresponding fourier transform

$$X_0(j\omega) = \frac{1-e^{-(1+j\omega)}}{1+j\omega}$$

(*i*) We know that

$$x_1(t) = x_0(t) - x_0(-t)$$

Using the linearity and time reversal property of the fourier transform,

We have

$$X_1(j\omega) = X_0(j\omega) + X_0(-j\omega) = \frac{2-2e^{-1}\cos\omega - 2\omega e^{-1}\sin\omega}{1+\omega^2}$$

(*ii*) We know that

$$x_2(t) = x_0(t) - x_0(-t)$$

Using the linearity and time reversal properties of the Fourier transform we have

$$X_2(j\omega) = X_0(j\omega) - X_0(j\omega) = j\left[\frac{-2\omega + 2e^{-1}\sin\omega + 2\omega e^{-1}\cos\omega}{1+\omega^2}\right]$$

(*iii*) We know that

$$x_3(t) = x_0(t) + x_0(t+1)$$

Using the linearity and time shifting properties of the fourier transform

We have

$$X_3(j\omega) = X_0(j\omega) + e^{j\omega}X_0(j\omega) = \frac{1+e^{j\omega} - e^{-1}\left(1+e^{-j\omega}\right)}{1+j\omega}$$

(*iv*) We know that

$$x_4(t) = t \cdot x_0(t)$$

Using the differentiation in frequency property

$$X_4(j\omega) = tx_0(t)$$

Using, the differentiation in frequency property

$$X_4(j\omega) = j\frac{d}{d\omega}X_0(j\omega)$$

Therefore $X_4(j\omega) = \dfrac{1-2e^{-1}e^{-j\omega} - j\omega e^{-1}e^{-j\omega}}{(1+j\omega)^2}$

Problem 4.22: (*a*) Determine which, if any, of the real signals depicted in Figure P. 4.22 have Fourier transforms that satisfy each of the following conditions:

(1) $Re\{X(j\omega)\} = 0$

(2) $Im\{X(j\omega)\} = 0$

(3) There exists a real α such that $e^{j\alpha\omega}X(j\omega)$ is real

(4) $X(j\omega)\, d\omega = 0$

(5) $\omega X(j\omega)\, d\omega = 0$

(6) $X(j\omega)$ is periodic

(*b*) Construct a signal that has properties (1), (4), and (5) and does *not* have the others.

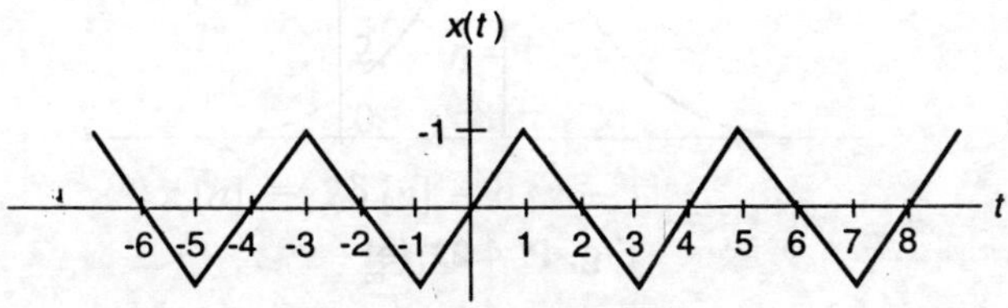

Fig. P. 4.22 (a)

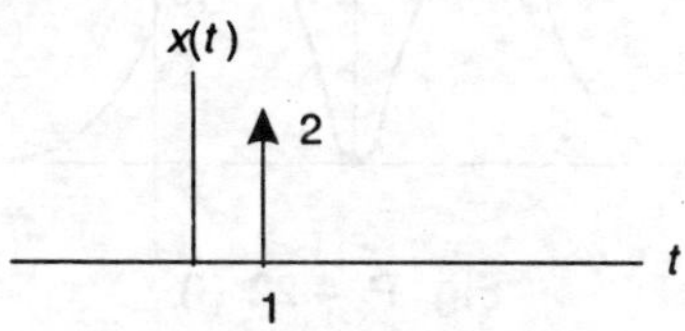

Fig. P. 4.22 (b)

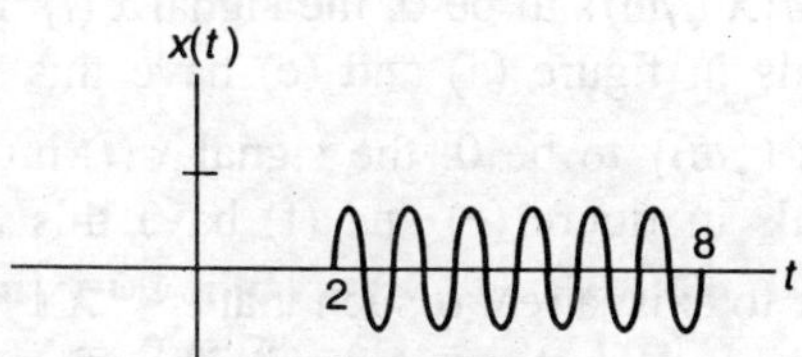

Fig. P. 4.22 (c)

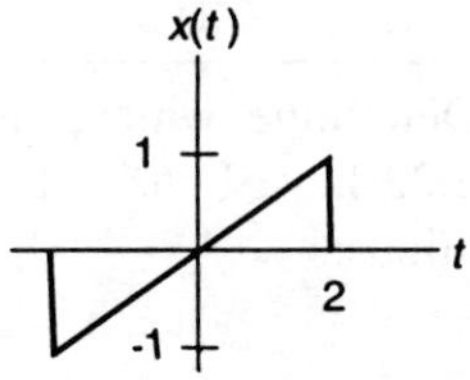

Fig. P. 4.22 (d)

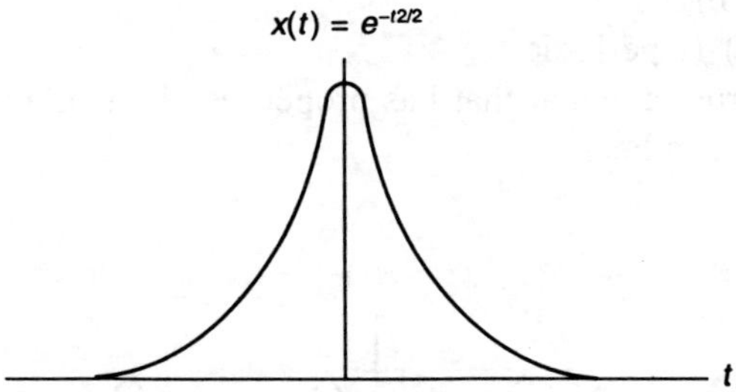

Fig. P. 4.22 (e)

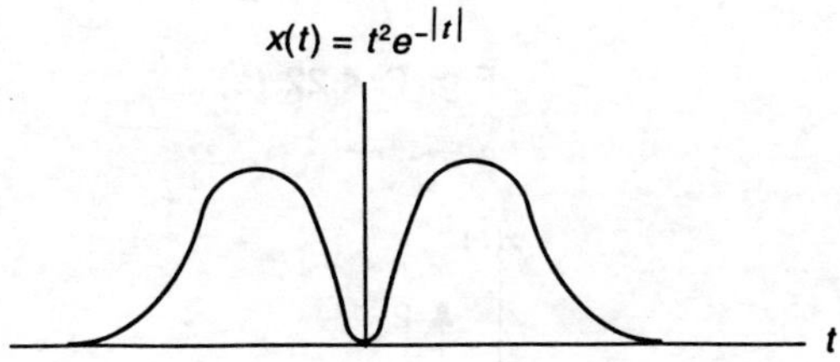

Fig. P. 4.22 (f)

Solution:

(*a*) (*i*) For $Re\{X(j\omega)\}$ to be 0, the signal $x(t)$ must be real and odd. Therefore signals in figure (a) and (c) have this property.

(*ii*) For $Im\{X(j\omega)\}$ to be 0, the signal $x(t)$ must be real and even. Therefore signals in figure (e) and (f) have this property.

(*iii*) For there to exist a real α such that $e^{j\alpha\omega} X(j\omega)$ is real. We require that $x(t - \alpha)$ be a real and even signal. Therefore signals in figure (*a*), (*b*), (*e*) and (*f*) have this property.

(*iv*) For this condition to be true $x(0) = 0$, therefore signals in figure (a), (b), (c), (d) and (f) have this property.

(*v*) For this condition to be true the derivative of $x(t)$ has to be zero at $t = 0$, therefore signals in figure (b), (c), (e) and (f), have this property.

(*vi*) For this to be true the signal $x(t)$ has to be periodic only the signal in figure (a) has this property.

(*b*) For a signal to satisfy only properties, (1), (4), and (5). It must be real and odd and

$$x(t) = 0, \; x'(0) = 0$$

The signal shown below is an example of that

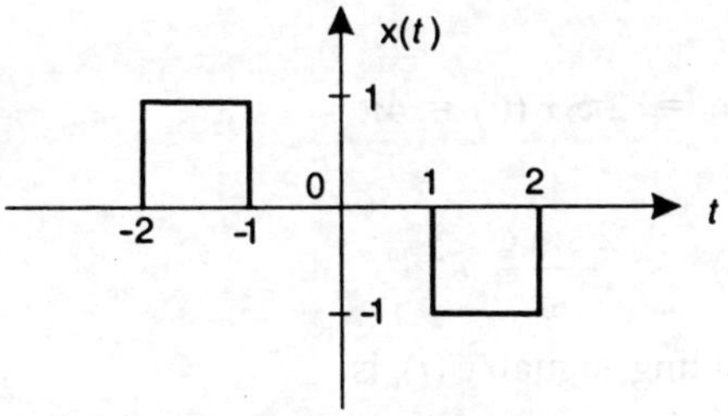

Fig. P. 4.22 (g)

Problem 4.23: Let $X(j\omega)$ denote the Fourier transform of the signal $x(t)$ depicted in Figure P4.23.

(*a*) Find $\measuredangle X(j\omega)$ (*b*) Find $X(j0)$

(*c*) Find $X\int_{-\infty}^{\infty} X(j\omega)\, d\omega$.

(*d*) Evaluate $\int_{-\infty}^{\infty} X(j\omega) \frac{2\sin\omega}{\omega} e^{j2\omega}\, d\omega$.

(*e*) Evaluate $\int_{-\infty}^{\infty} |X(j\omega)|^2\, d\omega$.

(*f*) Sketch the inverse Fourier transform of $(Re\{X(j\omega)\})$.

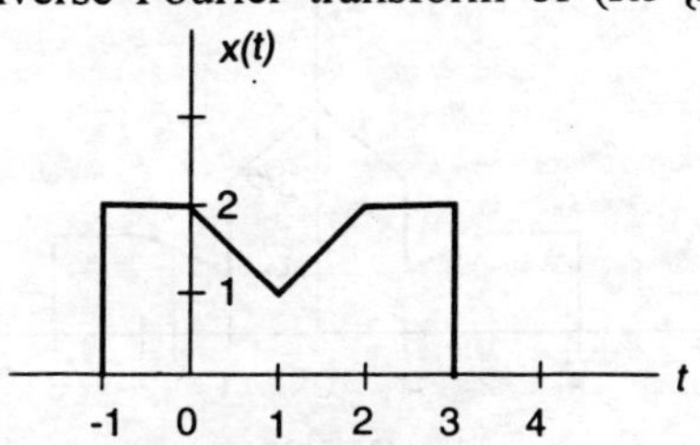

Fig. P. 4.23

Solution:

(*a*) Note that, $y(t) = x(t+1)$ is a real and even signal. therefore $y(j\omega)$ is also real and even. This implies that

$$\sphericalangle y(j\omega) = 0$$

Also, since $y(j\omega) = e^{j\omega} X(j\omega)$

We know that

$$\sphericalangle X(j\omega) = -\omega$$

(*b*) We have

$$X(j0) = \int_{-\infty}^{\infty} x(t)\, dt = 7$$

(*c*) We get

$$\int_{-\infty}^{\infty} X(j\omega)\, d\omega = 2\pi\, x(0) = 4\pi$$

(*d*) Let $\quad y(j\omega) = \dfrac{2\sin\omega}{\omega} e^{2j\omega}$

The corresponding signal $y(t)$ is

$$y(t) = \begin{cases} 1 & -3 < t < -1 \\ 0 & \text{otherwise} \end{cases}$$

Then the given integral is

$$\int_{-\infty}^{\infty} |X(j\omega)|^2\, d\omega = 2\pi \int_{-\infty}^{\infty} |X(t)|^2\, dt = 26\pi$$

(*f*)The inverse fourier transform at $Re\{X(j\omega)\}$ is the $\varepsilon v\{x(t)\}$

Which is $\dfrac{[x(t) + x(-t)]}{2}$

This is as shown in figure below

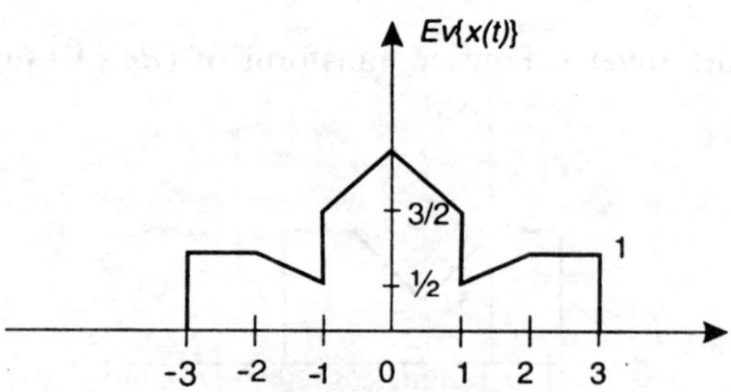

Fig. P. 4.23 (a)

Problem 4.24: (*a*) Compute the convolution of each of the following pairs of signals $x(t)$ and $h(t)$ by calculating $X(j\omega)$ and $H(j\omega)$, using the convolution property, and inverse transforming.

(*i*) $x(t) = te^{-2t}u(t),\ h(t) = e^{-4t}u(t)$

(*ii*) $x(t) = te^{-2t}u(t),\ h(t) = te^{-4t}u(t)$

(*iii*) $x(t) = e^{-t}u(t),\ h(t) = e^{t}u(-t)$

(*b*) Suppose that $x(t) = e^{-(t-2)}u(t-2)$ and $h(t)$ is as depicted in Figure P. 4.24 (a). Verify the convolution property for this pair of signals by showing that the Fourier transform of $y(t) = x(t) * h(t)$ equals $H(j\omega)\,X(j\omega)$.

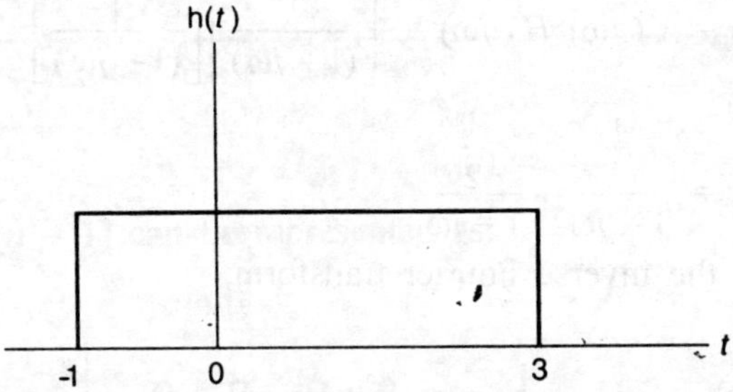

Fig. P. 4.24 (a)

Solution:

(*a*) (*i*) We know that

$$Y(j\omega) = X(j\omega)\,H(j\omega) = \left[\frac{1}{(2+j\omega)^2}\right]\left[\frac{1}{4+j\omega}\right]$$

$$= \frac{\frac{1}{4}}{4+j\omega} - \frac{\frac{1}{4}}{2+j\omega} + \frac{\frac{1}{2}}{(2+j\omega)^2}$$

Taking, the inverse Fourier transform
We obtain

$$y(t) = \frac{1}{4}e^{-4t}u(t) - \frac{1}{4}e^{-2t}u(t) + \frac{1}{2}te^{-2t}u(t)$$

(*ii*) We have

$$y(j\omega) = X(j\omega)\ H(j\omega) = \left[\frac{1}{(2+j\omega)^2}\right]\left[\frac{1}{(4+j\omega)^2}\right]$$

$$= \frac{\left(\frac{1}{4}\right)}{2+j\omega} + \frac{\left(\frac{1}{4}\right)}{(2+j\omega)^2} - \frac{\left(\frac{1}{4}\right)}{(4+j\omega)} + \frac{\left(\frac{1}{4}\right)}{(4+j\omega)^2}$$

Taking the inverse Fourier transform, we get

$$t(t) = \frac{1}{4}e^{-2t}u(t) + \frac{1}{4}te^{-2t}u(t) - \frac{1}{4}e^{-4t}u(t) + \frac{1}{4}te^{-4t}\cdot u(t)$$

(*iii*) We have

$$Y(j\omega) = X(j\omega)\ H(j\omega) = \left[\frac{1}{(1+j\omega)}\right]\left[\frac{1}{(1-j\omega)}\right]$$

$$= \frac{\frac{1}{2}}{1+j\omega} + \frac{\frac{1}{2}}{1-j\omega}$$

Taking, the inverse Fourier transform,
We get

$$y(t) = \frac{1}{2}e^{-|t|}$$

(*b*) By direct convolution of $x(t)$ with $h(t)$
We get

$$y(t) = \begin{cases} 0 & t<1 \\ 1-e^{-(t-1)} & 1<t\le 5 \\ e^{-(t-5)}-e^{-(t-1)} & t>5 \end{cases}$$

Taking the fourier transform of $y(t)$

$$y(j\omega) = \frac{2e^{-j3\omega}\sin(2\omega)}{\omega(1+j\omega)} = \left[\frac{e^{-j2\omega}}{1+j\omega}\right]\frac{e^{-j\omega}\cdot 2\sin(2\omega)}{\omega}$$

$$= X(j\omega)\ H(j\omega)$$

Problem 4.25: Consider the signals

$$x(t) = u(t-1) - 2u(t-2) + u(t-3)$$

and

$$\bar{x}(t) = \sum_{k=-\infty}^{\infty} x(t-kT)$$

where $T > 0$. let a_k denote the Fourier series coefficients of $\bar{x}(t)$, and let $X(j\omega)$ denote the Fourier transform of $x(t)$.

(*a*) Determine a closed-form expression for $X(j\omega)$

(*b*) Determine an expression for the Fourier coefficients a_k and verify

that $a_k = \frac{1}{T} X\left(j\frac{2\pi k}{T}\right)$

Solution:

(*a*) The fourier transform $X(j\omega)$ is

$$X(j\omega) = \int_{-\infty}^{\infty} x(t)\, e^{-j\omega t} \cdot dt = \int_{1}^{2} e^{-j\omega t} \cdot dt - \int_{2}^{3} e^{-j\omega t} \cdot dt$$

$$= \frac{2\sin\left(\frac{\omega}{2}\right)}{\omega} \left\{1 - e^{-j\omega t}\right\} e^{-j3\omega t}$$

(*b*) The fourier series coefficients a_k are

$$a_k = \frac{1}{T} \int_{<T>} x(t)\, e^{-j\frac{2\pi}{T}kt} = \frac{1}{2}\left\{\int_{1}^{2} e^{-j\frac{2\pi}{T}kt} dt - \int_{2}^{3} e^{-j\frac{2\pi}{T}kt} \cdot dt\right\}$$

$$= \frac{\sin k\frac{\pi}{2}}{k\pi}\left[1 - e^{-jk\pi}\right] e^{-j2k\frac{\pi}{2}}$$

Clearly the answer to part (a) and (b)
It is clear that

$$a_k = \frac{1}{T} X\left(j\frac{2\pi k}{T}\right) \quad \text{Where } T = 2$$

Problem 4.26: (*a*) Let $x(t)$ have the Fourier transform $X(j\omega)$, and let $p(t)$ be periodic with fundamental frequency ω_0 and Fourier series representation

$$p(t) = \sum_{n=-\infty}^{\infty} a_n e^{jn\omega_0 t}$$

Determine an expression for the Fourier transform of

$$y(t) = x(t)\,p(t) \qquad \text{...(1)}$$

(*b*) Suppose that $X(j\omega)$ is as depicted in Figure P. 4.26 (*a*). Sketch the spectrum of $y(t)$ in eq. (1) for each of the following choices of $p(t)$:

(*i*) $p(t) = \cos(t/2)$

(*ii*) $p(t) = \cos t$

(*iii*) $p(t) = \cos 2t$

(*iv*) $p(t) = (\sin t)(\sin 2t)$

(*v*) $p(t) = \cos 2t - \cos t$

(*vi*) $p(t) = \sum_{n=-\infty}^{+\infty} \delta(t - \pi n)$

(*vii*) $p(t) = \sum_{n=-\infty}^{+\infty} \delta(t - 2\pi n)$

(*viii*) $p(t) = \sum_{n=-\infty}^{+\infty} \delta(t - 4\pi n)$

(*ix*) $p(t) = \sum_{n=-\infty}^{+\infty} \delta(t - 2\pi n) - \frac{1}{2}\sum_{n=-\infty}^{+\infty} \delta(t - \pi n)$

(*x*) $p(t)$ = the periodic square wave shown in Figure P. 4.26 (*b*)

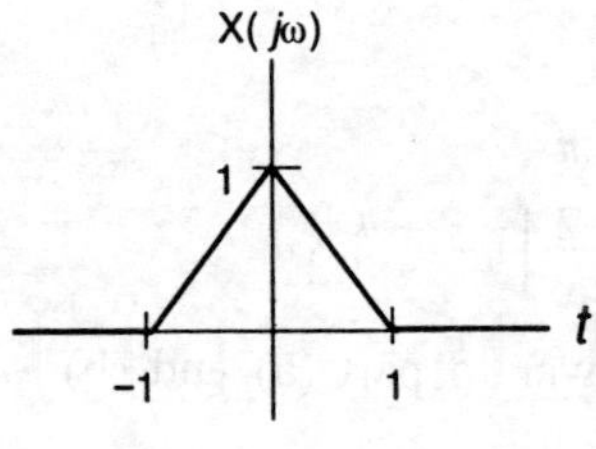

Fig. P. 4.26 (a)

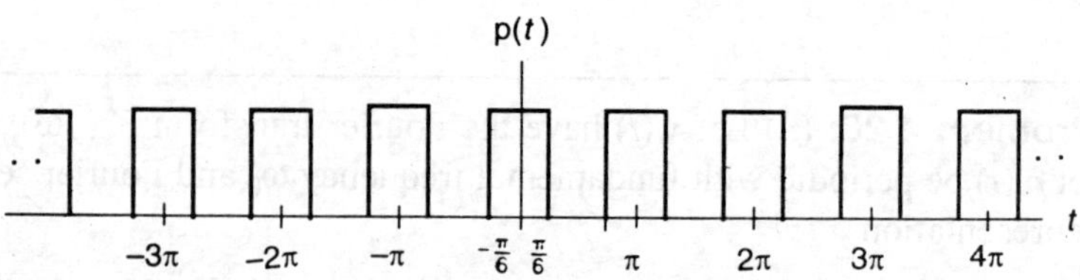

Fig. P. 4.26 (b)

Solution:

We know that

$$p(t) = \sum_{n=-\infty}^{\infty} a_n e^{jn\omega_0 t} \xleftrightarrow{FT} P(j\omega) = 2\pi \sum_{k=-\infty}^{\infty} a_k \,\delta(\omega - k\omega_0)$$

From this

$$y(j\omega) = \frac{1}{2\pi}\{X(j\omega) * H(j\omega)\} = \sum_{k=-\infty}^{\infty} a_k X\{j(\omega - k\omega_o)\}$$

(*b*)(*i*)

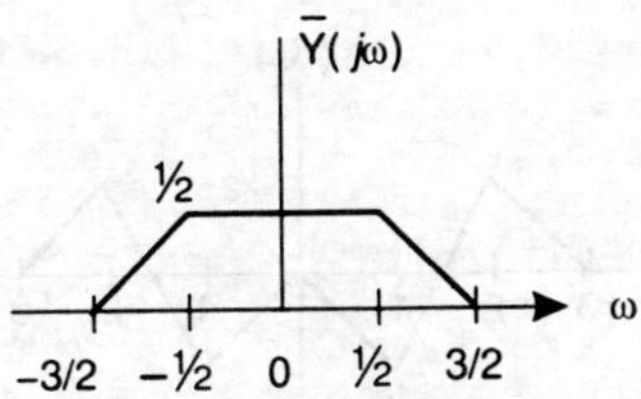

Fig. P. 4.26 (c)

(*ii*)

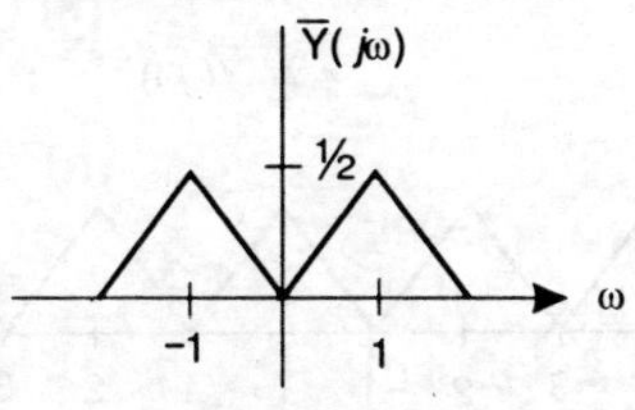

Fig. P. 4.26 (d)

(*iii*)

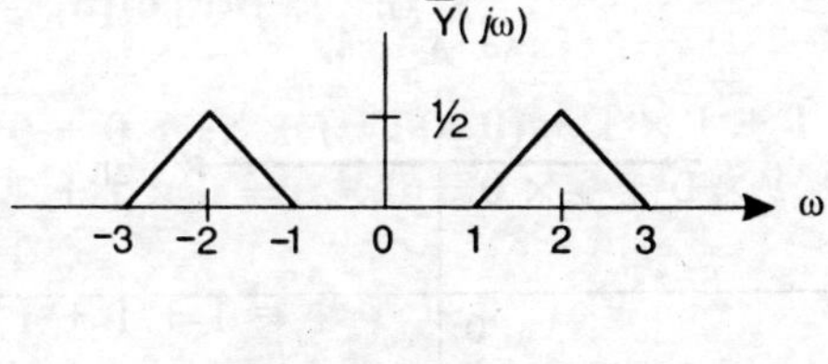

Fig. P. 4.26 (e)

(*iv*)

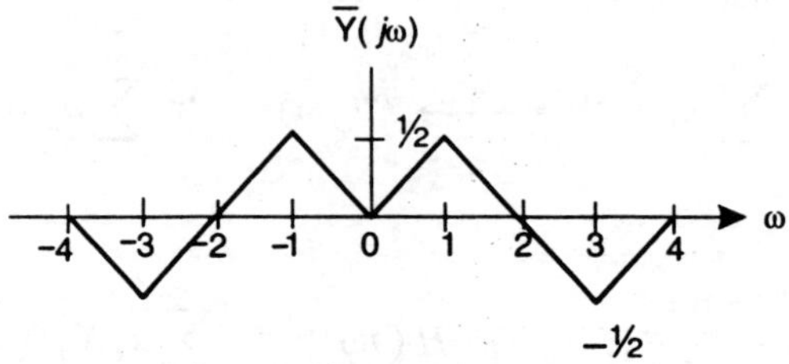

Fig. P. 4.26 (f)

(*v*)

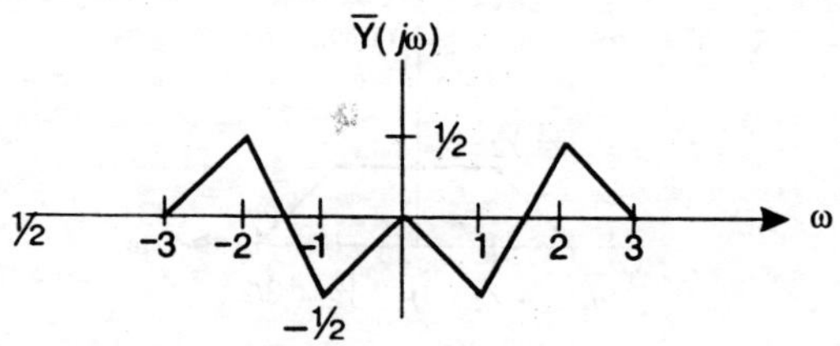

Fig. P. 4.26 (g)

(*vi*)

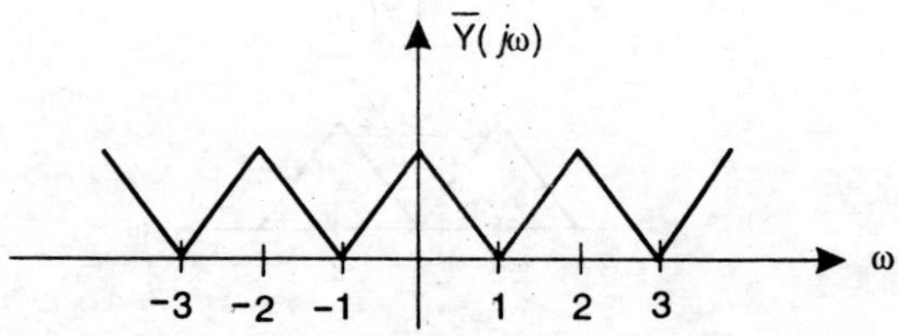

Fig. P. 4.26 (h)

(*vii*)

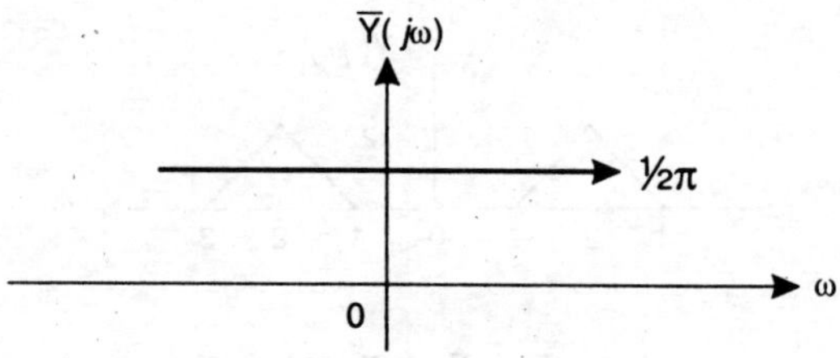

Fig. P. 4.26 (i)

(*viii*)

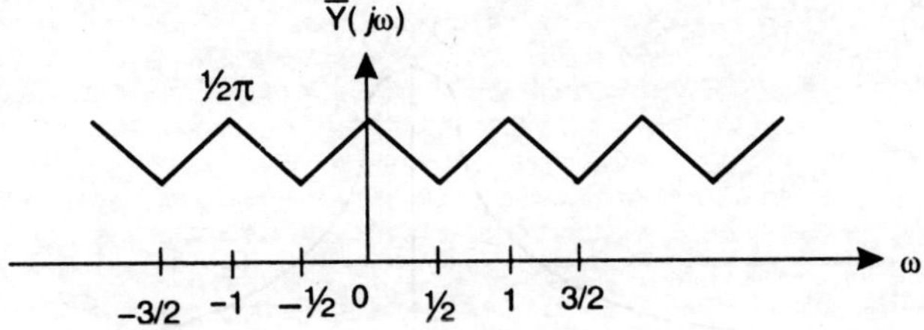

Fig. P. 4.26 (j)

(*ix*)

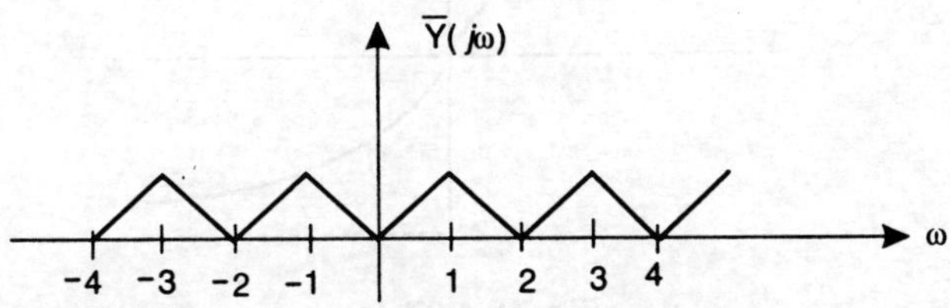

Fig. P. 4.26 (k)

(*x*)

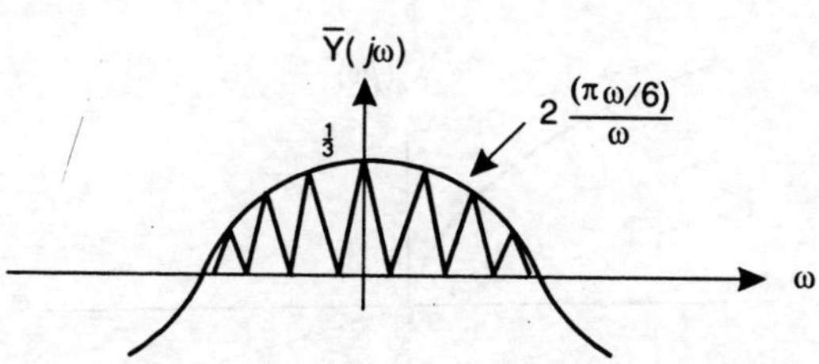

Fig. P. 4.26 (l)

Problem 4.27: A real-valued continuous-time function $x(t)$ has a Fourier transform $X(j\omega)$ whose magnitude and phase are as illustrated in Figure P4.27 (*a*).

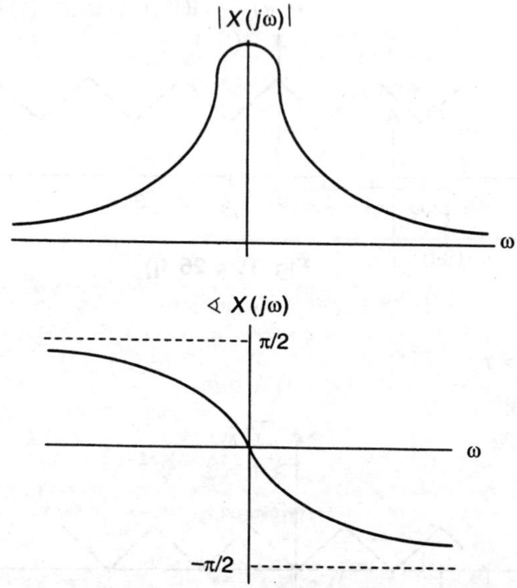

Fig. P. 4.27 (a)

The functions $x_a(t)$, $x_b(t)$, $x_c(t)$ and $x_d(t)$ have Fourier transforms whose magnitudes are identical to $X(j\omega)$, but whose phase functions differ, as shown in Figures P4.27 (*b*)–(*e*). The phase functions $\measuredangle X_a(j\omega)$ and $\measuredangle X_b(j\omega)$ are formed by adding a linear phase to $\measuredangle X(j\omega)$.

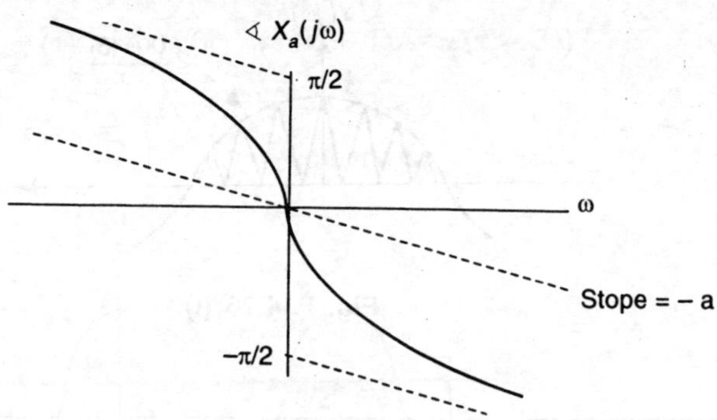

Fig. P. 4.27 (b)

The function $\measuredangle X_c(j\omega)$ is formed by reflecting $\measuredangle X(j\omega)$ about $\omega = 0$, and $\measuredangle X_d(j\omega)$ is obtained by a combination of a reflection and an addition of a linear phase. Using the properties of Fourier transforms, determine the expressions for $x_a(t)$, $x_b(t)$, $x_c(t)$, and $x_d(t)$ in terms of $x(t)$.

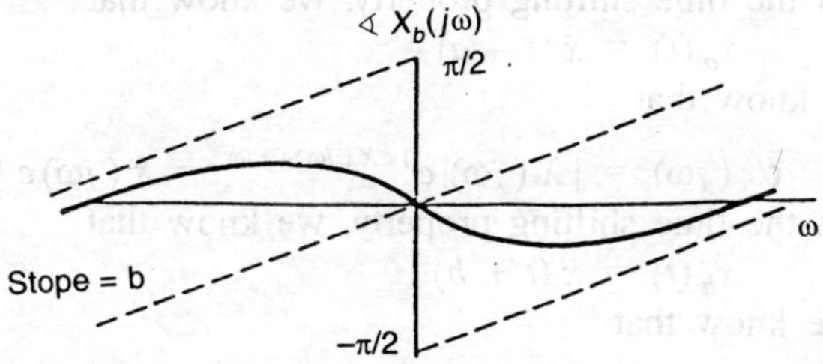

Fig. P. 4.27 (c)

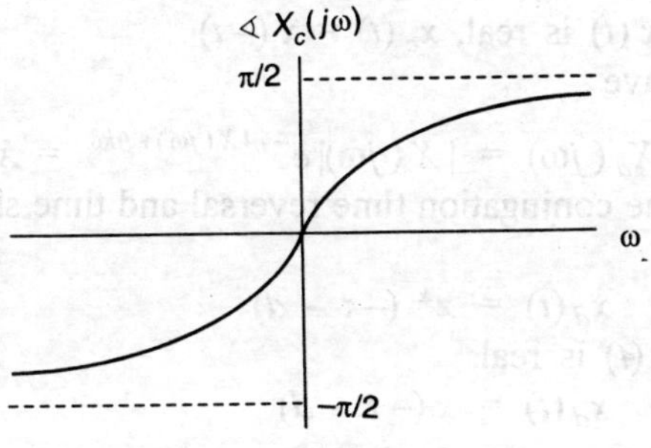

Fig. P. 4.27 (d)

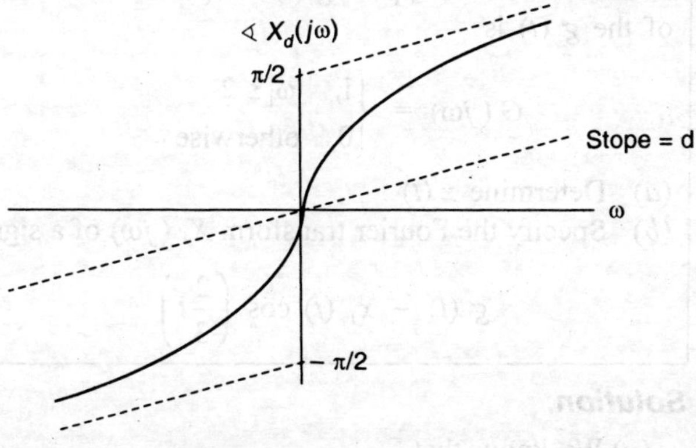

Fig. P. 4.27 (e)

Solution:

(*i*) We have

$$X_a(j\omega) = |X(j\omega)|\ e^{jX(j\omega)-j\omega a} = X(j\omega)\,e^{-j\omega a}$$

From the time shifting property, we know that

$$x_a(t) = x(t-a)$$

(*ii*) We know that

$$\alpha_b(j\omega) = |X(j\omega)|\,e^{j<X(j\omega)+jb\omega} = X(j\omega)\,e^{jb\omega}$$

From the time shifting property, we know that

$$x_b(t) = x(t+b)$$

(*iii*) We know that

$$X_c(j\omega) = |X(j\omega)|\,e^{-j\measuredangle X(j\omega)} = X^*(j\omega)$$

From the conjugation and time reversal propery, we know that

$$x_c(t) = x^*(-t)$$

Since, $x(t)$ is real, $x_c(t) = x(-t)$

(*d*) We have

$$X_d(j\omega) = |X(j\omega)|\,e^{-j4X(j\omega)+jd\omega} = X^*(j\omega)\,e^{j\omega d}$$

From the conjugation time reversal and time shifting properties, we know that

$$x_d(t) = x^*(-t-d)$$

since $x(t)$ is real,

$$x_d(t) = x(-t-d)$$

Problem 4.28: Suppose $g(t) = x(t)\cos t$ and the Fourier transform of the $g(t)$ is

$$G(j\omega) = \begin{cases} 1, & |\omega| \le 2 \\ 0, & \text{otherwise} \end{cases}$$

(*a*) Determine $x(t)$

(*b*) Specify the Fourier transform $X_1(j\omega)$ of a signal $x_1(t)$ such that

$$g(t) = x_1(t)\cos\left(\frac{2}{3}t\right)$$

Solution:

(*a*) We know that

$$\omega(t) = \cos t \xleftrightarrow{FT} W(j\omega) = \pi[\delta(\omega-1) + \delta(\omega+1)]$$

and

$$g(t) = x(t)\cos t \xleftrightarrow{FT} G(j\omega) = \frac{1}{2\pi}[X(j\omega) * W(j\omega)]$$

Therefore,

$$G(j\omega) = \frac{1}{2}X[j(\omega-1)] + \frac{1}{2}X(j(\omega+1))$$

Since $G(j\omega)$ is as shown in figure below. It is clear that the above equation that $X(j\omega)$ is as shown below.

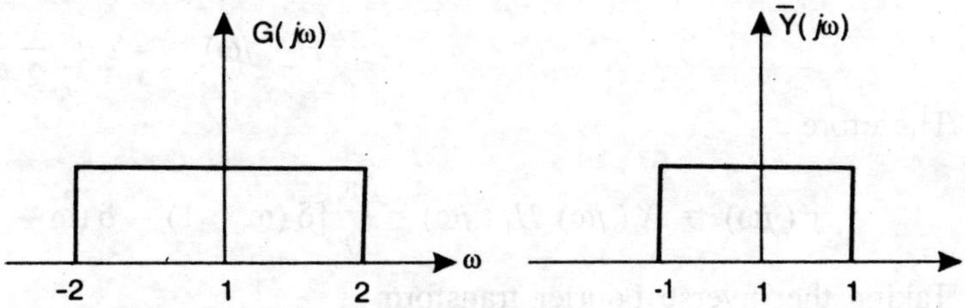

Fig. P. 4.28 (a)

(*b*) $X_1(j\omega)$ is as shown in above figure.

Problem 4.31: (*a*) Show that the three LTI systems with impulse responses

$$h(t) = u(t)$$
$$h_2(t) = -2\delta(t) + 5e^{-2t}u(t),$$

and

$$h_3(t) = 2te^{-t}u(t)$$

all have the same response to $x(t) = \cos t$.

(*b*) Find the impulse response of another LTI system with the same response to $\cos t$.

This problem illustrates the fact that the response to $\cos t$ cannot be used to specify an LTI system uniquely.

Solution:

We know that

$$x(t) = \cos t \xleftrightarrow{FT} X(j\omega) = \pi[\delta(\omega+1) + \delta(\omega-1)]$$

(*i*) We have

$$h_1(t) = u(t) \xleftrightarrow{FT} H_1(j\omega) = \frac{1}{j\omega} + \pi\delta(\omega)$$

Therefore

$$Y(j\omega) = X(j\omega)\, H_1(j\omega) = \frac{\pi}{j}[\delta(\omega+1) - \delta(\omega-1)]$$

Taking the inverse Fourier transform,
Then, We get

$$y(t) = \sin(t)$$

(*ii*) We have

$$h_2(t) = -2\delta(t) + 5e^{-2t}u(t) \xleftrightarrow{FT} H_2(j\omega) = -2 + \frac{5}{2+j\omega}$$

Therefore

$$y(j\omega) = X(j\omega)\, H_1(j\omega) = \frac{\pi}{j}[\delta(\omega+1) - \delta(\omega-1)]$$

Taking the inverse Fourier transform
We get $y(t) = \sin(t)$

(*iii*) We have

$$h_3(t) = 2\, te^{-t}u(t) \xleftrightarrow{FT} H_2(j\omega) = \frac{2}{(1+j\omega)^2}$$

Therefore

$$Y(j\omega) = X(j\omega)\, H_1(j\omega) = \frac{\pi}{j}[\delta(\omega+1) - \delta(\omega-1)]$$

Taking, the inverse fourier transofrm, We get

$$y(t) = \sin(t)$$

(*b*) An *LTI* system with impulse response

$$h_4(t) = \frac{1}{2}[h_1(t) + h_2(t)]$$

will have the same response of $x(t) = \cos(t)$. We can find other such impulse responses by suitably scaling and linearity combining $h_1(t)$, $h_2(t)$ and $h_3(t)$.

Problem 4.29: Consider an LTI system S with impulse response

$$h(t) = \frac{\sin(4(t-1))}{\pi(t-1)}$$

Determine the output of S for each of the following inputs:

(*a*) $x_1(t) = \cos\left(6t + \frac{\pi}{2}\right)$	(*b*) $x_2(t) = \sum_{k=0}^{\infty}\left(\frac{1}{2}\right)^k \sin(3kt)$
(*c*) $x_3(t) = \frac{\sin(4(t+1))}{\pi(t+1)}$	(*d*) $x_4(t) = \left(\frac{\sin 2t}{\pi t}\right)^2$

Solution:

Note that

$$h(t) = h_1(t-1), \text{ where}$$

$$h_1(t) = \frac{\sin 4t}{\pi t}$$

The Fourier transform $H_1(j\omega)$ of $h_1(t)$ is as shown in Fig. P. 4.29.

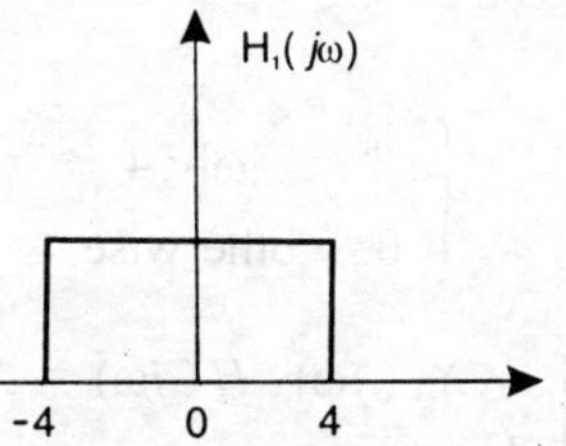

Fig. P. 4.29 (a)

From fig. it is clear that $h_1(t)$ is the impulse response of an ideal low pass filter pass bend is in the range $|\omega| < 4$.

Therefore, $h(t)$ is the impulse response of an ideal low pass filter shifted by one to the right.

Using the shifting property

$$H(j\omega) = \begin{cases} e^{-j\omega} & |\omega| < 4 \\ 0 & \text{otherwise} \end{cases}$$

(*a*) We have

$$X_1(j\omega) = \pi e^{j\frac{\pi}{12}}\delta(\omega - 6) + \pi e^{j\frac{\pi}{12}}\delta(\omega + 6)$$

It is clear that

$$y_1(j\omega) = X_1(j\omega)\, H(j\omega) = 0$$

$$\Rightarrow \quad y_1(t) = 0$$

This result is equivalent to saying that $X_1(j\omega)$ is zero, in the passband of $H(j\omega)$.

(*b*) We know that

$$X_2(j\omega) = \frac{\pi}{j}\left[\sum_{k=-\infty}^{\infty}\left(\frac{1}{2}\right)^k \{\delta(\omega - 3k) - \delta(\omega + 3k)\}\right]$$

Therefore

$$Y_2(j\omega) = X_2(j\omega)\, H(j\omega) = \frac{\pi}{j}\left[\left(\frac{1}{2}\right)\{\delta(\omega - 3) - \delta(\omega + 3)\}\, e^{-j\omega}\right]$$

This implies that

$$y_2(t) = \frac{1}{2}\sin(3t - 1)$$

We may have obtained the same result by noting that only the sinusoidal with frequency 3 in $X_2(j\omega)$ lies in the passband of $H(j\omega)$.

(*c*) We have

$$X_3(j\omega) = \begin{cases} e^{jw} & |\omega| < 4 \\ 0 & \text{otherwise} \end{cases}$$

$$Y_3(j\omega) = X_3(j\omega) \cdot H(j\omega) = X_3(j\omega)\, e^{-j\omega}$$

This implies that

$$y_3(t) = x_3(t - 1) = \frac{\sin(4t)}{\pi t}$$

We may have obtained the same result by noting that $X_3(j\omega)$ lies entirely in the passband of $H(j\omega)$.

(*d*) $X_4(j\omega)$ is as shown in figure below

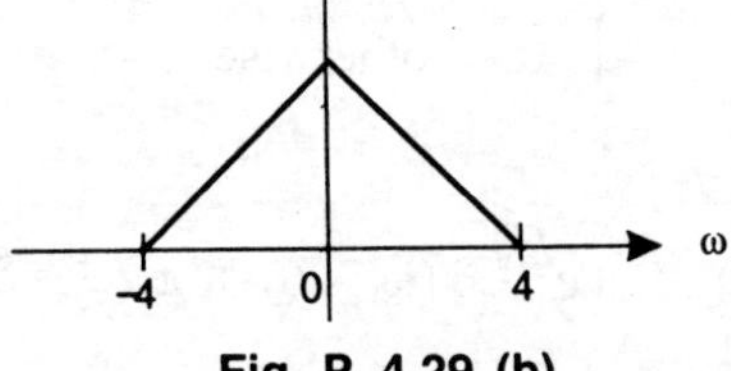

Fig. P. 4.29 (b)

Therefore

$$y_4(j\omega) = X_4(j\omega)\, H(j\omega) = X_4(j\omega)\, e^{-j\omega}$$

This implies that

$$y_4(t) = x_4(t-1) = \left[\frac{\sin(2(t-1))}{\pi(t-1)}\right]^2$$

We may have obtained the same result by noting that $X_4(j\omega)$ lies entirely in the passband of $H(j\omega)$.

Problem 4.30: The input and the output of a causal LTI system are related by the differential equation

$$\frac{d^2y(t)}{dt^2} + 6\frac{dy(t)}{dt} + 8y(t) = 2x(t)$$

(*a*) Find the impulse response of this system.

(*b*) What is the response of this system if $x(t) = te^{-2t}u(t)$?

(*c*) Repeat part (*a*) for the causal LTI system described by the equation

$$\frac{d^2y(t)}{dt^2} + \sqrt{2}\frac{dy(t)}{dt} + y(t) = 2\frac{d^2x(t)}{dt^2} - 2x(t)$$

Solution:

(*a*) Taking the fourier transform of both sides of the given differential equation,

We get

$$H(j\omega) = \frac{Y(j\omega)}{X(j\omega)} = \frac{2}{-\omega^2 - 2j\omega + 8}$$

Using the partial fraction
We get

$$H(j\omega) = \frac{1}{j\omega + 2} - \frac{1}{j\omega + 4}$$

Taking the inverse fourier transform

$$h(t) = e^{-2t}u(t) - e^{-4t}u(t)$$

(*b*) For the given signal $x(t)$,

We have $X(j\omega) = \dfrac{1}{(2 + j\omega)^2}$

$$X(j\omega) = X(j\omega)\,H(j\omega) = \frac{2}{(-\omega^2 + 2j\omega + 8)\,(2 + j\omega)^2}$$

Using the partial fraction expansion,
We get

$$y(j\omega) = \frac{\frac{1}{4}}{j\omega + 2} - \frac{\frac{1}{2}}{(j\omega + 2)^2} + \frac{1}{(j\omega + 2)^3} - \frac{\frac{1}{4}}{(j\omega + 4)}$$

Taking the inverse Fourier transform

$$y(t) = \frac{1}{4}e^{-2t}u(t) - \frac{1}{2}te^{-2t} \cdot u(t) + t^2e^{-2t} \cdot u(t) - \frac{1}{4}e^{-4t} \cdot u(t)$$

(c) Taking the fourier transform of both sides of the given differential equation, we get

$$H(j\omega) = \frac{Y(j\omega)}{X(j\omega)} = \frac{2(-\omega^2 - 1)}{-\omega^2 + \sqrt{2}j\omega + 1}$$

Using the partial fraction expansion, we get

$$H(j\omega) = 2 + \frac{-\sqrt{2} - 2\sqrt{2}j}{j\omega - \frac{\left(-\sqrt{2} - j\sqrt{2}\right)}{2}} + \frac{-\sqrt{2} + 2\sqrt{2}j}{\left(j\omega - \frac{-\sqrt{2} - j\sqrt{2}}{2}\right)}$$

Taking the inverse Fourier transform

$$h(t) = 2\delta(t) - \sqrt{2}(1 + 2j)\,e^{-(1+j)\frac{t}{\sqrt{2}}}u(t) -$$

$$\sqrt{2}(1 - 2j)\,e^{-(1-j)\frac{t}{\sqrt{2}}}u(t).$$

Problem 4.31: A causal and stable LTI system S has the frequency response

$$H(j\omega) = \frac{j\omega + 4}{6 - \omega^2 + 5j\omega}$$

(a) Determine a differential equation relating the input $x(t)$ and output $y(t)$ of S.

(b) Determine the impulse response $h(t)$ of S.

(*c*) What is the output of S when the input is

$$x(t) = e^{-4t}u(t) - te^{-4t}u(t)?$$

Solution:

(*a*) We know that

$$\frac{Y(j\omega)}{X(j\omega)} = \frac{j\omega + 4}{6 - \omega^2 + 5j\omega}$$

Cross multiplication and taking the inverse Fourier transform, we get

$$\frac{d^2y(t)}{dt^2} + \frac{5dy(t)}{dt} + 6y(t) = \frac{dx(t)}{dt} + 4x(t)$$

(*b*) We have

$$H(j\omega) = \frac{2}{2 + j\omega} - \frac{1}{3 + j\omega}$$

Taking the inverse Fourier transform, we get

$$h(t) = 2\,e^{-2t} \cdot u(t) - e^{-3t} \cdot u(t)$$

(*c*) We have

$$X(j\omega) = \frac{1}{1 + j\omega} - \frac{1}{(4 + j\omega)^2}$$

Therefore

$$y(j\omega) = X(j\omega)\,H(j\omega) = \frac{1}{(4 + j\omega)(2 + j\omega)}$$

Finding the partial fraction expansions of $y(j\omega)$ and taking the inverse fourier transform.

$$y(t) = \frac{1}{2}e^{-2t}u(t) - \frac{1}{2}e^{-4t} \cdot u(t)$$

Problem 4.32: In this problem, we provide examples of the effects of nonlinear changes in phase.

(*a*) Consider the continuous-time LTI system with frequency response

$$H(j\omega) = \frac{a - j\omega}{a + j\omega}$$

where $a > 0$. What is the magnitude of $H(j\omega)$? What is $\sphericalangle H(j\omega)$? What is the impulse response of this system?
(*b*) Determine the output of the system of part (*a*) with $a = 1$ when the input is

$$\cos\left(t/\sqrt{3}\right) + \cos t + \cos\sqrt{3}t$$

Roughly sketch both the input and the output.

Solution:

(*a*) From the given information

$$|H(j\omega)| = \frac{\sqrt{a^2+\omega^2}}{\sqrt{a^2+\omega^2}} = 1$$

$$\text{Also,}\quad H(j\omega) = -\tan^{-1}\frac{\omega}{a} - \tan^{-1}\frac{\omega}{a} = -2\tan^{-1}\left(\frac{\omega}{a}\right)$$

$$\text{Also,}\quad H(j\omega) = -1 + \frac{2a}{a+j\omega}$$

$$\Rightarrow \qquad h(t) = -\delta(t) + 2ae^{-at}u(t)$$

(*b*) If $a = 1$,

We have,

$$|H(j\omega)| = 1 \sphericalangle H(j\omega) = -2\tan^{-1}\omega$$

Therefore

$$y(t) = \cos\left(\frac{t}{\sqrt{3}} - \frac{\pi}{3}\right) - \cos\left(t - \frac{\pi}{2}\right) + \cos\left(\sqrt{3}t - \frac{2\pi}{3}\right)$$

Problem 4.33: Consider an LTI system whose response to the intput

$$x(t) = [e^{-t} + e^{-3t}]\,u(t)$$

is

$$y(t) = [2e^{-t} - 2e^{-4t}]\,u(t)$$

(*a*) Find the frequency response of this system.
(*b*) Determine the system's impulse response
(*c*) Find the differential equation relating the input and the output of this system.

Solution:

(*a*) The frequency response is

$$H(j\omega) = \frac{Y(j\omega)}{X(j\omega)} = \frac{3(3+j\omega)}{(4+j\omega)(2+j\omega)}$$

(*b*) Finding the partial fraction expansion of answer in part (*a*), and taking its inverse Fourier transform,

We obtain

$$h(t) = \frac{3}{2}[e^{-4t} + e^{-2t}]\,u(t)$$

(*c*) We have

$$\frac{Y(j\omega)}{X(j\omega)} = \frac{(9+3\,j\omega)}{8+6\,j\omega-\omega^2}$$

Cross multiplying and taking the inverse fourier transform We get

$$\frac{d^2y(t)}{dt^2} + \frac{6dy(t)}{dt} + 8y(t) = \frac{3\,dx(t)}{dt} + 9x(t)$$

Problem 4.34: Consider the signal $x(t)$ in Figure P. 4.34 (*a*)

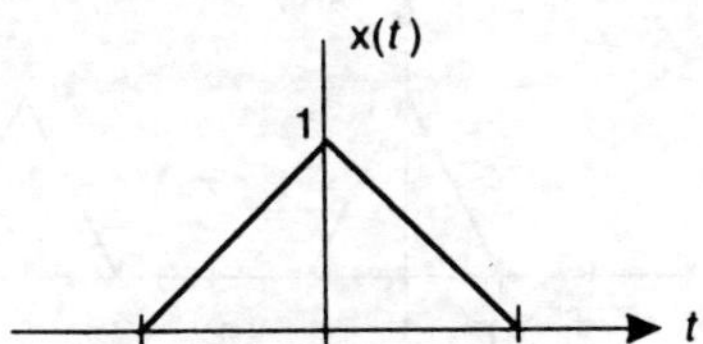

Fig. P. 4.34 (a)

(*a*) Find the Fourier transform $X(j\omega)$ of $x(t)$

(*b*) Sketch the signal

$$\bar{x}(t) = x(t) * \sum_{k=-\infty}^{\infty}\delta(t-4k)$$

(*c*) Find another signal $g(t)$ such that $g(t)$ is not the same as $x(t)$ and

$$\bar{x}(t) = g(t) * \sum_{k=-\infty}^{\infty} \delta(t-4k)$$

Solution:

(*a*) Note that

$$x(t) = x_1(t) * x_1(t)$$

Where, $x_1(t) = \begin{cases} 1 & |\omega| < \frac{1}{2} \\ 0 & \text{otherwise} \end{cases}$

Also, the fourier transform $X_1(j\omega)$ of $x_1(t)$ is

$$X_1(j\omega) = 2\frac{\sin\left(\frac{\omega}{2}\right)}{\omega}$$

Using the convolution property
We have

$$X(j\omega) = X_1(j\omega)\, X_1(j\omega) = \left(\frac{2\sin\left(\frac{\omega}{2}\right)}{\omega}\right)^2$$

(*b*) The signal $x(t)$ is as shown in figure

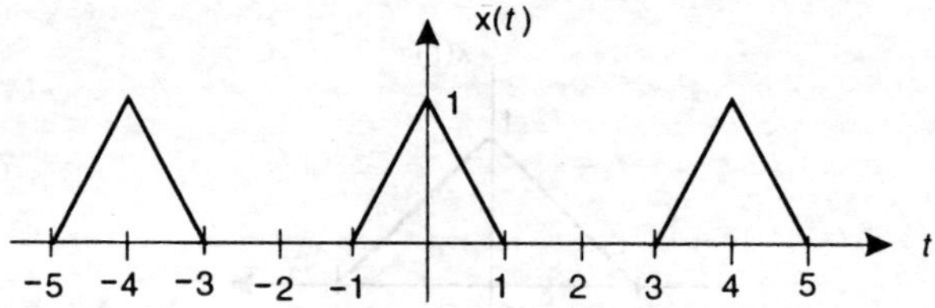

Fig. P. 3.34 (b)

(*c*) One possible of $g(t)$ as shown in figure (b).

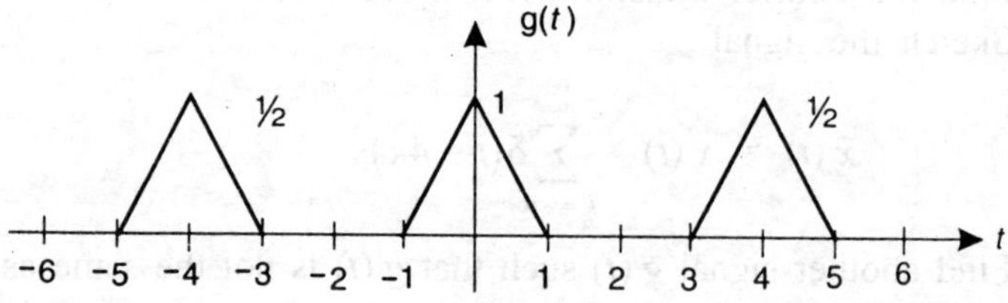

Fig. P. 3.34 (c)

Problem 4.35: The output of a system in response to an input $x(t) = e^{-2t}u(t)$ is $y(t) = e^{-t}u(t)$. Find the frequency response and the impulse response of this system.

Solution:

Take the *FT* of $x(t)$ and $y(t)$, obtaining

$$X(j\omega) = \frac{1}{j\omega + 2} \quad \text{and} \quad Y(j\omega) = \frac{1}{j\omega + 1}$$

Now use the definition

$$H(j\omega) = \frac{Y(j\omega)}{X(j\omega)}$$

to obtain the system frequency response

$$H(j\omega) = \frac{j\omega + 2}{j\omega + 1}$$

which may be reqritten as

$$H(j\omega) = \left(\frac{j\omega + 1}{j\omega + 1}\right) + \frac{1}{j\omega + 1} = 1 + \frac{1}{j\omega + 1}$$

Take the inverse *FT* of each term to obtain the impulse response of the system:

$$h(t) = \delta(t) + e^{-t}u(t)$$

Problem 4.36: Find the frequency response and impulse response of the system described by the differential equation

$$\frac{d^2}{dt^2}y(t) + 3\frac{d}{dt}y(t) + 2y(t) = 2\frac{d}{dt}x(t) + x(t)$$

Solution:

Here we have

$$N = 2, \qquad M = 1.$$

Now, we obtain the frequency response

$$H(j\omega) = \frac{2j\omega + 1}{(j\omega)^2 + 3j\omega + 2}$$

The impulse response is given by the iverse *FT* of $H(j\omega)$. Rewrite $H(j\omega)$ using the partial fraction expansion:

$$\frac{2j\omega + 1}{(j\omega)^2 + 3j\omega + 2} = \frac{A}{j\omega + 1} + \frac{B}{j\omega + 2}$$

Solving for A and B we obtain $A = -1$, $B = 3$. Hence

$$H(j\omega) = \frac{-1}{j\omega + 1} + \frac{3}{j\omega + 2}$$

The inverse *FT* gives the impulse response

$$h(t) = 3e^{2t}u(t) - e^{-t}u(t)$$

Problem 4.36: The mechanical system depicted in Fig. 4.36 (*a*) has the applied force $x(t)$ as its input and position $y(t)$ as its output. The relationship between $x(t)$ and $y(t)$ is governed by the differential equation

$$m\frac{d^2}{dt^2}y(t) + f\frac{d}{dt}y(t) + ky(t) = x(t)$$

Find the frequency response of the system and plot the magnitude response in dB for $m = 0.5$ kg, $f = 0.1$ N . s/m, and $k = 50$ N/m.

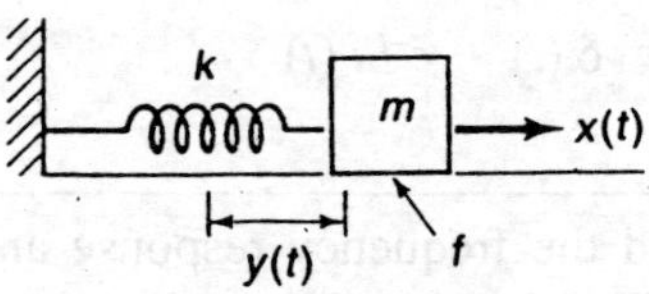

Fig. P. 4.36 (a)

(*a*) Mechanical system with input $x(t)$ and output $y(t)$

Solution:

The transfer function,

$$H(j\omega) = \frac{1}{m(j\omega)^2 + fj\omega + k}$$

Figure P. 4.36 (b) depicts the magnitude response in *dB* for the specified values, of *m*, *f*, and *k*. This system tends to attenuate frequencies both below and above 10 rads/s. hence it has a very narrow bandpass filtering characteristic and effectively only responds to input requency components near 10 rad/s.

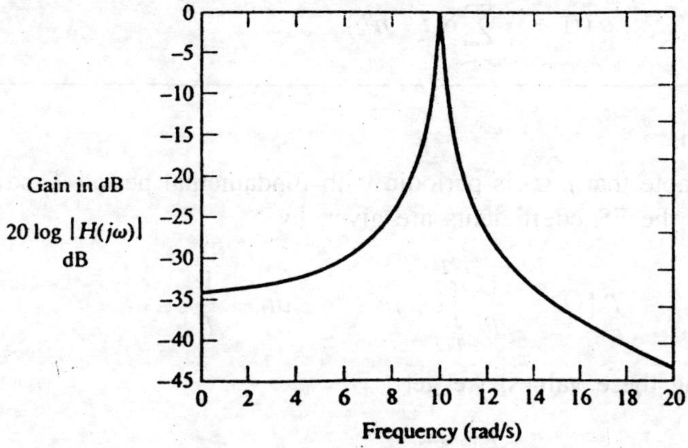

Fig. P. 4.36 (b)

Problem 4.37: Find the FT representation for $x(t) = \cos(\omega_0 t)$

Solution:

The *FS* representation for $x(t)$ is

$$\cos(\omega_0 t) \xleftrightarrow{FT} X[k] = \begin{cases} \frac{1}{2}, & k = \pm 1 \\ 0, & k \neq \pm 1 \end{cases}$$

Using these coefficients we have

$$\cos(w_0 t) \xleftrightarrow{FT} X(j\omega) = \pi\delta(\omega - \omega_0) + \pi\delta(\omega + \omega_0)$$

The pair is depicted graphically in Fig. 4.37 (a).

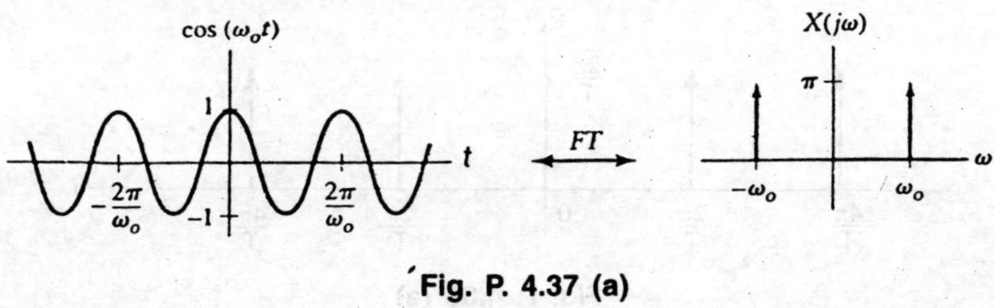

Fig. P. 4.37 (a)

Problem 4.38: Find the FT of the impulse train

$$p(t) = \sum_{n=-\infty}^{\infty} \delta(t - nT)$$

Solution:

We note that $p(t)$ is periodic with fundamental period T, so $w_0 = 2\pi/T$ and the FS coefficients are given by

$$P[k] = \frac{1}{T}\int_{-T/2}^{T/2} \delta(t) e^{-jk\omega_0 t} \cdot dt = \frac{1}{T}$$

Using these values, we get

$$P(j\omega) = \frac{2\pi}{T}\sum_{k=-\infty}^{\infty} \delta(\omega - k\omega_0)$$

Hence the *FT* of an impulse train is another impulse train. The spacing between the impulses in frequency is inversely related to the spacing between impulses in time. This *FT* pair is depicted in Fig. 4.38 (a).

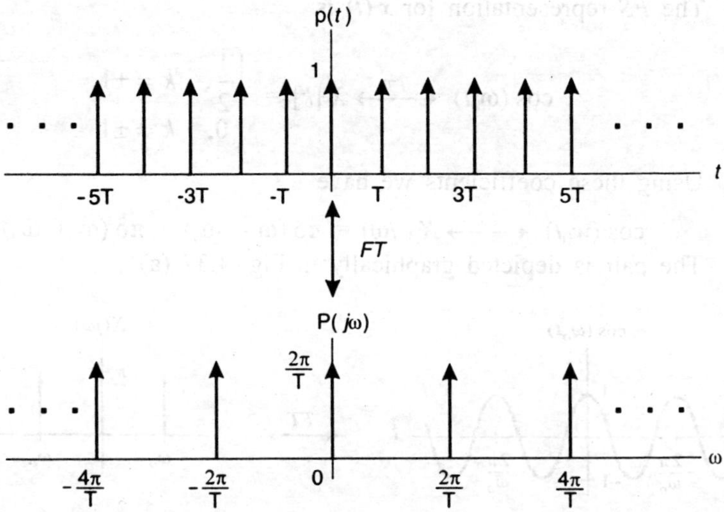

Fig. P. 4.38 (a)

Problem 4.39: Let the input to a system with impulse response $h(t) = (1/\pi t)\sin(\pi t)$ be the periodic square wave. Use the convolution property to find the output of this system.

Solution:

The frequency response of the system is

$$h(t) \xleftrightarrow{FT} H(j\omega) = \begin{cases} 1, & |\omega| \leq \pi \\ 0, & \text{otherwise} \end{cases}$$

The *FT* of the periodic input signal was obtained in Drill Problem 4.4 and is given by

$$X(j\omega) = \sum_{k=-\infty}^{\infty} \frac{2\sin\left(k\dfrac{\pi}{2}\right)}{k}\,\delta\left(\omega - k\frac{\pi}{2}\right)$$

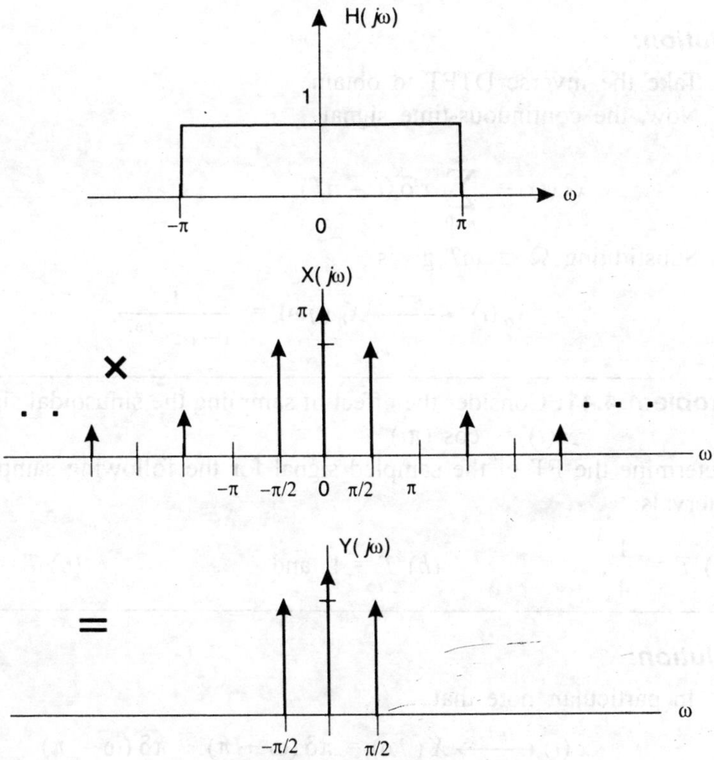

Fig. P. 4.39 (a)

The *FT* of the system output is $Y(j\omega) = H(j\omega)$. This product is depicted in Fig. 4.39, where we see that

$$Y(j\omega) = 2\delta\left(\omega+\frac{\pi}{2}\right) + \pi\delta(\omega) + 2\delta\left(\omega-\frac{\pi}{2}\right)$$

In effect, the system described by $H(j\omega)$ acts as a lowpass filter, passing the discrete-frequency components at $-\frac{\pi}{2}$, 0, and $\frac{\pi}{2}$, while suppressing all others. Taking the inverse *FT* of $Y(j\omega)$ gives the output

$$y(t) = \frac{1}{2}+\frac{1}{\pi}\cos\left(\omega\frac{\pi}{2}\right)$$

> **Problem 4.40:** Determine the FT pair associated with a signal whose
>
> $$X(e^{j\omega}) = \frac{1}{1-ae^{-j\omega}}$$

Solution:

Take the inverse DTFT to obtain
Now, the continuous-time signal

$$x_\delta(t) = \sum_{n=0}^{\infty} a^n\delta(t-nT)$$

Substituting $\Omega = \omega T$ gives

$$x_\delta(t) \xleftrightarrow{FT} X_\delta(j\omega) = \frac{1}{1-ae^{-j\omega T}}$$

> **Problem 4.41:** Consider the effect of sampling the sinusoidal signal
>
> $$x(t) = \cos(\pi t)$$
>
> Determine the FT of the sampled signal for the following sampling intervals:
>
> (*a*) $T = \frac{1}{4}$, (*b*) $T = 1$, and (*c*) $T = \frac{1}{2}$.

Solution:

In particular, note that

$$x(t) \xleftrightarrow{FT} X(j\omega) = \pi\delta(\omega+\pi) + \pi\delta(\omega-\pi)$$

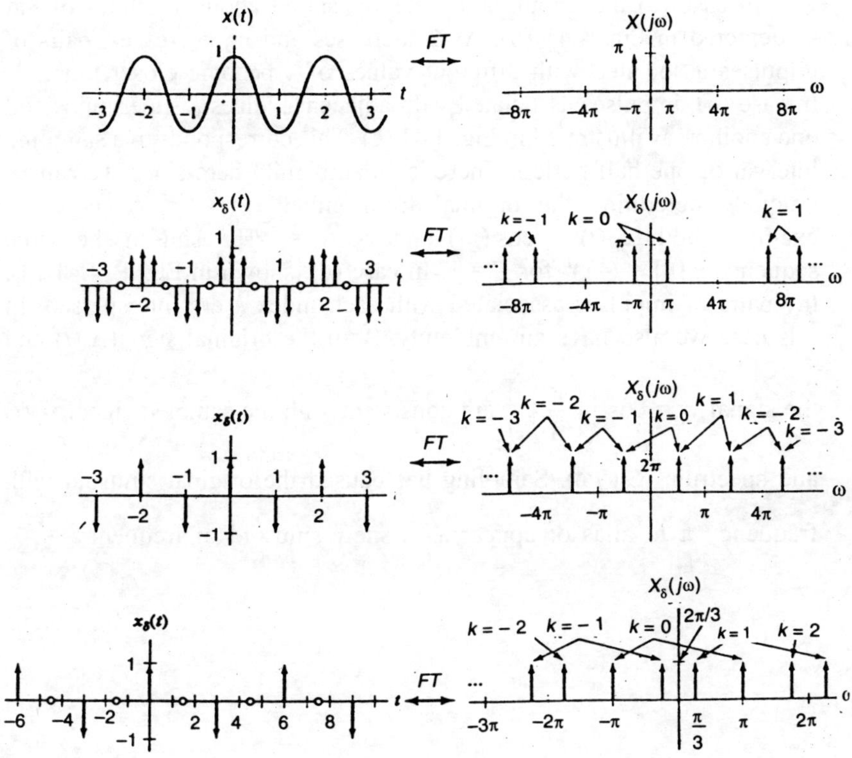

Fig. P. 4.41

Now,

$$X_\delta(j\omega) = \frac{\pi}{T}\sum_{k=-\infty}^{\infty} \delta(\omega + \pi - k\omega_s) + \delta(\omega - \pi - k\omega_s)$$

Hence $X_\delta(j\omega)$ consists of pairs of impulses separated by 2π centered on integer multiples of the sampling frequency, ω_s. The sampling frequency is different in each case. Using $\omega_s = 2\frac{\pi}{T}$ gives (*a*) $\omega_s = 8\pi$, (*b*) $\omega_s = 2\pi$, and (*c*) $\omega_s = 4\frac{\pi}{3}$, respectively. the sampled continuous-time signals and their *FT*s are depicted in Fig. 4.41.

In case (a) the impulses are clearly paired about multiples of 8π, as depicted in Fig. 4.41 (b). As T increases and w_s decreases, pairs of impulses associated with different values of k become closer together. In case (*b*), impulses associated with adjacent indices k superimpose on one another, as illustrated in Fig. 4.41 (c). This corresponds to a sampling interval of one-half period. There is an ambiguity here since we cannot uniquely determine the original from either $x_\delta(t)$ or $X_\delta(j\omega)$. For example, both $x_1(t) = \cos(\pi t)$ and $x_2(t) = e^{j\pi t}$ result in the same sequence $x[n] = (-1)^n$ for $T = 1$. In case (c), shown in Fig. P. 4.41 (d), the pairs of impulses associated with each index k are interspersed. In this case we also have an ambiguity. Both the original signal $x(t)$ and the signal $x_1 = \cos\left(\left(\frac{\pi}{3}\right)t\right)$ are consistent with the sampled signal $x_\delta(t)$ and spectrum $X_\delta(j\omega)$. Sampling has caused the original sinusoid with frequency π to alias or appear as a snew sinusoid of frequency $\frac{\pi}{3}$.

Chapter 5

Communication Systems

Problem 5.1: You are given a nonlinear device whose input–output relation is described by

$$i_o = a_1 v_i + a_2 v_i^2$$

where a_1 and a_2 are constants, v_i is the input voltage, and i_o is the output current. Let

$$v_i\,(t) = A_c \cos(\omega_c t) + A_m \cos(\omega_m t)$$

where the first term represents a sinusoidal carrier and the second term represents a sinusoidal modulating signal.

(*a*) Determine the frequency content of $i_o\,(t)$.

(*b*) The output current $i_o\,(t)$ contains an AM signal produced by the two components of $v_i\,(t)$. Describe the specification of a filter that extracts this AM signal from $i_o\,(t)$.

Solution:

(*a*) $i_o = a_1[A_c \cos(\omega_c t) + A_m \cos(\omega_m t)] + [A_c \cos(\omega_c t) + A_2 \cos(\omega_m t)]^2 \cdot a_2$

$= a_1\,[A_c \cos(\omega_c t) + A_m \cos(\omega_m t)] + \dfrac{a_2 A_c^2}{2}\,(1 + \cos 2\,\omega_c t)$

$+ \dfrac{a_2 A_2^2}{2}\,(1 + \cos 2\,\omega_m t) + 2\,a_2 A_2 A_c \times \cos(\omega_c t)\,\cdot$

$\cos(\omega_m t)$

$$i_o(t) = a_1[A_c \cos(\omega_c t) + A_m \cos(\omega_m t)] + \frac{a_2 A_c^2}{2}[1 + \cos(2\omega_c t)]$$

$$+ \frac{a_2 A_2^2}{2}[1 + \cos(2\omega_m t)] + a_2 A_2 A_c [\cos(\omega_c - \omega_m)t + \cos(\omega_c + \omega_m)t]$$

The Fourier transform, $i_o(\omega)$ is then,

$$i_o(w) = a_1 A_c \pi[\delta(\omega - \omega_c) + \delta(\omega - \omega_c)] + a_1 A_m \pi[\delta(\omega - \omega_m) + \delta(\omega + \omega_m)] + \frac{a_2 A_c^2}{2} + \frac{a_2 A_c^2}{2}\pi[\delta(\omega - 2\omega_c) + \delta(\omega + \omega_c)]$$

$$+ \frac{a_2 A_2^2}{2} + \frac{a_2 A_2^2}{2}\pi[\delta(\omega - 2\omega_m) + \delta(\omega + 2\omega_m)] + a_2 A_2 A_c \pi [\delta(\omega - \omega_c + \omega_m) + \delta(\omega + \omega_c - \omega_m)] + a_2 A_2 A_c \pi [\cos(\omega - \omega_c - \omega_m) + \delta(\omega + \omega_c + \omega_m)]$$

Fourier spectra is shown below

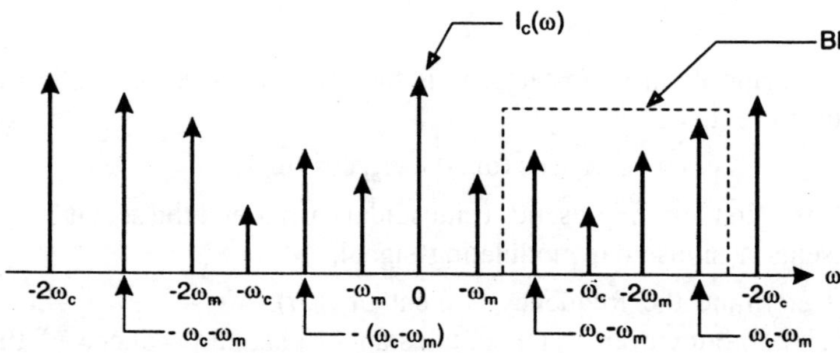

Fig. P. 5.1 (a)

(*b*) The *AM* signal is expressed as,

$$i_{AM}(t) = a_1 A_c \cos(\omega_c t) + 2a_2 A_2 A_c \cos(\omega_c t) \cdot \cos(\omega_m t)$$

Using Band-pass filter having frequency greater than ω_m and less than $2\omega_c$ as illustrated in figure above.

Problem 5.2: Figure P5.2 shows the circuit diagram of a balanced modulator. The input applied to the top AM modulator is $m(t)$ whereas that applied to the lower AM modulator is $-m(t)$ these two modulators have the same amplitude sensitivity. Show that the output $s(t)$ of the balanced modulator consists of a DSB-SC modulated signal.

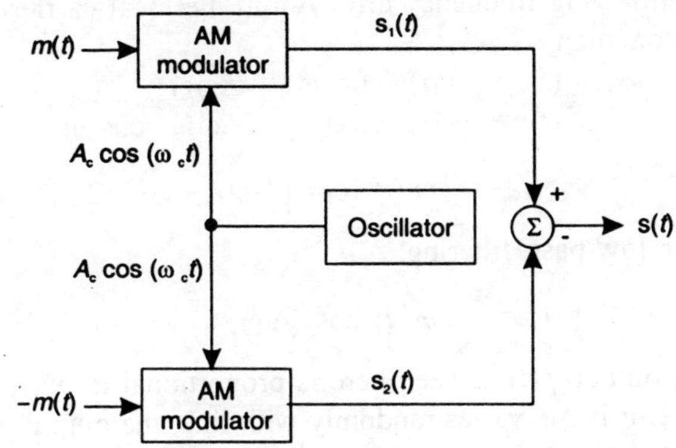

Fig. P. 5.2

Solution:

From the figure,

$$s_1(t) = A_c[1 + k_a\, m(t)] \cdot \cos(2\pi f_c t)$$
$$= A_c[1 + k_a m(t)] \cdot \cos \omega_c t$$

where k_a is the amplitude sensitivity

and, $\quad s_2(t) = A_c[1 - k_a m(t)] \cdot \cos \omega_c t$

Then, $\quad s(t) = s_1(t) - s_2(t) = 2\, A_c k_a m(t) \cdot \cos \omega_c t$

Since, singal $s(t)$ consists of product of $m(t)$ and carrier signal, it must be a DSB-SC modulated signal.

Problem 5.3: A DSB-SC modulated signal is demodulated by applying it to a coherent detector. Evaluate the effect of a frequency error $\Delta\omega$ in the local carrier frequency of the detector, measured with respect to the carrier frequency of the incoming DSB-SC signal.

Solution:

A coherent detector is shown below, having an input of DSB-SC modulated signal.

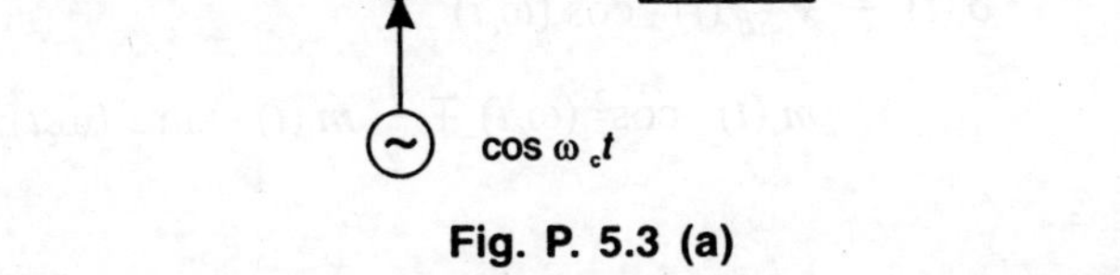

Fig. P. 5.3 (a)

$$x_{DSB}(t) = m(t) \cdot \cos \omega_c(t)$$

When having frequency error $\Delta\omega$ in the local carrier frequency of the detector, then

$$x_{DSB}(t) = m(t) \cdot \cos(\omega_c + \Delta\omega)\, t$$

and

$$\delta(t) = m(t) \cdot \cos(\omega_c + \Delta\omega)\, t \cdot \cos \omega_c t$$

$$= \frac{1}{2} m(t) \left[\cos(\Delta\omega t) + \cos(2\omega_c + \Delta\omega)\, t\right]$$

After low-pass filtering,

$$\gamma(t) = \frac{1}{2} m(t) \cos(\Delta\omega t)$$

The, output $\gamma(t)$ as seen here is proportional to $m(t)$, when Δw is constant. But if Δw varies randomly with line, the output will vary also which is undesirable.

Problem 5.4: A single sideband modulated wave $s(t)$ is applied to the coherent detector shown in Fig. P5.4. The cutoff frequency of the lowpass filter is set equal to the highest frequency component of the messag signal. Using frequency-domain ideas, show that this detector produces an output that is a scaled version of the original message signal. You may assume that the carrier frequency ω_c satisfies the condition $\omega_c > \omega_m$.

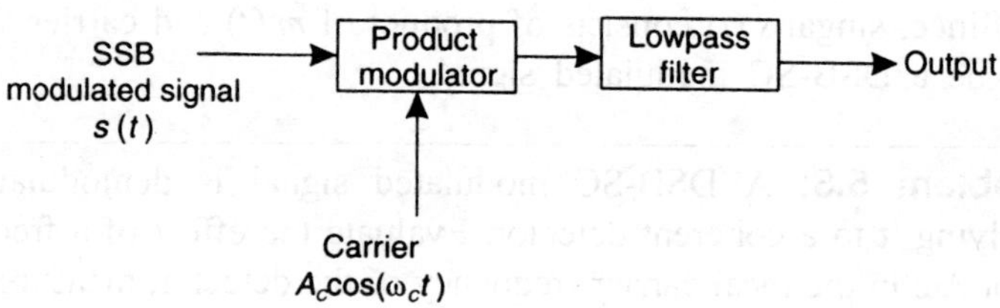

Fig. P. 5.4

Solution:

The SSB modulated signal $s(t)$ is expressed as,

$$x_{SSB}(t) = m(t) \cdot \cos(\omega_c t) \mp \hat{m}(+) \sin(\omega_c t)$$

Then, the output of the product modulator is,

$$\delta(t) = x_{SSB}(t) \cdot \cos(\omega_c t)$$

$$= m(t) \cdot \cos^2(\omega_c t) \mp \frac{1}{2} \hat{m}(t) \cdot \sin z(\omega_c t)$$

$$= \frac{1}{2} m(t) \cdot [1 + \cos 2\omega_c t] \mp \frac{1}{2} \hat{m}(t) \cdot \sin z\, \omega_c t$$

After low-pass filtering, the output is

$$y(t) = \frac{1}{2} m(t).$$

Which is a scaled version of the message signal $m(t)$.

Problem 5.5: In this problem we study the idea of mixing used in a *superheterodyne receiver*. To be specific, consider the block diagram of the *mixer* shown in Fig. P. 5.5 (a) that consists of a product modulator with a local oscillator of variable frequency, followed by a bandpass filter. The input signal is an AM wave of bandwidth 10 kHz of carrier frequency that may lie anywhere in the range 0.535–1.605 MHz; these parameters are typical of AM radio broadcasting. It is required to translate this signal to a frequency band centered at a fixed *intermediate frequency* (IF) of 0.455 MHz. Find the range of tuning that must be provided in the local oscillator in order to achieve this requirement.

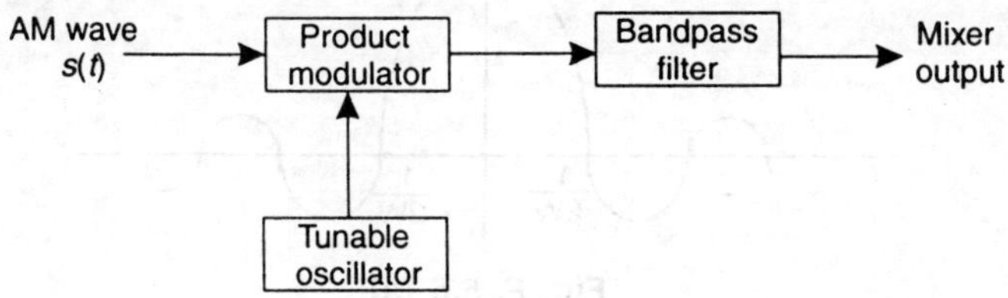

Fig. P. 5.5 (a)

Solution:

Case I:

When $f_{Lo} > f_c$

$$535 < f_c < 1605 \text{ kHz}$$

$$f_{Lo} - f_c = 455$$

$$f_{Lo} = f_c + 455$$

When $f_c = 535$ kHz, $f_{Lo} = 990$ kHz

When $f_c = 1605$ kHz, $f_{Lo} = 2060$ kHz

Thus, the required tuning range is 990 – 2060 kHz

Case II:

When $\qquad f_{Lo} < f_c,$

$$f_c - f_{Lo} = 455$$

When $\qquad f_c = 535 \text{ kHz}, f_{Lo} = 80 \text{ kHz}$

When $\qquad f_c = 1605 \text{ kHz}, f_{Lo} = 1150 \text{ kHz}$

Thus, required tuning range for this case is

80 kHz – 1150 kHz

> **Problem 5.6:** Specify the Nyquist rate for each of the following signals:
> (*a*) $g(t) = \sin c\ (200\ t)$
> (*b*) $g(t) = \sin c^2\ (200\ t)$
> (*c*) $g(t) = \sin c\ (200\ t) + \sin c^2\ (200\ t)$

Solution:

(*a*) $\qquad g(t) = \sin c\ (200\ t)$

The $g(t)$ is drawn below,

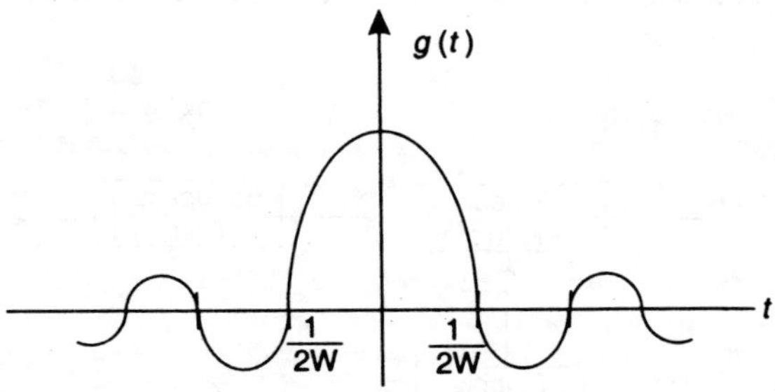

Fig. P. 5.6 (a)

The Fourier transform of $g(t)$ is,

$$G(f) = \frac{A}{2W} \text{rect}\left(\frac{f}{2W}\right)$$

and is drawn as shown in Fig. P. 5.6 (b).

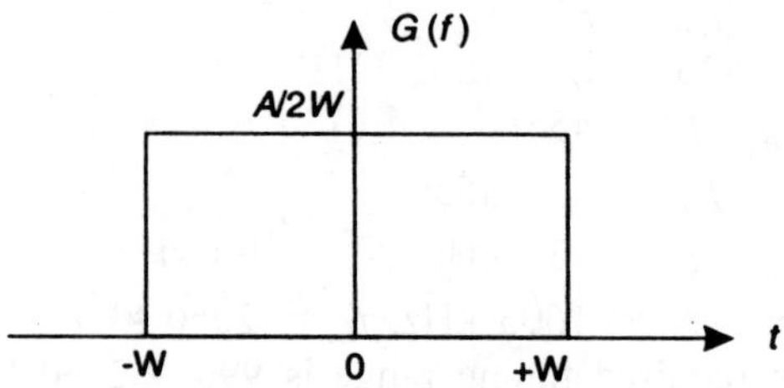

Fig. P. 5.6 (b)

Here, we see that $m(t)$ is a band-limited signal with $f_M = 100$ Hz. Then, the Nyquist rate is 200 Hz.

(*b*) From the frequency convolutional theorem, the signal $g(t) = \sin c^2 (200\, t)$ is also band-limited signal and its bandwidth is twice that of signal in part (*a*), i.e., 200 Hz. Hence, the Nyquist rate is 400 Hz.

(*c*) $$g(t) = \sin c\,(200\, t) + \sin c^2 (200\, t)$$

Here, $\sin c\,(200\, t)$ is band-limited with 100 Hz and $\sin c^2 (200\, t)$ is band-limited with 200 Hz. Then $g(t)$ is band-limited with 200 Hz and then the Nyquist rate would be 400 Hz.

Problem 5.7: Assume that the 10.8 MHz signal in the figure below is derived from the 200-kHz oscillator (multiplicated by 54) and that the 200-kHz oscillator drift is 0.1 Hz.

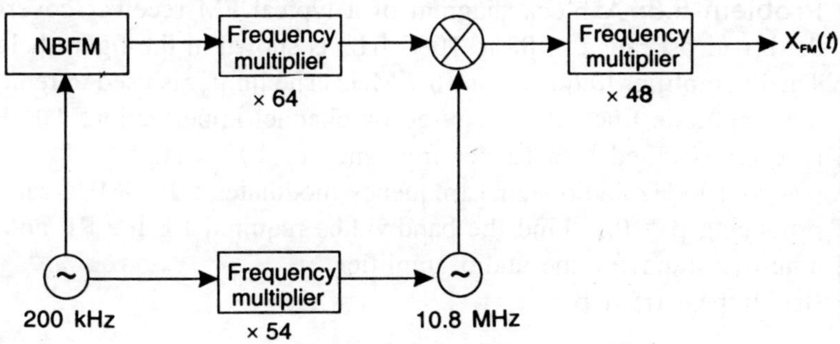

Fig. P. 5.7 (a)

(*i*) Find the drift in the 10.8 MHz signal.
(*ii*) Find the drift in the carrier of the resulting FM signal.

Solution:

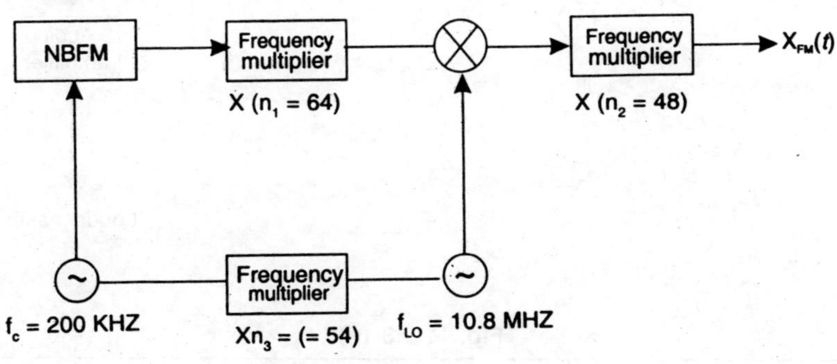

Fig. P. 5.7 (b)

(i) From the figure, $f_{Lo} = f_c H_3$

Then, drift in 10.8 MHz signal,

$$\Delta f_{Lo} = H_3 \Delta f_c = 54 \times (\pm 0.1 \text{ Hz}) = \pm 5.4 \text{ Hz}$$

(ii) Frequency of resulting FM signal,

$$f = (f_c n_l \pm f_{LO}/n_2)$$

then, drift in this signal,

$$\Delta f = (\Delta f_c n_1 \pm \Delta f_{LO}) n_2 = (6.4 \pm 5.4)\, 48$$

$$= \begin{cases} 11.8 \times 48 \text{ Hz} \\ 1 \times 48 \end{cases} = \begin{cases} 566.4 \text{ Hz} \\ 48 \text{ Hz} \end{cases}$$

Problem 5.8: A block diagram of a typical FM receiver covering the broadcast range of 88 to 108 MHz, is shown in the figure below. The IF amplifier frequency is 10.7 MHz. The limiter is used to remove the amplitude fluctuations caused by channel imperfection. The FM receiver is tuned to a carrier frequency of 100 MHz.

(i) A 10 kHz audio signal frequency modulates a 100-MHz carrier, producing $\beta = 0.2$. Find the bandwidths required for the RF and IF amplifiers and for the audio amplifier.

(ii) Repeat (i) if $\beta = 5$.

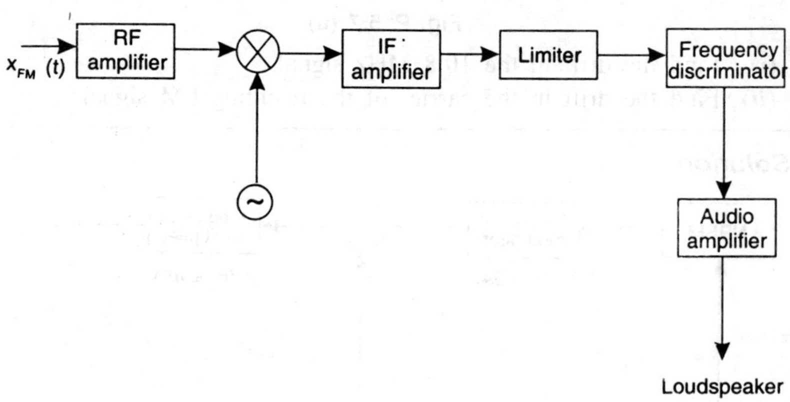

Fig. P. 5.8 (a)

Solution:

(*i*) Bandwidth for RF and IF amplifiers: 24 kHz; audio amplifier: 10 kHz.

(*ii*) Bandwidth for RF and IF amplifiers: 120 kHz; audio amplifier: 10 kHz.

Problem 5.9: Figure given below shows the circuit diagram of a square-law modulator. The signal applied to the nonlinear device is relatively weak, such that it can be represented by a square law:

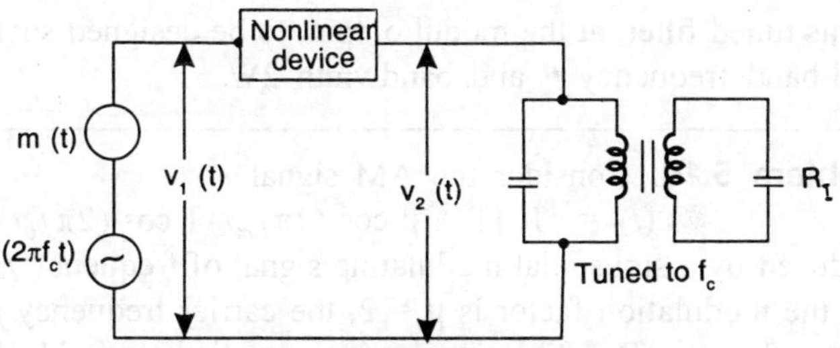

Fig. P. 5.9 (a)

$$v_2(t) = a_1 v_1(t) + a_2 v_1^2(t)$$

where a_1 and a_2 are constants, $v_1(t)$ is the input voltage, and $v_2(t)$ is the output voltage. The input voltage is defined by

$$v_1(t) = a_c \cos(2\pi f_c t) + m(t)$$

where $m(t)$ is a message signal and $A_c \cos(2\pi f_c t)$ is the carrier wave.

(*i*) Evaluate the output voltage $v_2(t)$

(*ii*) Specify the frequency response that the tuned circuit in given in the figure below must satisfy in order to generate an AM signal with f_c as the carrier frequency.

(*iii*) What is the amplitude sensitivity of this AM signal?

Solution:

As, input voltage $v_1(t)$ is given by,

$$v_1(t) = A_c \cos(2\pi f_c t) + m(t)$$

then, output voltage,

$$v_2(t) = a_1 [A_c \cos(2\pi f_c t) + m(t)] + a_2 [A_c \cos(2\pi f_c t) + m(t)]^2$$

(i) $$v_2(t) = a_1 A_c\left[1+\frac{2a_2}{a_1}m(t)\right]\cos(2\pi f_c t) + a_1 m(t) + a_2 m^2(t) + a_2 A_c^2 \cos^2(2\pi f_c t)$$

The required term for the AM signal is the term $a_1 A_c\left[1+\frac{2a_2}{a_1}m(t)\right] \times \cos(2\pi f_c t)$

Then, the amplitude sensitivity, $k_a = \frac{2a_2}{a_1}$

(ii) The remaining terms after AM signal can be removed by filtering. For this tuned filter, at the modul output, to be designed such that it has a mid-band frequency f_c and bandwidth 2W.

Problem 5.10: Consider the AM signal

$$s(t) = A_c\,[1 + \mu \cos(2\pi f_m t)]\cos(2\pi f_c t)$$

Produced by a sinusoidal modulating signal of frequency f_m. Assume that the modulation factor is $\mu = 2$, the carrier frequency f_c is much greater than f_m. The AM signal $s(t)$ is applied to an ideal envelope determine producing the output $v(t)$.

(i) Determine the Fourier series representation of $v(t)$.

(ii) What is the ratio of second-harmonic amplitude to fundamental amplitude in $v(t)$?

Solution:

(i) Envelope detector output,

$$v(t) = A_c\,|1 + \mu \cos(2\pi f_m t)|$$

which is illustrated below for the case when $\mu = 2$

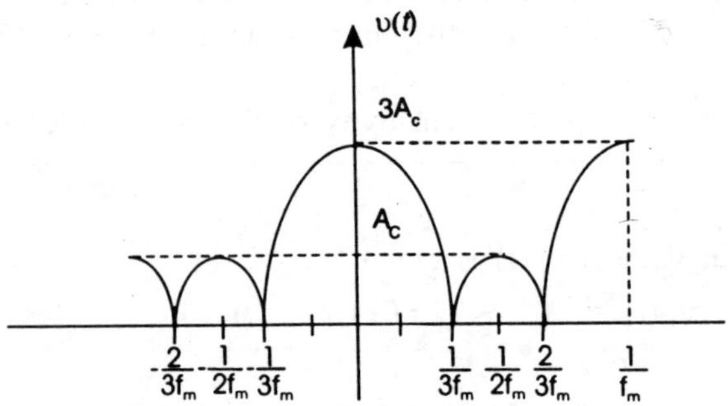

Fig. P. 5.10 (a)

From the figure, we see that, $v(t)$ is periodic with a period equal to f_m, and an even function of t, and so only consine terms will be present in its fourier series representation i.e.,

$$v(t) = a_0 + 2\sum_{n=1}^{\infty} a_n \cos(2n\pi f_m t) \quad (1)$$

where $$a_0 = 2f_m \int_0^{1/2f_m} v(t)\, dt$$

$$= 2A_c f_m \int_0^{1/3f_m} [1 + 2\cos(2n\pi f_m t)]\, dt +$$

$$2A_c f_m \int_{1/3f_m}^{1/2f_m} [-1 - 2\cos(2\pi f_m t)]\, dt$$

$$= \frac{A_c}{3} + \frac{4A_c}{\pi}\sin\left(\frac{2\pi}{3}\right)$$

$$a_n = 2f_m \int_0^{1/2f_m} v(t)\cos(2n\pi f_m t)\, dt$$

$$= 2A_c f_m \int_0^{1/3f_m} [1 + 2\cos(2n\pi f_m t)\cos(2\pi f_m t)\, dt +$$

$$2A_c f_m \int_{1/3f_m}^{1/2f_m} [1 - 2\cos(2n\pi f_m t)\cos(2n\pi f_m t)\, dt$$

$$= \frac{A_c}{n\pi}\left[2\sin\left(\frac{2n\pi}{3}\right) - \sin(n\pi)\right] +$$

$$\frac{A_c}{(n+1)\pi}\left\{2\sin\left[\frac{2\pi}{3}(n+1)\right] - \sin[\pi(n+1)]\right\} +$$

$$\frac{A_c}{(n-1)\pi}\left\{2\sin\left[\frac{2\pi}{3}(n-1)\right] - \sin[\pi(n-1)]\right\} \quad \ldots(2)$$

For $n = 0$, equation (2) reduces to that shown in equation (1)

(*ii*) For $n = 1$, equation (2) yields

$$a_1 = A_c\left(\frac{\sqrt{3}}{2\pi} + \frac{1}{3}\right)$$

For $\qquad n = 2$, it yields $a_2 = \dfrac{A_c\sqrt{3}}{2\pi}$

Therefore, the required ratio of second-harmonic amplitude to fundamental amplitude in $v(t)$ is

$$\frac{a_2}{a_1} = \frac{3\sqrt{3}}{2\pi + 3\sqrt{3}} = 0.452$$

Problem 5.11: Consider a square-law detecter, using a nonlinear device whose transfer characteristic is defined by

$$v_2(t) = a_1 v_2(t) + a_2 v_1^2(t)$$

where a_1 and a_2 are constants, $v_1(t)$ is the input, and $v_2(t)$ is the output. The input consists of the Am wave

$$v_1(t) = A_c[1 + k_a m(t)]\cos(2\pi f_c t)$$

(*i*) Evaluate the output $v_2(t)$

(*ii*) Find the conditions for which the message signal $m(t)$ may be recovered from $v_2(t)$.

Solution:

Given, $\quad v_1(t) = A_c[1 + k_a m(t)]\cos 2\pi f_c t$

Then, output of the square-law device will be,

(*i*) $$v_2(t) = a_1 v_1(t) + a_2 v_1^2(t)$$

$$= a_1 A_c[1 + k_a m(t)]\cos 2\pi f_c t + \frac{1}{2}a_2 A_c^2$$

$$[1 + k_a m(t) + a + k_a^2 m^2(t)][1 + \cos(4\pi f_c t)]$$

(*ii*) The required signal $a_2 A_c^2 k_a m(t)$ can be extracted by a low pass filter. Also the term $\frac{1}{2}a_2 A_c^2 k_a^2 m^2$ will give rise to plurality of similar frequency components within the baseband spectrum. Thus ratio wanted signal to distortion is $\dfrac{2}{k_a m(t)}$. To make this ration large, $k_a m(t)$ should be small compared unity.

Problem 5.12: The AM signal

$$s(t) = A_c[1 + k_a m(t) \cos(2\pi f_c t)$$

is applied to the system shown in the figure below. Assuming that $|k_a m(t) < 1$ for all t and the message signal $m(t)$ is limited to the interval $-W \leq f \leq W$, and that the carrier frequency $f_c > 2W$, show that $m(t)$ be obtained from the square-rooter output $v_3(t)$.

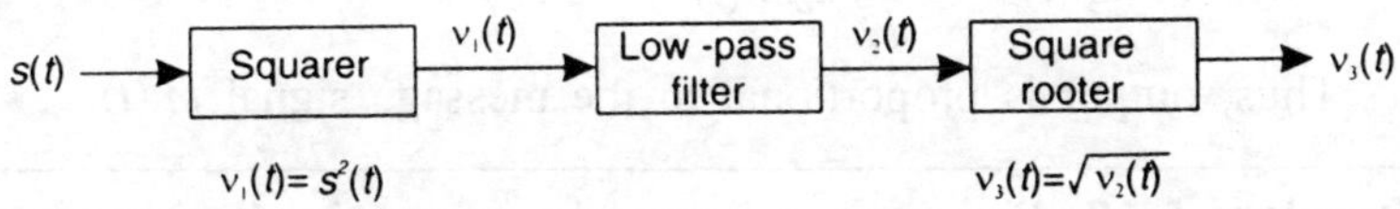

Fig. P. 5.12 (a)

Solution:

Squarer output is

$$v_t(t) = s_1^2(t)$$

or $$v_1(t) = A_c^2[1 + k_a m(t)]^2 \cos^2(2\pi f_c t)$$

$$= \frac{A_c^2}{2}[1 + 2k_a m(t)] + m^2(t)[1 + \cos(4\pi f_c t)]$$

The amplitude spectrum of $v_1(t)$ is therefore as follows, assuming that $m(t)$ is limited to the interval $-W \leq f \leq W$:

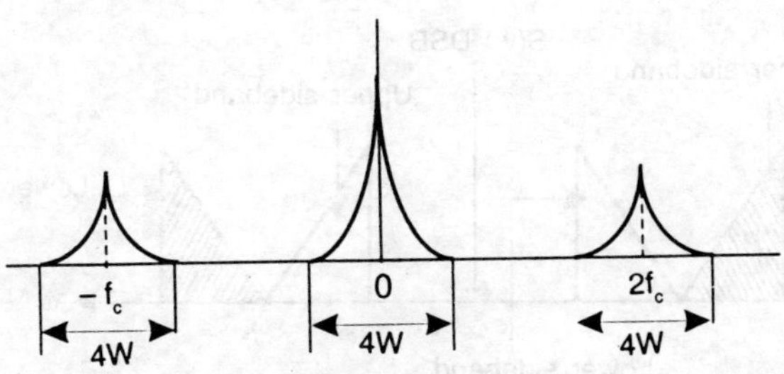

Fig. P. 5.12 (b)

Since $f_c > 2W$, then $2f_c - 2W > 2W$. Therefore, by choosing the cutoff frequency of the low–pass filter greater than $2W$, but less than $2f_c - 2W$. The output will be

$$v_2(t) = \frac{A_c^2}{2}[1 + k_a m(t)]^2$$

$\therefore$ output of square rooter,

$$v_3(t) = \sqrt{v_2(t)} = \frac{A_c}{\sqrt{2}}[1 + k_a m(t)]$$

Now, the blocking capacitor will suppress the dc term present, thus,

$$v_3(t) = \frac{A_c}{\sqrt{2}} k_a m(t)$$

Thus, output is proportional to the message signal $m(t)$.

Problem 5.13: Figure given below shows the circuit diagramm of a balanced modulator. The input applied to the top AM modulator is $m(t)$, whereas that applied to the lower AM modulator is $-m(t)$; these two modulators have the same amplitude sensitivity. Show that the outputs $s(t)$ of the balanced modulator consists of a DAB-SC modulated signal.

Solution:

The two AM modulator outputs will be

$$s_1(t) = A_c[1 + k_a m(t)] \cos(2\pi f_c t) \quad \ldots(1)$$

$$s_2(t) = A_c[1 - k_a m(t)] \cos(2\pi f_c t) \quad \ldots(2)$$

Subtracting equation (2) from equation (1), we have

$$s(t) = s_2(t) - s_1(t) = 2k_a A_c m(t) \cos(2\pi f_c t)$$

Since $s(t)$ consists only of message signal $m(t)$ and the carrier wave $c(t)$, i.e. $s(t)$ is proportional to $m(t)$. Hence $s(t)$ is DSB-SC modulated signal. Figure shown below is the spectrum of $s(t)$.

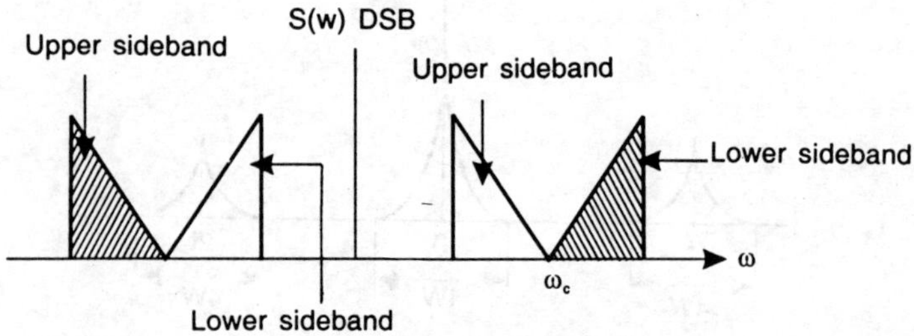

Fig. P. 5.13 (a)

As seen from the spectrum, there is no identifiable carrier it S_{DSB} (ω), so called Double sideband suppress carrier modulation.

Problem 5.14: A DSB-SC modulated signal is demodulated by applying it to a coherent detector.
(*i*) Evaluate the effect of a frequency error Δf in the local carrier frequency of the detector, measure with respect to the carrier frequency of the incoming DSB-SC signal.
(*ii*) For the case of a sinusoidal modulating wave, show that because of this frequency error, the demodulated signal exhibits beats at the error frequency. Give your answer with a sketch of this demodulated signal.

Solution:

(*i*) Output of product modulator,

$$v(t) = A_c m(t) \cos(2\pi f_c t) \cos[2\pi (f_c + \Delta t) t]$$

$$= \frac{A_c}{2} m(t) \{\cos(2\pi \Delta f_c t) + \cos[2\pi (2f_c + \Delta f) t]\}$$

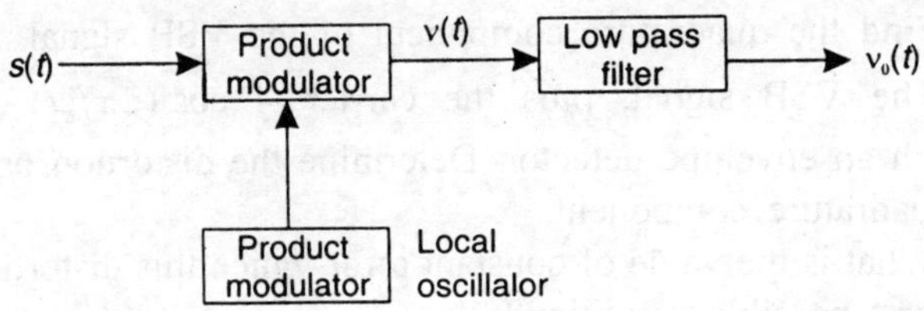

then after low pass filtering,

$$v_0(t) = \frac{A_c}{2} m(t) \cos(2\pi f_c t)$$

Thus, the output signal is the message signal modulated by a sinusoid of frequency Δf.

(*ii*) If $\quad m(t) = \cos(2\pi f_m t)$

then $\quad s_2(t) = \dfrac{A_c}{2} \cos(2\pi f_m t) \cos(2\pi \Delta f t)$

$$= \frac{A_c}{4} \{\cos[2\pi (f_m + \Delta f) t] + \cos[2\pi (f_m - \Delta f) t]$$

the wave form of $s_2(t)$ is shown below,

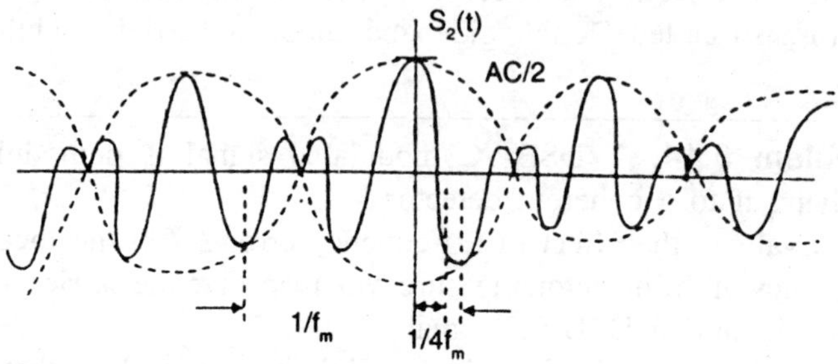

Fig. P. 5.14 (a)

Problem 5.15: The single tone modulating signal $m(t) = A_m \cos(2\pi f_m t)$ is used to generate the VSB signal.

$$s(t) = \frac{1}{2} aA_mA_c \cos 2\pi (f_c + f_m)t] + \frac{1}{2} A_mA_c$$

$$(1 - a) \cos [2\pi (f_c - f_m)t]$$

where a is a constant, less than unity, representing the attenuation of the upper side frequency.

(*i*) Find the quadrature component of the VSB signal $s(t)$.

(*ii*) The VSB signal, plus the carrier $A_c \cos(2\pi f_c t)$, is passed through an envelope detector. Determine the distortion produced by the quadrature component.

(*iii*) What is the value of constant a for which this distortion reaches its worst possible condition?

Solution:

(*i*) Any signal $s(t)$ can be expressed in terms of in phase and quardative component as,

$$s(t) = s_I(t) \cos(2\pi f_c t) - s_Q(t) \sin(2\pi f_c t) \quad ...(1)$$

Now, expanding the given signal $s(t)$,

$$s(t) = \frac{1}{2} aA_mA_c \cos(2\pi f_c t) \cos(2\pi f_m t) - \frac{1}{2} aA_mA_c \sin(2\pi f_c t)$$

$$\sin(2\pi f_m t) + \frac{1}{2}(1 - a) A_cA_m \cos(2\pi f_c t) \cos(2\pi f_m)t +$$

$$\frac{1}{2}(1 - a) A_mA_c \sin(2\pi f_c t) \sin(2\pi f_m t)$$

$$s(t) = \frac{1}{2} A_m A_c \cos(2\pi f_c t) \cos(2\pi f_m t) + \frac{1}{2} A_m A_c (1 - 2a)$$

$$\sin(2\pi f_c t) \sin(2\pi f_m t) \qquad \ldots(2)$$

Comparing equation (2) with equation (1), we have quadrative component,

$$s_Q(t) = -\frac{1}{2} A_c A_m (1 - 2a) \sin(2\pi f_m t)$$

(*ii*) Now, the VSB signal after adding carrier signal $A_c \cos(2\pi f_c t)$ is,

$$s(t) = A_c \left[1 + \frac{A_m}{2} \cos(2\pi f_m t)\right] \cos(2\pi f_c t) +$$

$$\frac{1}{2} A_c A_m (1 - 2a) \sin(2\pi f_m t) \sin(2\pi f_c t)$$

Then, envelope equal to

$$a(t) = \sqrt{s_Q^2(t) + s_1^2(t)}$$

$$= A_c \sqrt{\left[1 + \frac{1}{2} A_m \cos(2\pi f_m t)\right]^2 + \left[\frac{1}{2} A_m (1 - 2a) \sin(2\pi f_m t)\right]^2}$$

$$= A_c \left[1 + \frac{1}{2} A_m \cos(2\pi f_m t)\right] \sqrt{1 + \left[\frac{\frac{1}{2} A_m (1 - 2a) \sin(2\pi f_m t)}{1 + \frac{1}{2} A_m \cos(2\pi f_m t)}\right]^2}$$

$$= A_c \left[1 + \frac{1}{2} A_m \cos(2\pi f_m t)\right] d(t)$$

where distortion,

$$\delta(t) = \sqrt{1 + \left[\frac{\frac{1}{2} A_m (1 - 2a) \sin(2\pi f_m t)}{1 + \frac{1}{2} A_m \cos(2\pi f_m t)}\right]^2}$$

(*ii*) Distortion will be maximum when,

$$a = 0.$$

Problem 5.16: Consider the modulated wave

$$s(t) = A_c \cos(2\pi f_c t) + m(t)\cos(2\pi f_c t) - \hat{m}(t)\sin(2\pi f_c t)$$

which represents a carrier plus an SSB signal, with $m(t)$ denoting the message signal and $\hat{m}(t)$ its Hilbert transform. Determine the conditions for which an ideal envelope detector, with $s(t)$ as input, would produce a good approximation to the message signal $m(t)$.

Solution:

Given the modulate wave

$$s(t) = [A_c + m(t)\cos 2\pi f_c(t) - \hat{m}(t)\sin(2\pi f_c t).$$

then, in phase component $s_I(t) = A_c + m(t)$

and, quadrative component $s_Q(t) = +\hat{m}(t)$

then, envelope detector output

$$a(t) = \sqrt{s_I^2(t) + s_Q^2(t)} = \sqrt{[A_c + m(t)^2] + [\hat{m}(t)]^2}$$

If carrier amplitude is such that,

$A_c \gg |m(t)|$ and $A_c \gg |\hat{m}(t)|$, for all values of t.

Then $$a(t) \approx \sqrt{A_c^2 + 2A_c\, m(t)} \approx A_c\sqrt{1 + \frac{2}{A_c}m(t)}$$

$$\approx A_c\left[1 + \frac{m(t)}{A_c}\right] = A_c + m(t)$$

It is seen that except for dc bias A_c, the envelope $a(t)$ is approximately equal to $m(t)$.

Problem 5.17: Consider the SSB wave

$$s(t) = m(t)\cos(2\pi f_c t) - \hat{m}(t)\sin(2\pi f_c t)$$

where f_c is the carrier frequency. $m(t)$ is the message signal, and $\hat{m}(t)$ is its Hilbert transform. This modulated wave is applied to a square-law device characterized by

$$y(t) = s^2(t)$$

Show that the output $y(t)$ contains a frequency component at twice the carrier frequency but that it has a time-varying phase, which makes it impractical to recover the carrier by squaring.

Solution: The square law device output, with SSB wave $s(t)$ as input, is given by

$$y(t) = s^2(t)$$

$$= [m(t)\cos(2\pi f_c t) - \hat{m}(t)\sin(2\pi f_c t)]^2$$

$$= m^2(t)\cos^2(2\pi f_c t) + \hat{m}^2(t)\sin^2(2\pi f_c t) -$$

$$2m(t)\,\hat{m}(t)\cos(2\pi f_c t)\sin(2\pi f_c t)$$

$$= \frac{1}{2}m^2(t)\,[1 + \cos(4\pi f_c t)] + \frac{1}{2}\hat{m}^2(t)$$

$$[1 - \cos(4\pi f_c t)] - m(t)\,\hat{m}(t)\sin(4\pi f_c t)$$

$$y(t) = \frac{1}{2}[m^2(t) + \hat{m}^2(t)] + \left\{\frac{1}{4}\left[m^2(t) + \hat{m}^2(t)\right]^2 + m^2(t)\hat{m}^2(t)\right\}^{\frac{1}{2}} -$$

$$\cos[4\pi f_c t + Q(t)]$$

where $\quad Q(t) = \tan^{-1}\left[\dfrac{2m(t)\,\hat{m}(t)}{m^2(t) + \hat{m}^2(t)}\right]$

Since, $Y(t)$ is a mixed combination of amplitude and frequency modulation. The AM can be eliminated by filtering but the phase modulation is unaffected by such an operation, which makes it impractical to recover the carrier wave by squaring.

Problem 5.18: In this problem we study the idea of mixing in a superheterodyne receiver. To be specific, consider the block diagram of the mixer shown in the figure below that consists of a product modulator with a local oscillator of variable frequency f_t followed by a band-pass filter. The input signal is an AM wave of band-width 10 kHz and carrier frequency that may lie anywhere in the range 0.535–1.605 MHz; these parameters are typical of AM radio broadcasting. It is required to translate this signal to a frequency band centered at a fixed intermediate frequncy (IF) of 0.455 MHz. Find the range of tuning that must be provided in the local oscillator in order to achieve this requirement.

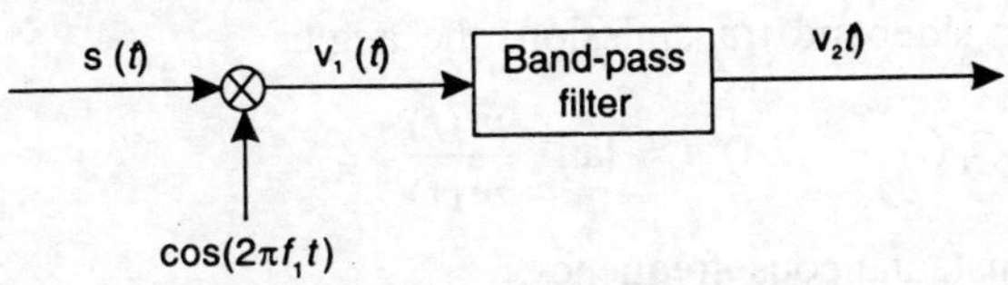

Solution:

Let f_c be the mid-brand frequency of the incoming signal and f_c denote the local oscillator frequency.

Then, as given,

$$0.535 < f_c < 1.605$$

and $$f_c - f_l = 0.455$$

where f_c and f_t are expressed in MHz.

or $$f_l = f_c - 0.455.$$

Now, when $f_c = 0.535$ MHz,

$$f_l = 0.535 - 0.455 = 0.08 \text{ MHz}$$

and when $f_c = 1.605$ MHz,

$$f_l = 1.605 - 0.455 = 1.15 \text{ MHz}.$$

Hence, the required range of tuning of the local oscillator is 0.08 1.15 MHz.

Problem 5.19: Single-sideband modulation may be viewed as a hybrid form of amplitude modulation and frequency modulation. Evaluat the envelope and instantaneous frequency of an SSB wave for the following two cases:

(*i*) When only the upper sideband is transmitted.

(*ii*) When only the lower sideband is transmitted.

Solution:

(*i*) For SSB modulation, the modulated wave is,

$$s(t) = \frac{A_c}{2}[m(t)\cos(2\pi f_c t) \pm \hat{m}(t)\sin(2\pi f_c t)]$$

where +ve and –ve sign corresponds to upper and lower sideband of the modulated wave respectively.

Now, the envelope for the SSB wave is given by,

$$a(t) = \frac{A_c}{2}\sqrt{m^2(t) + \hat{m}^2(t)}$$

which is same for both transmitted upper sideband and lower sideband.

For upper sideband transmission, the angle,

$$Q_i(t) = 2\pi f_c t + \tan^{-1}\frac{\hat{m}(t)}{m(t)}$$

and the instantaneous frequency,

$$f_i(t) = \frac{1}{2\pi}\frac{d\,\theta_i(t)}{dt} = f_c + \frac{m(t)\,\hat{m}'(t) - \hat{m}(t)\,m'(t)}{2\pi\,[m^2(t) + \hat{m}^2(t)]}$$

where $'$ denotes time derivative.

(*ii*) For lower sideband transmission, the angle

$$\theta_i(t) = 2\pi f_c t + \tan^{-1}\left(-\frac{\hat{m}(t)}{m(t)}\right)$$

and, the instantaneous frequency

$$f_i(t) = f_c + \frac{m'(t)\,\hat{m}(t) - \hat{m}'(t)\,m(t)}{2\pi\,[m^2(t) + \hat{m}^2(t)]}$$

Problem 5.20: The sinusoidal modulating wave

$$m(t) = A_m \cos(2\pi f_m t)$$

As applied to a phase modulator with phase sensitivity k_p. The unmodulated carrier wave has frequency f_c band amplitude A_c.

(*i*) Determine the spectrum of the resulting phase-modulated signal, assuming that the maximum phase deviation $\beta_p = k_p A_m$ does not exceed 0.3 radians.

(*ii*) Construct a phasor diagram for this modulated signal, and compare it with that of the corresponding narrow-band FM signal.

Solution:

(*i*) A phase-modulated wave expressed as,

$$s(t) = A_c \cos[w_c t + k_p m(t)] \qquad \ldots(1)$$

where, $m(t) = A_m \cos(2\pi f_m t)$

then, $s(t) = A_c \cos[2\pi f_c t + k_p A_m \cos(2\pi f_m t)]$

$= A_c \cos[2\pi f_c t + \beta_p \cos(2\pi f_m t)]$

where $\beta_p = k_p A_m$ known as modulation index for PM.

$= A_c \cos(2\pi f_c t) \,.\, \cos[\beta_p \cos(2\pi f_m t) -$

$A_c \sin(2\pi f_c t)] + \sin[\beta_p \cos(2\pi f_m t)]$

Now, if $\beta_p \le 0.5$, then

$$\cos[\beta_p \cos(2\pi f_m t)] \approx 1$$

$$\sin[\beta_p \cos(2\pi f_m t)] \approx \beta_p \cos(2\pi f_m t)$$

Then, we can write equation (i) as,

$$s(t) \approx A_c \cos(2\pi f_c t) - \beta_p A_c \sin(2\pi f_m t)$$

$$= A_c \cos(2\pi f_c t) - \frac{1}{2}\beta_p A_c \sin[2\pi(f_c + f_m)t] -$$

$$\frac{1}{2}\beta_p A_c \sin[2\pi(f_c - f_m)t] \qquad ...(2)$$

$\therefore$ spectrum of $s(t)$

$$\cos(2\pi f_c t) \rightleftharpoons \frac{1}{2}[\delta(f - f_c) + \delta(f + f_c)]$$

and $\sin 2\pi f_c t \rightleftharpoons \frac{1}{2\int}[\delta(f - f_c) + \delta(f + f_c)]$

Using, Fourier-transform pairs as,

$$S(f) \approx \frac{1}{2} A_c[\delta(f - f_c) + \delta(f + f_c)] - \frac{1}{4j}\beta_p A_c$$

$$[\delta(f - f_c - f_m) - \delta(f + f_c + f_m)]$$

$$-\frac{1}{4j}\beta_p A_c [\delta(f - f_c + f_m) - \delta(f + f_c + f_m)]$$

(*ii*) The phasor diagram for $s(t)$ is deduced from equation (2) to be as follows:

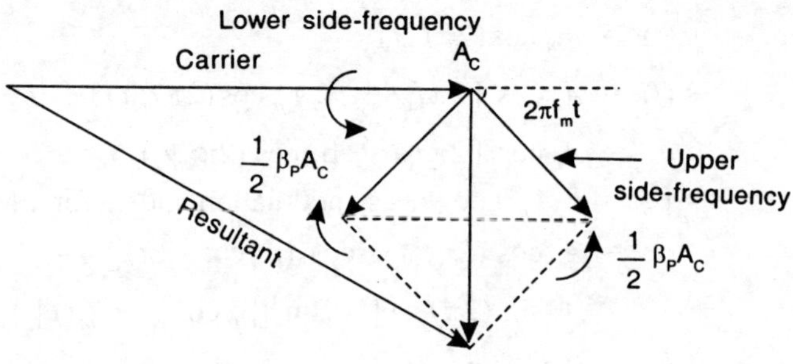

Fig. P. 5.20 (a)

Phasor diagram for narrow band FM is shown below,

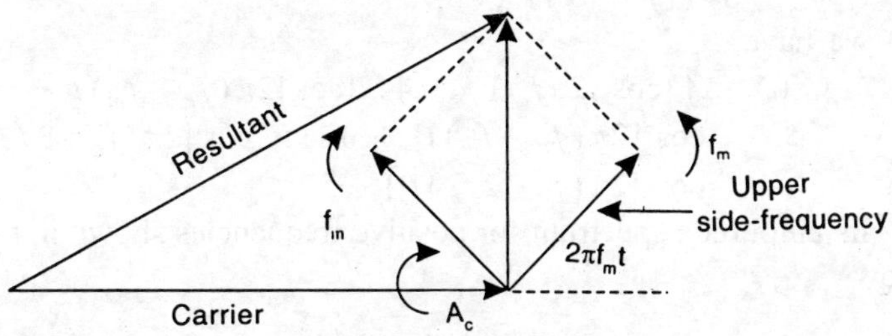

Fig. P. 5.20 (b)

Comparing these two phasor diagrams in Fig. 5.20 (a) and Fig. 5.20 (b) we see that, except for a phase difference, the narrow-band PM and FM waves have the same form.

Problem 5.21: An FM signal with modulation inxed $\beta = 1$ is transmitted through an ideal band-pass filter with mid-band frequency f_c and bandwidth $5f_m$ where f_c is the carrier frequency and f_m is the frequency of the sinusoidal modulating wave. Determine the amplitude spectrum of the filter output.

Solution:

An FM signal is also represented as,

$$s_0(t) = A \sum_{n=-\infty}^{\infty} J_n(\beta) \cos(\omega_c + n\omega_m) t$$

where $J_n(\beta)$ is the Bessel function of the first kind of order n and argument β.

Given $\beta = 1$,

then, $$s_0(t) = A \sum_{n=-\infty}^{\infty} J_n(1) \cos(\omega_c + n\omega_m) t$$

Since, the FM signal is passed through ideal band apss filter with mid frequency f_c and bandwidth $5f_m$, then only FM signal components which are at frequencies, $\omega_c \pm n\omega_m$ $[n = 1, 2]$ can pass.

Then, $$S_0(t) = AJ_0(1) \cdot \cos(\omega_c t) + J_1(1) \cdot \cos(\omega + \omega_m) t + AJ_2(1) \cdot \cos(\omega_c + 2\omega_m) t$$

Now, as $J_0(1) = 0.765$

$$J_1(1) = 0.44$$
$$J_2(1) = 0.115,$$

we have,

$$s_0(t) = A[\cos(2\pi f_c t) + 0.44\{\cos[2\pi(f_c + f_m)t + \cos[2\pi(f_c - f_m)t\} + 0.115\{\cos[2\pi(f_c + 2f_m)t] + \cos[2\pi(f_c - 2f_m)t]\}$$

Its amplitude spectrum for positive frequencies shown in the figure below.

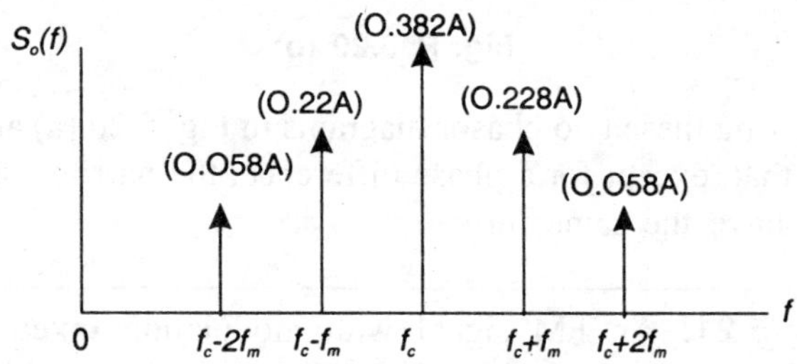

Fig. P. 5.21 (a)

Problem 5.22: Consider a wide-band PM signal produced by a sinusoidal modulating wave $A_m \cos(2\pi f_m t)$, using a modulator with a phase sensitivity equal to k_p radians per volt.
(*i*) Show that if the maximum phase deviation of the PM signal is large compared with one radian, the bandwidth of the PM signal varies linearly with the modulation frequency f_m.
(*ii*) Compare this characteristic of a wide-band PM signal with that of a wide-band FM signal.

Solution:

(*i*) A PM signal expressed as,

$$x_{PM}(t) = A\cos[W_c t + k_P m(t)]$$

where $m(t)$ is the message signal and k_P is phase deviation constant. then the angle of the PM wave is,

$$\theta_i(t) = 2\pi f_c t + k_P m(t) = 2\pi f_c t + k_P A_m \cos(2\pi f_m t)$$
$$= 2\pi f_c t + \beta_P \cos(2\pi f_m t)$$

where $\beta_P = k_P A_m$

The instantaneous frequency of the PM wave is therefore

$$f_i(t) = \frac{1}{2\pi}\frac{d\theta_i(t)}{dt} = f_c - \beta_P f_m \sin(2\pi f_m t)$$

Here, we see that the maximum frequency deviation in a PM wave varies linearly with the modulation frequency f_m.

Using Carson's rule, we find that the transmission bandwidth of the PM wave is approximately (for the case when $\beta_P \gg 1$)

$$B_T = 2(f_m + \beta_P f_m) = 2f_m(1+\beta_P) \approx 2f_m \beta_P$$

This shows that B_T varies linearly with f_m.

(*ii*) A FM signal given by

$$x_{FM}(t) = A\cos[\omega_c t + k_f \int^{t} (\lambda)\, d\lambda]$$

Its angle is,

$$\theta_t(t) = \omega_c t + k_f \int_{-\infty}^{t} m(\lambda) \cdot d\lambda$$

or $$f_i(t) = \frac{i}{2\pi}\cdot\frac{d\theta_i(t)}{dt}$$

$$= f_c(t) + k_f m(t)$$

then, maximum, frequency deviation,

$$\Delta f = |f_i(t) - f_c(t)|_{\max} = |k_f m(t)|_{\max} = k_f A_m$$

Then,

$$\beta_f(\text{modulation index}) = \frac{\Delta f}{fm} = \frac{k_f A_m}{fm}$$

Hence, Bandwidth,

$$W_B = 2(\beta_f + 1) f_m = 2\beta_f f_m \text{ (When } \beta_f \gg 1) = k_f A_m$$

Hence, we see that for FM wave, bandwidth is independent of the modulation frequency f_m.

Problem 5.23: Figure given below shows the block diagram of a real-time spectrum analyzer working on the principle of frequency modulation. The given signal $g(t)$ and a frequency-modulated signal $s(t)$ are applied to a multiplier and the output $g(t)\, s(t)$ is fed into a filter of impulse response $h(t)$. The $s(t)$ and $h(t)$ are linear.

$$g(t) \rightarrow \otimes \rightarrow \boxed{\text{Filter}} \rightarrow \text{Output}$$
$$\uparrow$$

FM signals whose instantaneous frequencies vary at opposite rates, as shown by

$$s(t) = \cos(2\pi f_c t - \pi k t^2),$$

$$h(t) = \cos(2\pi f_c t + \pi k t^2)$$

where k is a constant, Show that the envelope of the filter output is proportional to the amplitude spectrum of the input signal $g(t)$ with kt playing hte role of frequency f.

Solution:

The input at the filter is,

$$v_1(t) = g(t)\, s(t) = g(t) \cdot \cos(2\pi f_c t - \pi k t^2)$$

Then, the complex envelope of $v_1(t)$ is,

$$\tilde{v}_1(t) = g\,[v_1(t) + j\tilde{v}_1(t)] \cdot \exp(-j2\pi f_c t)$$

$$= g(t) \cdot [\cos(2\pi f_c t - \pi k t^2) + j\sin(2\pi f_c t - \pi k t^2)]$$

$$\cdot \exp(-j\pi f_c t)$$

$$= g(t) \cdot e^{\left(j2\pi f_c t - \pi k t^2\right)} \cdot e^{(-j2\pi f_c t)} = g(t) \cdot e^{-\pi k t^2} \qquad \text{...(1)}$$

The impulse response can be expressed as,

$$h(t) = Re\left[\tilde{h}(t) \cdot \exp(j2\pi_c t)\right]$$

where, $$\tilde{h}(t) = [h(t)\, j\tilde{h}(t)] \cdot \exp(-j\pi f_c t)$$

$$= [\cos(2\pi f_c t + \pi K t^2 + j\sin(2\pi f_c t + \pi K t^2)] \,.$$
$$\exp(-j2\pi f_c t)$$

$$= e^{\left(j2\pi f_c t + \pi k t^2\right)} \cdot e^{-j2\pi f_c t} = \exp(j\pi k t^2) \qquad \text{...(2)}$$

Now, the complex envelope filter of the output is

$$\tilde{v}(t) = \frac{1}{2}\,\tilde{h}(t) * \tilde{v}_1(t) = \frac{1}{2}\int_{-\infty}^{\infty} \tilde{h}(\tau) \,.\, \tilde{v}_1(t-\tau)\, d\tau,$$

Using equations (1) and (2), we get

$$\tilde{v}_o(t) = \frac{1}{2}\int_{-\infty}^{\infty} g(\tau)\exp(-j\pi k\tau^2)\exp[j\pi k(t-\tau)]^2\, d\tau$$

$$= \frac{1}{2}(j\pi k\tau^2)\int_{-\infty}^{\infty} g(\tau)\exp(-j2\pi kt\tau)\, d\tau$$

$$= \frac{1}{2}\exp(j\pi kt^2)\, G(kt)$$

$$\therefore \quad |\tilde{v}_o(t)| = \frac{1}{2}|G(kt)|$$

This shows that the envelope of the filter output is, except for the scale factor of $\frac{1}{2}$, equal to the magnitude of the Fourier transform of the input signal $g(t)$, with kt in place of frequency f.

Problem 5.24: An FM signal is applied to a square-law device with output voltage v_2 related to input voltage v_1 by

$$v_2 = a v_1^2$$

where a is a constant. Explain how such a device can be used to obtain an FM signal with a greater frequency deviation than that available at the input.

Solution:

An FM signal can be expressed as,

$$v_1(t) = A_c \cos[2\pi f_1 t + 2\pi k_{f1}\int_0^t m(t)\, dt]$$

where $m(t) = a_m \cos\omega_m t.$

Then, the resulting output at the square-law device is,

$$v_2(t) = a v_1^2(t)$$

$$= a A_c^2 \cos^2[2\pi f_1 t + 2\pi k_{f1}\int_0^t m(t)\, dt]$$

$$= \frac{1}{2} a A_c^2 \{1 + \cos[4\pi f_1 t + 4\pi k_{f1}\int_0^t m(t)\, dt]\}$$

Now, by passing $v_2(t)$ through a band-pass filter centered at $2f_1$ and with a bandwidth just large enough to pass the FM wave component of $v_2(t)$, we have

$$v_0(t) = \frac{1}{2} a A_c^2 \cos\left[2\pi f_2 t + 2\pi k_{f2} \int_0^t m(t)\, dt\right],$$

where $f_2 = 2f_1$ and $K_{f2} = 2k_{f1}$

We know $k_{f1} = \frac{\Delta f}{a_m}$,

where a_m is the amplitude of message signal.

Then, $k_{f2} = \frac{2\Delta f}{a_m}$

Hence, obtained FM wave have twice the frequency deviation as that of input FM wave.

Problem 5.25: An FM signal with a frequency deviation of 10 kHz at a modulation frequency of 5 kHz is applied to two frequency multipliers connected in cascade. The first multiplier doubles the frequency and the second multiplier triples the frequency. Determine the frequency deviation and the modulation index of the FM signal obtained at the second multiplier output. What is the frequency separation of the adjacent side frequencies of this FM signal.

Solution:

Overall frequency multiplication ratio,

$$n = 2 \times 3 = 6$$

Instantaneous frequency of the FM wave at the input of the first frequency multiplier

$$f_{i1}(t) = f_c + \Delta f \cos(2\pi f_m t)$$

Then, the instantaneous frequency of the resulting FM wave at the output of the second frequency multiplier is,

$$f_{i2}(t) = n f_c + n\Delta f \cos(2\pi f_m t)$$

Thus, the frequency deviation of this FM wave

$$= n\Delta f = 6 \times 10 = 60 \text{ kHz}$$

$$\therefore \text{Modulation index} = \frac{n\Delta f}{f_m} = \frac{60}{5} = 12$$

The frequency separation of the adjacent side-frequencies of this FM wave is unchanged at $f_m = 5$ kHz.

Chapter 6

Laplace Transform

Problem 6.1: Determine the bilateral Laplace transform and ROC for each of the following signals.

Solution:

(*a*)
$$x(t) = u(t-2)$$

$$x(t) \overset{L}{\longleftrightarrow} X(s) = \int_{-\infty}^{\infty} x(t)\, e^{-st}\, dt$$

$$X(s) = \int_{-\infty}^{\infty} u(t-2)\, e^{-st}\, dt = \int_{-\infty}^{\infty} e^{-st}\, dt = \left[\frac{-1}{s} \cdot e^{-st}\right]_2^{\infty} = \left[0 + \frac{e^{-2s}}{s}\right]$$

$$X(s) = \frac{e^{-2s}}{s}$$

ROC: Pole is at $s = 0$, ROC cannot contain any poles.

The ROC consists of those values of $s = \sigma + jw$ for which the Fourier transform of $x(t)\, e^{-\sigma t}$ converges

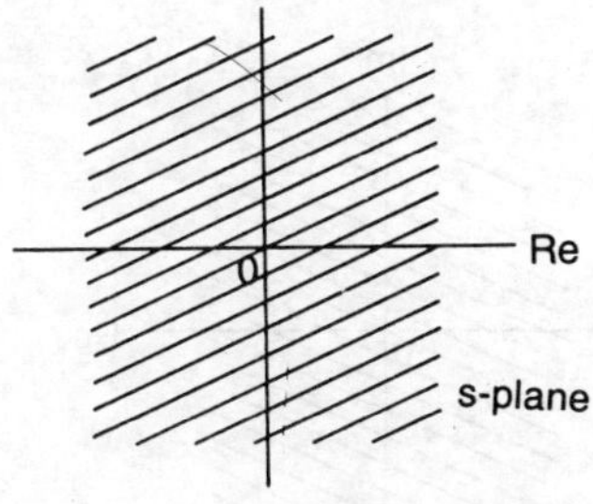

Fig. P. 6.1 (a)

(*b*) $x(t) = e^{2t} u(-t+2)$

$$x(t) \overset{L}{\longleftrightarrow} X(s) = \int_{-\infty}^{\infty} x(t) e^{-st} dt = \int_{-\infty}^{\infty} e^{2t} u(-t+2) e^{-st} dt$$

$$= \int_{-\infty}^{2} e^{(2-s)t} dt = \left[\frac{1e^{(2-s)t}}{(2-s)} \right]_{-\infty}^{2} = \frac{e^{(2-s)2}}{2-s} - 0$$

$$X(s) = \frac{e^{(4-2s)}}{2-s}$$

ROC:

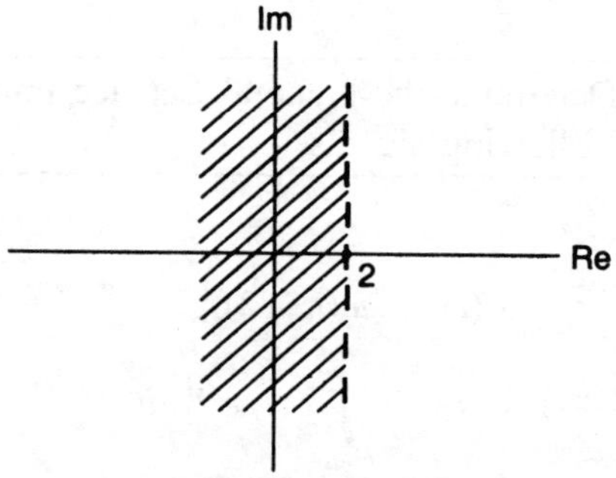

Fig. P. 6.1 (b)

(*c*) $x(t) = s(t - t_o)$

$$x(t) \overset{L}{\longleftrightarrow} X(s) = \int_{-\infty}^{\infty} x(t) e^{-st} dt$$

$$X(s) = \int_{-\infty}^{\infty} s(t - t_o) e^{-st} dt = 1$$

Which is valid for any value of s.
That is, the ROC of $L[x(t)]$ is the entire s-plane.

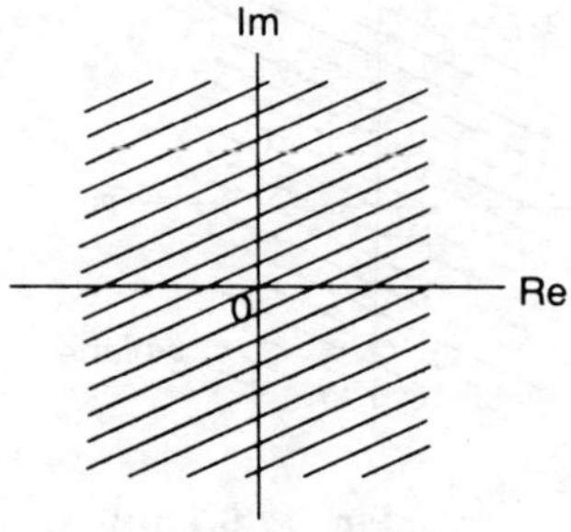

Fig. P. 6.1 (c)

(*d*) $x(t) = \cos(2t)\,u(t)$

$$x(t) \xleftrightarrow{L} X(s) = \int_{-\infty}^{\infty} x(t)\, e^{-st}\, dt = \int_{-\infty}^{\infty} \cos(2t)\; u(t)\; e^{-st}\, dt$$

$$x(t) = \frac{1}{2}(e^{-2tj} + e^{2tj})\, u(t)$$

$$X(s) = \frac{1}{2}\int_{-\infty}^{\infty} [e^{-2tj} + e^{2tj}]\, u(t) \cdot e^{-st}\, dt$$

$$= \frac{1}{2}\int_{-\infty}^{\infty} e^{(-2tj)}\, e^{(-st)}\, u(t)\; dt + \frac{1}{2}\int_{-\infty}^{\infty} e^{2tj} \cdot e^{-st} \cdot u(t)\, dt$$

$$= \frac{1}{2}\left(\frac{1}{s+2j}\right) + \frac{1}{2}\left(\frac{1}{s-2j}\right)$$

$$X(s) = \frac{1}{2}\frac{2s}{(s^2+4)} = \frac{s}{(s^2+4)}$$

Problem 6.2: Determine the unilateral Laplace transform of each of the following signals using the defining equation.

Solution:

(*a*) $x(t) = u(t-1)$

$$X(s) = \int_{0}^{\infty} x(t) \cdot e^{-st}\, dt = \int_{0}^{\infty} u(t-1)\, e^{-st}\, dt$$

$$= \int_{0}^{\infty} e^{-st}\, dt = \left[\frac{-e^{-st}}{s}\right]_1^{\infty} = \frac{e^{-s}}{s}$$

(*b*) $x(t) = u(t+1)$

$$X(s) = \int_{0}^{\infty} x(t) \cdot e^{-st}\, dt = \int_{0}^{\infty} u(t+1)\, e^{-st}\, dt = \int_{0}^{\infty} e^{-st}\, dt = \frac{1}{s}$$

(*c*) $x(t) = e^{-t+2}\, u(t)$

$$X(s) = \int_0^{\infty} x(t) \cdot e^{-st}\, dt = \int_0^{\infty} e^{-t+2}\, u(t) \cdot e^{-st}\, dt$$

$$= e^2 \int_0^{\infty} e^{-(1+s)t}\, dt = \frac{e^2}{.}\left[\frac{e^{-(1+s)t}}{-(1+s)}\right]_0^{\infty}$$

$$X(s) = \frac{e^2}{(1+s)}$$

(*d*) $x(t) = \cos(w_o t)\, u(t-3)$

$$X(s) = \int_0^{\infty} \cos(w_o t)\, u(t-3)\, e^{-st}\, dt$$

$$\cos(w_o t) = \frac{1}{2}[e^{-w_o jt} + e^{w_o jt}] = \frac{1}{2}\int_3^{\infty} (e^{-(w_o j+s)t} + e^{-(s-w_o j)t})\, dt$$

$$= \frac{1}{2}\left[\frac{e^{(w_o j+s)t}}{-(w_o j+s)}\right]_3^{\infty} + \frac{1}{2}\left[\frac{e^{-(s-w_o j)t}}{-(s-w_o j)}\right]_3^{\infty}$$

$$= \frac{1}{2}\left[\frac{e^{-(w_o j+s)3}}{(w_o j+s)} + \frac{e^{-(s-w_o j)3}}{(s-w_o j)}\right]$$

$$= \frac{e^{-3s}}{2}\frac{\left[(s-w_o j)e^{-w_o j} + (s+jw_o)\cdot e^{w_o j}\right]}{(s^2+w_o^2)}$$

$$= \frac{e^{-3s}}{(s^2+w_o^2)}\left[\frac{s\,(e^{-w_o j}+e^{w_o j})}{2} + \frac{j \times jw_o\,(e^{w_o j}-e^{-w_o j})}{2j}\right]$$

$$= \frac{e^{-3s}}{(s^2+w_o^2)}[s\cos w_o - w_o \sin w_o]$$

(*e*) $x(t) = \sin(w_o t)(t+2)$

$$X(s) = \int_0^{\infty} x(t) \cdot e^{-st}\, dt$$

$$= \int_0^\infty \sin(w_o t) \cdot u(t+2) \cdot e^{-st}\, dt$$

$$\sin(w_o t) = \frac{1}{2j}\left(e^{jw_o t} - e^{-jw_o t}\right) = \frac{1}{2j}\int_0^\infty \left[e^{-(s-jw_o)t} - e^{-(s+jw_o)t}\right] dt$$

$$= \frac{1}{2j}\left[\frac{e^{-(s-jw_o)t}}{-(s-jw_o)} - \frac{e^{-(s+jw_o)t}}{-(s+jw_o)}\right]_0^\infty$$

$$= \frac{1}{2j}\left[\frac{1}{(s-jw_o)} - \frac{1}{(s+jw_o)}\right]$$

$$= \frac{1}{2j}\left[\frac{s+jw_o-s+jw_o}{(s^2+w_o^2)}\right] = \frac{w_o}{(s^2+w_o^2)}$$

(*f*) $x(t) = e^{2t}[u(t) - u(t-4)]$

$$X(s) = \int_0^\infty x(t) \cdot e^{-st}\, dt = \int_0^\infty e^{2t}\,[u(t) - u(t-4)] \cdot e^{-st}\, dt$$

$$= \int_0^\infty e^{(2-s)t}\, u(t)\, dt - \int_0^\infty e^{(2-s)t} \cdot u(t-4)\, dt$$

$$= \int_0^\infty e^{-(-2+s)t}\, dt - \int_4^\infty e^{-(s-2)t}\, dt = \left.\frac{e^{-(s-2)t}}{-(s-2)}\right|_0^\infty - \left.\frac{e^{-(s-2)t}}{-(s-2)}\right|_4^\infty$$

$$= \frac{1}{(s-2)} - \frac{e^{-4(s-2)}}{(s-2)} = \frac{1-e^{-4(s-2)}}{(s-2)}$$

(*g*) $x(t) = \begin{cases} \sin(\pi t), & 0 < t < 1 \\ 0, & \text{otherwise} \end{cases}$

$$X(s) = \int_0^\infty x(t) \cdot e^{-st}\, dt$$

$$= \int_0^1 \sin(\pi t) \cdot e^{-st}\, dt \qquad \left[\begin{array}{ll} x(t) = \sin(\pi t), & 0 < t < 1 \\ \quad\;\; = 0, & \text{otherwise} \end{array}\right]$$

$$\sin(\pi t) = \frac{1}{2j}(e^{j\pi t} - e^{-j\pi t}) = \frac{1}{2j}\int_0^1 (e^{j\pi t} - e^{-j\pi t}) \cdot e^{-st} \cdot dt$$

$$= \frac{1}{2j}\int_0^1 [e^{-(s-j\pi)t} - e^{-(s+j\pi)t}] \cdot dt$$

$$= \frac{1}{2j}\left[\frac{e^{-(s-j\pi)t}}{-(s-j\pi)} - \frac{e^{-(s+j\pi)t}}{-(s+j\pi)}\right]_0^1$$

$$= \frac{1}{2j}\left[\left(\frac{e^{-(s-j\pi)}}{-(s-j\pi)} + \frac{1}{(s-j\pi)}\right) - \left(\frac{e^{-(s+j\pi)}}{-(s+j\pi)} + \frac{1}{(s+j\pi)}\right)\right]$$

$$= \frac{1}{2j}\left[\frac{s+j\pi - s + j\pi}{(s^2+\pi^2)}\right] + \frac{1}{2j}\left[\frac{e^{-(s+j\pi)}}{(s+j\pi)} - \frac{e^{-(s-j\pi)}}{(s-j\pi)}\right]$$

$$= \frac{\pi}{(s^2+\pi^2)} - \frac{e^{-s}}{2j}\left[\frac{-se^{-j\pi} + j\pi \cdot e^{-j\pi} + s \cdot e^{j\pi} + j\pi \cdot e^{j\pi}}{(s^2+\pi^2)}\right]$$

$$= \frac{\pi}{(s^2+\pi^2)} - e^{-s}\left[\frac{\pi\cos\pi + s\sin\pi}{s^2+\pi^2}\right]$$

Problem 6.3: Use the Laplace transform properties and the table of basic Laplace transforms to determine the unilateral Laplace transform of each of the following signals.

Solution:

(*a*) $x(t) = t^2e^{-2t}u(t)$

Now,

$$L\{t^2e^{-2t}u(t)\} = L\{t \cdot te^{-2t}u(t)\}$$

consider $L[t \cdot e^{-2t}u(t)] = \frac{-d}{ds}[L\{e^{-2t}u(t)\}]$

where $f(t) = e^{-2t}u(t)$

$\therefore$ $F(s) = \frac{1}{s+2}$

$$\therefore\ L\,[t \cdot e^{-2t}u\,(t)] = \frac{-d}{ds}\left[\frac{1}{S+2}\right] = -\left[-\frac{1}{(s+2)^2}\right] = \frac{1}{(s+2)^2}$$

Now

$$[t \cdot te^{-2t}u\,(t)] = \frac{-d}{ds}F\,(s)$$

where $F\,(s) = L\,[t \cdot e^{-2t}u\,(t)] = \dfrac{1}{(s+2)^2}$

$$\therefore\ L\,[t \cdot te^{-2t}u\,(t)] = \frac{-d}{ds}\left[\frac{1}{(s+2)^2}\right] = -\left[\frac{(s+2)^2 \cdot 0 - 1 \times 2(s+2)}{(s+2)^4}\right]$$

$$= \frac{2\,(s+2)}{(s+2)^4} = \frac{2}{(s+2)^3}$$

(*b*) $x\,(t) = e^{-t}u\,(t) * \sin\,(3\pi t)\,u\,(t)$

Convolution in time corresponds to multiplication of Laplace transforms.

$x'\,(t) * y'\,(t) \xleftrightarrow{L} X'\,(s)\ Y'\,(s)$

Thus $x'\,(t) = e^{-t}u\,(t),$

$y'\,(t) = \sin\,(3\pi t)\ u\,(t)$

$$X'\,(s) = \frac{1}{s+1}$$

$$Y'\,(s) = \frac{3\pi}{s^2 + (3\pi)^2}$$

$$\therefore\quad X\,(s) = X'\,(s) \cdot Y'\,(s) = \frac{1}{(s+1)} \times \frac{3\pi}{(s^2 + 9\pi^2)}$$

(*c*) $x\,(t) = \dfrac{d}{dt}\,[tu\,(t)]$

$$L\left[\frac{df\,(t)}{dt}\right] = sF\,(s) - f(0^-)$$

value of $f\,(t)$ at $t = 0^-$ i.e. $f(0^-) = 0$

$$f(t) = tu(t)$$

$$\therefore \quad F(s) = \frac{1}{s^2}$$

$$\therefore \quad X(s) = s \cdot \frac{1}{s^2} = \frac{1}{s}$$

(*d*) $$x(t) = tu(t) - (t-1)\,u(t-1) - (t-2)\,u(t-2) + (t-3)\,u(t-3)$$

From Linearity property

$$X(s) = X_1(s) - X_2(s) - X_3(s) + X_4(s)$$

From time shifting property

$$x(t - t_o) = e^{-st0}X(s)$$

$$X_1(s) = \frac{1}{s^2}$$

$$X_2(s) = \frac{e^{-s}}{s^2}$$

$$X_3(s) = \frac{e^{-2s}}{s^2}$$

$$X_4(s) = \frac{e^{-3s}}{s^2}$$

$$\therefore X(s) = \frac{1}{s^2} - \frac{e^{-s}}{s^2} - \frac{e^{-2s}}{s^2} + \frac{e^{-3s}}{s^2} = \frac{1 - e^{-s} - e^{-2s} + e^{-3s}}{s^2}$$

(*e*) $$x(t) = \int_0^t e^{-2\tau} \cos(3\tau)\, d\tau$$

$$x(\tau) = e^{-2\tau} \cos(3\tau)\, d\tau$$

$$x(\tau) = \frac{s+2}{(s+2)^2 + 3^2}$$

From $\int_{-\infty}^{t} x(\tau)\, d(\tau) = \frac{1}{s} X(s)$

$$X(s) = \frac{1}{s}\left(\frac{s+2}{(s+2)^2 + 3^2}\right)$$

(*f*) $$x(t) = 2t\frac{d}{dt}(e^{-t}\sin(t)\,u(t))$$

Let $$y(t) = e^{-t}\sin(t)\,u(t)$$

then $$Y(s) = \frac{1}{(s+1)^2+1}$$

Now we have $\frac{d}{dt}[x(t)] \xleftrightarrow{L} sX(s)$

$$\therefore\ L\left(\frac{d}{dt}(e^{-t}\sin(t)\,u(t))\right) = \frac{s}{(s+1)^2+1} = \frac{s}{(s+1)^2+1}$$

Now we have $ty(t) \xleftrightarrow{L} \frac{d}{ds}F(s)$

$$\therefore\ -2\frac{d}{ds}\left(\frac{s}{(s+1)^2+1}\right) = -2\left[\frac{\left[(s+1)^2+1\right]-2s(s+1)}{\left[(s+1)^2+1\right]^2}\right]$$

$$= \frac{-2}{\left[(s+1)^2+1\right]} + \frac{4s(s+1)}{\left[(s+1)^2+1\right]^2}$$

$$= \frac{-2(s+1)^2-2+4s^2+4s}{\left[(s+1)^2+1\right]^2} = \frac{2(s^2-2)}{\left[(s+1)^2+1\right]^2}$$

Problem 6.4: Given the transform pair $\cos(2t)\,u(t) \xleftrightarrow{L} X(s)$, determine the time signals corresponding to the following Laplace transform.

Solution:

(*a*) $sX(s) - 1$

We have $\frac{d}{dt}x(t) \longleftrightarrow sX(s)$

$$\therefore\ \frac{d}{dt}x(t) = \frac{d}{dt}\cos(2t)\,u(t) = -2\sin 2t$$

Also

$$s(t) \xleftrightarrow{L} 1$$

Thus the time signal corresponding to $[sX(s) - 1]$

$$= s(t) - 2\sin 2t$$

(*b*) $X(2s)$

We have

$$x(at) \xleftrightarrow{L} \frac{1}{|a|} X\left(\frac{s}{a}\right)$$

$$X(2s) = \frac{2}{2} X\left(\frac{s}{\frac{1}{2}}\right)$$

put $a = \frac{1}{2}$

$$\text{Thus } 2X\left(\frac{s}{\frac{1}{2}}\right) \xleftrightarrow{L^{-1}} x - (\frac{1}{2}t)$$

$$X(2s) \longleftrightarrow \frac{1}{2}\cos tu\left(\frac{t}{2}\right)$$

(*c*) $X(s + 1)$

We have

$$e^{s_o t} x(t) \longleftrightarrow X(s - s_o)$$

Put $s_o = -1$

$$X(s + 1) \longleftrightarrow e^{-1t} x(t)$$

We have $x(t) = \cos(2t) \cdot u(t)$

$$\therefore \quad X(s + 1) \longleftrightarrow e^{-t} \cdot \cos(2t) \cdot u(t)$$

(*d*) $s^{-1} X(s)$

We have

$$\int_{-\infty}^{t} x(\tau)\, d\tau \xleftrightarrow{L} \frac{X(s)}{s}$$

We have $x(t) = \cos 2\, tu(t)$

$$s^{-1} X(s) \longleftrightarrow \int_{-\infty}^{t} \cos(2\tau)\, u(\tau)\, d\tau$$

$$= \int_{-\infty}^{t} \cos(2\tau)\, d\tau \left[\frac{\sin 2\tau}{2}\right]_0^t \quad s^{-1}X(s) \longleftrightarrow \frac{\sin 2t}{2}$$

Problem 6.5: Given the transform pair $x(t) \stackrel{L}{\longleftrightarrow} \dfrac{2s}{(s^2-2)}$, determine the Laplace transform of each of the following time signals.

Solution:

(*a*) $x(2t)$

We have, $x(at) \longleftrightarrow \dfrac{1}{|a|} X\left(\dfrac{s}{a}\right)$

Here $a = 2$

Thus, $x(2t) \stackrel{L}{\longleftrightarrow} \dfrac{1}{2}\left[\dfrac{2\frac{s}{2}}{\left(\frac{s}{2}\right)^2 - 2}\right] = \dfrac{1}{2} \times \dfrac{4s}{s^2-8} = \dfrac{2s}{s^2-8}$

(*b*) $x(t-3)$

We have, $x(t-\tau) \stackrel{L}{\longleftrightarrow} e^{-s\tau}X(s)$

here $\tau = 3$

$$x(t-3) \stackrel{L}{\longleftrightarrow} e^{-3s} \cdot \frac{2s}{(s^2-2)} = \frac{e^{-3s} \cdot 2s}{(s^2-2)}$$

(*c*) $x(t) * tx(t)$

We have $x(t) * y(t) \longleftrightarrow -\dfrac{d}{ds} X(s)$

Let $y(t) = tx(t)$

$$tx(t) \stackrel{L}{\longleftrightarrow} -\frac{d}{ds}\left(\frac{2s}{s^2-2}\right) = \left[\frac{(s^2-2)\cdot 2 - 2s \cdot 2s}{(s^2-2)^2}\right]$$

$$= -\left[\frac{2s^2 - 4 - 4s^2}{(s^2-2)^2}\right] = \frac{2s^2+4}{(s^2-2)^2}$$

Thus,

$$x(t) * tx(t) \longleftrightarrow \frac{2s}{(s^2-2)} \times \frac{(2s^2+4)}{(s^2-2)^2} = \frac{4s(s^2+2)}{(s^2-2)^3}$$

(*d*) $e^{-2t}\, x(t)$

We have,

$$e^{s_o t}\, x(t) \overset{L}{\longleftrightarrow} X(s - s_o)$$

here, we have

$$s_o = -2$$

Thus,

$$e^{-2t}\, x(t) \overset{L}{\longleftrightarrow} \frac{2(s+2)}{(s+2)^2 - 2}$$

(*e*) $2\frac{d}{dt}x(t)$

We have, $\frac{d}{dt}x(t) \overset{L}{\longleftrightarrow} sX(s)$

also

$$x(t) \overset{L}{\longleftrightarrow} \frac{2s}{s^2-2}$$

Thus,

$$2\frac{d}{dt}\, x(t) \overset{L}{\longleftrightarrow} \frac{4s^2}{s^2-2}$$

(*f*) $\int_0^t x(2\tau)\, d\tau$

We have,

$$\int_{-\infty}^{t} x(\tau)\, d\tau \overset{L}{\longleftrightarrow} \frac{X(s)}{s}$$

also, $x(at) \overset{L}{\longleftrightarrow} \frac{1}{|a|} X\left(\frac{s}{a}\right)$

Thus,

$$\int_0^t x(2\tau)\, d\tau \overset{L}{\longleftrightarrow} \frac{1}{s} \cdot \frac{2s}{s^2-8} = \frac{2}{s^2-8}$$

Problem 6.6: Use the *s*-domain shift property and the transform pair $e^{-at}u(t) \overset{L}{\longleftrightarrow} \dfrac{1}{(s+a)}$ to derive the unilateral Laplace transform of $x(t) = e^{-at}\cos(w_1 t)\, u(t)$.

Solution:

We have

$$\cos(w_1 t) = \frac{1}{2}[e^{jw_1 t} + e^{-jw_1 t}]$$

Thus $$x(t) = \frac{e^{-at}}{2} u(t)\,[e^{+jw_1 t} + e^{-jw_1 t}]$$

also, we have,

$$e^{-at}u(t) \overset{L}{\longleftrightarrow} \frac{1}{(s+a)}$$

$$x(t) = \frac{1}{2}[e^{-at}u(t)\cdot e^{jw_1 t} + e^{-at}u(t)\cdot e^{-jw_1 t}]$$

using shifting property in the *s* domain

$$x(t) \overset{1}{\longleftrightarrow} \frac{1}{2}\left[\frac{1}{s - jw_1 + a} + \frac{1}{(s + jw_1 t + a)}\right]$$

$$\longleftrightarrow \frac{1}{2}\left[\frac{2(s+a)}{(s+a)^2 + w_1^2}\right] = \frac{(s+a)}{(s+a)^2 + w_1^2}$$

Problem 6.7: Prove the following properties of the unilateral Laplace transform:

Solution: (*a*) Linearity:

The transform of a finite sum of time functions is the sum of the Laplace transforms of the individual functions.

$$L[x_1(t) + x_2(t) + \ldots + x_n(t)] = X_1(s) + X_2(s) + \ldots + X_n(s)$$

Proof:

Let us find the Laplace transform of the sum from the basic definition

$$L[x_1(t) + x_2(t) + \ldots + x_n(t)]$$

$$= \int_0^\infty [x_1(t) + x_2(t) + \ldots + x_n(t)]\, e^{-st}\, dt$$

$$= \int_0^\infty x_1(t)\, e^{-st}\, dt + \int_0^\infty x_2(t)\, e^{-st}\, dt + \ldots + \int_0^\infty x_n(t)\, e^{-st}\, dt$$

$$= X_1(s) + X_2(s) + \ldots + X_n(s)$$

The property can be further extended if the time functions are multiplied by the constants i.e.

$$L[a_1 x_1(t) + a_2 x_2(t) + \ldots + a_n x_n(t)]$$

$$= a_1 X_1(s) + a_2 X_2(s) + a_2 X_2(s) + \ldots + a_n X_n(s)$$

where a_1, a_2 ... an are constants.

(*b*) Time shift

If $X(s)$ is the Laplace transform of $x(t)$ the the Laplace transform of the function delayed by time to is

$$L[x(t - t_o)] = e^{-t_o s} X(s)$$

Proof: By the definition

$$L[x(t - t_o)] = \int_0^\infty e^{-st} x(t - t_o)\, dt$$

As function is shifted by time t_0, the lower limit of integration can be changed from o^- to t_o.

$$L[x(t - t_o)] = \int_{t_o}^\infty e^{-st} x(t - t_o)\, dt$$

Let $\tau = t - t_o$

$t = t_o, \tau = 0$ and $t = \infty, \tau = \infty$

$$\therefore L[x(t - t_o)] = \int_0^\infty e^{-s(\tau + t_o)} x(\tau)\, d\tau$$

$$= \int_0^\infty e^{-s\tau} e^{-st_o} \cdot x(\tau)\, d\tau = e^{-st_o} \int_0^\infty e^{-s\tau} \cdot x(\tau)\, d\tau$$

Now $\int_0^\infty e^{-s\tau} x(\tau)\, d\tau = X(s)$ from the defnition

$$\therefore \qquad L\,[x\,(t - t_o)] = e^{-st_o}X(s)$$

(*c*) *S-domain shift*

Proof: The Laplace transform of $e^{s_o t}x\,(t)$ is

$$L\,[e^{st}x\,(t)] = \int_0^\infty e^{s_o t}\,x\,(t)\cdot e^{-st}\,dt = \int_0^\infty x\,(t)\cdot e^{-(s\,-\,s_o)t}\,dt = X\,(s - s_o)$$

Thus, $\qquad L\,[e^{s_o t}\,x\,(t)] = X\,(s - s_o)$

(*d*) *Differentiation in the s-domain*

If $X\,(s)$ is the Laplace transform of $x\,(t)$ then the differentiation by s in the complex frequency domain corresponds to the multiplication by t in the time domain

$$L\,[tx\,(t)] = \frac{dX\,(s)}{ds}$$

Proof: From the definition of the Laplace transform

$$X\,(s) = \int_0^\infty x\,(t)\;e^{-st}\;dt$$

Differentiating both sides with respect to s

$$\frac{dX\,(s)}{ds} = \int_0^\infty x\,(t)\,\frac{d}{ds}(e^{-st})\,dt = \int_0^\infty x\,(t) - t\cdot e^{-st}\,dt$$

(*e*) Convolution:

$$L\,[x_1\,(t)\,*\,x_2\,(t)] = X_1\,(s)\cdot X_2\,(s)$$

Proof:

Let $\qquad X\,(s) = X_1\,(s)\cdot X_2\,(s)$

where $\quad X_1\,(s) = \int_0^\infty x_1\,(m)\,e^{-sm}\,dm$

and $X_2\,(s) = \int_0^\infty x_2\,(n)\cdot e^{-sn}\,dn \qquad$ *m* and *n* are dummy variables.

$$\therefore \qquad X\,(s) = \int_0^\infty x_1\,(m)\;e^{-sm}\,dm\cdot\int_0^\infty x_2\,(n)\cdot e^{-sn}\,dn$$

As m and n are independent variables we can write

$$X(s) = \int_0^{\infty}\int_0^{\infty} x_1(m)\, x_2(n) \cdot e^{-s(m+n)}\, dm\, dn$$

Consider the new variable t and τ such that

$$t = m + n \text{ and } \tau = m$$

and $\quad dmdn = d\tau dt$

Let us consider limits of integration in terms of t and τ

$$t = m + n = \tau + n$$

The smallest value of n is zero hence $t \geq \tau$. The equation $t = \tau$ is straight line in $t - \tau$ plane. So, to integrate area between the line $t = \tau$ and $\tau = 0$. We get the limits for t as $0 \to \infty$. While limits of τ as 0 to t

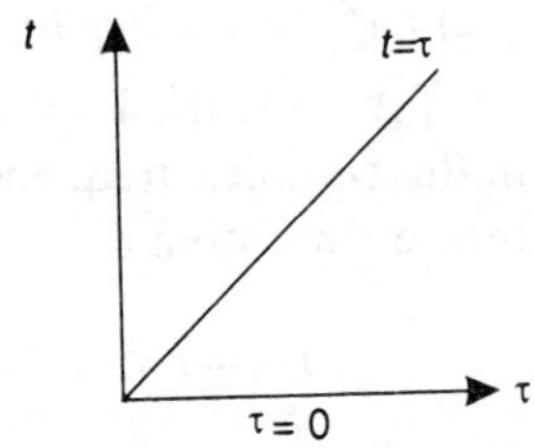

Fig. P. 6.7 (a)

Now

$m = \tau$ and $n = t - \tau$

$$X(s) = \int_0^{\infty}\left[\int_0^{t} x_1(\tau)\, x_2(t-\tau)\, d\tau\right] e^{-st}\, dt \qquad \ldots(1)$$

From the definition of Laplace transform the equation (1) represents

Laplace transform of integral $\int_0^{t} x_1(\tau)\, x_2(t-\tau)\, d\tau$

which is called convolution integral, so in the equation (1)

$$x(t) = \int_0^{t} x_1(\tau)\, x_2(t-\tau)\, d\tau$$

$$= x_1(t) * x_2(t) \ldots \text{ convolution integral}$$

$\therefore\ L[x_1(t) * x_2(t)] = X_1(s) \cdot X_2(s)$ **Proved**

$$= \int_0^{\infty} [tx(t)]\, e^{-st}\, dt = -L[tx(t)] \text{ as per definition}$$

$$\therefore \quad L[tx(t)] = \frac{-dX(s)}{ds} \quad \textbf{Proved}$$

(*f*) Scaling:

$$x(at) = \xleftrightarrow{L} \frac{1}{|a|} X\left(\frac{s}{a}\right)$$

Proof:

$$L[x(at)] = \int_0^{\infty} e^{-st} x(at)\, dt \text{ as per definition}$$

$$\begin{bmatrix} \text{Now put} & at = u \\ & dt = \dfrac{du}{a} \end{bmatrix}$$

Thus,

$$L[x(at)] = \int_0^{\infty} e^{-su/a} x(u) \cdot \frac{du}{a} = \frac{1}{a}\int_0^{\infty} e^{-su/a} x(u)\, du = \frac{1}{a} X\left(\frac{s}{a}\right)$$

Problem 6.8: Determine the initial value $x(0^+)$ given the following Laplace transforms.

Solution:

(*a*) $$X(s) = \frac{3}{s^2 + 5s - 1}$$

We may apply the imitial value theorem to obtain

$$x(0^+) = \lim_{s\to\infty} s \cdot \frac{3}{s^2 + 5s - 1} = \lim_{s\to\infty} \frac{3}{s + 5 - \frac{1}{s}} = \frac{1}{\infty} = 0$$

(*b*) $$X(s) = \frac{2s + 3}{s^2 + 5s - 6}$$

We may apply the initial value theorem to obtain

$$x(0^+) = \lim_{s\to\infty} s \cdot \frac{2s + 3}{s^2 + 5s - 6} = \lim_{s\to\infty} \frac{2 + \frac{3}{s}}{1 + \frac{5}{s} - \frac{6}{s^2}} = \frac{2}{1} = 2.$$

(*c*) $$X(s) = e^{-5s} \cdot \frac{3s^2 + 2s}{s^2 + s - 1}$$

We may apply the initial value theorem to obtain

$$x(0^+) = \lim_{s\to\infty} \frac{s\cdot e^{-5s}\cdot s(3s+2)}{s^2+s-1} = \lim_{s\to\infty} \frac{e^{-5s}(3s+2)}{1+\frac{1}{s}-\frac{1}{s^2}}$$

$$= 0\cdot\infty = 0$$

Problem 6.9: Determine the final value $x(\infty)$ given the following Laplace transforms.

Solution:

(*a*) $$X(s) = \frac{2s^2+3}{s^2+5s+1}$$

The final value theorem is applicable since poles of $x(s)$ are in the left half of s-plance.

We have

$$x(\infty) = \lim_{S\to 0} s\cdot\frac{2s^2+3}{s^2+5s+1} = 0$$

(*b*) $$X(s) = \frac{2s+3}{s^3+5s^2+6s}$$

The final value theorem is applicable since $X(s)$ has only a single pole at $s = 0$ and the remaining poles are in the left of the s-plane

We have,

$$x(\infty) = \lim_{S\to 0} s\cdot\frac{(2s+3)}{s^3+5s^2+6s} = \lim_{S\to 0}\frac{s(2s+3)}{s\left(s^2+5s+6\right)} = \frac{3}{6} = \frac{1}{2}$$

(*c*) $$X(s) = e^{-3s}\frac{3s^2+2}{s(s+1)(s+2)}$$

The final value theorem is applicable. Since $X(s)$ has only a single pole at $s = 0$ and the remaining poles are in the left half of the s-plane we have

$$x(\infty) = \lim_{S\to 0} s\cdot e^{-3s}\frac{\left(3s^2+2\right)}{s(s+1)(s+2)} = \frac{1\times 2}{1\times 2} = 1$$

Problem 6.10: Use the method of partial fractions to find the time signals corresponding to the following unilateral Laplace transforms.

Solution:

(*a*) $X(s) = \dfrac{-(s-4)}{(s^2+3s+2)} = \dfrac{-(s+4)}{(s+1)^2(s+2)} = \dfrac{-(s+4)}{(s+1)(s+2)}$

Use a partial fraction expansion

$$X(s) = \frac{A}{(s+1)} + \frac{B}{(s+2)}$$

Solving for A and B

$$X(s) = \frac{-3}{(s+1)} + \frac{2}{(s+2)}$$

Taking inverse Laplace

$$x(t) = -3e^{-t} + 2e^{-2t}$$

(*b*) $X(s) = \dfrac{s}{s^2+5s+6} = \dfrac{s}{(s+2)^2+s+2} = \dfrac{s}{(s+2)(s+3)}$

Use a partial fraction expansion

$$= \frac{A}{(s+2)} + \frac{B}{(s+3)}$$

Solving for A and B

$$X(s) = \frac{-2}{(s+2)} + \frac{3}{(s+3)}$$

Taking inverse Laplace

$$x(t) = -2e^{-2t} + 3e^{-3t}$$

(*c*) $X(s) = \dfrac{2s-1}{s^2+2s+1} = \dfrac{(2s-1)}{(s+1)^2}$

Use a partial fraction expansion

$$= \frac{A}{(s+1)^2} + \frac{B}{(s+1)}$$

Solving for A and B

$$X(s) = \frac{-3}{(s+1)^2} + \frac{2}{(s+1)}$$

$$x(t) = L^{-1}[X(s)]$$

$$x(t) = -3te^{-t} + 2e^{-t}$$

(*d*) $X(s) = \dfrac{5s+4}{s^3+6s^2+2s} = \dfrac{5s+4}{s(s^2+3s+2)} = \dfrac{5s+4}{s(s+1)(s+2)}$

Use a partial fraction expansion

$$X(s) = \frac{A}{s} + \frac{B}{(s+1)} + \frac{C}{(s+2)}$$

Solving for A, B and C

$$X(s) = \frac{2}{s} + \frac{1}{(s+1)} - \frac{3}{(s+2)}$$

$$x(t) = L^{-1}[X(s)]$$

$$x(t) = 2 + e^{-t} - 3 \cdot e^{-2t}$$

(*e*) $X(s) = \dfrac{3s^2+8s+5}{(s+2)(s^2+2s+1)^2}$

$$X(s) = \frac{3s^2+8s+5}{(s+2)(s+1)^2} = \frac{(3s+5)(s+1)}{(s+2)(s+1)^2} = \frac{3s+5}{(s+2)(s+1)}$$

Use a partial fraction expansion

$$X(s) = \frac{A}{(s+2)} + \frac{B}{(s+1)}$$

Solving for A and B

$$A = 1,$$

$$B = 2$$

$$X(s) = \frac{1}{(s+2)} + \frac{2}{(s+1)}$$

$$x(t) = L^{-1}[X(s)]$$

$$x(t) = e^{-2t} + 2e^{-t}$$

(*f*) $$X(s) = \frac{3s+2}{s^2+4s+5}$$

This may be written as

$$X(s) = \frac{3(s+2)-4}{(s+2)^2+1^2} = \frac{3(s+2)}{(s+2)^2} - \frac{4.1}{(s+2)^2+1^2}$$

$$A_2 L^{-1}\left[\frac{(s+a)}{(s+a)^2+w^2}\right] = [e^{-at} \cos wt] \text{ and}$$

$$L^{-1}\left[\frac{w}{(s+a)^2+w^2}\right] = [e^{-at} \sin wt]$$

$$x(t) = L^{-1}[X(s)] = 3 \cdot e^{-2t} \cos t - 4 \cdot e^{-2t} \sin t$$

$$\therefore \quad x(t) = 3e^{-2t} \cos t - 4\, e^{-2t} \sin t.$$

(*g*) $$X(s) = \frac{4s^2+8s+10}{(s+2)(s^2+2s+5)}$$

$$X(s) = \frac{As+B}{s^2+2s+5} + \frac{C}{s+2}$$

Solving for A, B and C

$$X(s) = \frac{2s}{(s^2+2s+5)} + \frac{2}{(s+2)} = \frac{2s}{(s+1)^2+2^2} + \frac{2}{(s+2)}$$

$$= 2\left[\frac{s+1}{(s+1)^2+2^2} - \frac{1}{2}\left(\frac{2}{(s+1)^2+2^2}\right)\right] + \frac{2}{(s+2)}$$

$$x(t) = L^{-1}[X(s)] = 2\, e^{-t} \cos 2t - e^{-t} \sin 2t + 2\, e^{-2t}$$

$$x(t) = 2\, e^{-t} \cos 2t - e^{-t} \sin 2t + 2\, e^{-2t}$$

(*h*) $$X(s) = \frac{-9}{(s+1)(s^2+2s+10)}$$

$$X(s) = \frac{As+B}{(s^2+2s+10)} + \frac{C}{(s+1)}$$

Solving for A, B and C

$$A = 1, \quad B = 1, \quad C = -1$$

$$\therefore X(s) = \frac{s+1}{(s^2+2s+10)} - \frac{1}{(s+1)} = \frac{s+1}{(s+1)^2+3^2} - \frac{1}{(s+1)}$$

$$x(t) = L^{-1}[X(s)]$$

$$\therefore \quad x(t) = e^{-t}\cos(3t) - e^{-t}$$

Problem 6.11: Determine the forced and natural responses for the systems described by the following differential equations with the specified input and initial conditions.

Solution:

$$(a) \; 5\frac{d}{dt}t(t) + 10y(t) = 2x(t), y(0^+) = 2, x(t) = u(t)$$

$$5\frac{d}{dt}y(t) + 10y(t) = 2x(t),$$

Taking the Laplace transform of both sides of the differential equation

We obtain,

$(5s + 10)\,Y(s) - 5y(0^+) = 2X(s)$

Solving for $Y(s)$, we get

$$Y(s) = \frac{2\,X(s)}{(5s+10)} + \frac{5y(0^+)}{(5s+10)}$$

The first term is associated with the forced response of the system. The second term corresponds to the natural response.

Substituting for $X(s) \cdot y(0^+)$

$$Y^{(f)}(s) = \frac{2}{s(5s+10)}$$

$$Y^{(n)}(s) = \frac{5.2}{5(s+2)} = \frac{2}{(s+2)}$$

Expanding both of these in a partial fraction expansion yields

$$Y^{(f)}(s) = \frac{1}{5s} - \frac{1}{(5s+10)}$$

$$Y^{(n)}(s) = \frac{2}{(s+2)}$$

Next, taking the inverse unilateral Laplace transforms of $Y^{(f)}(s)$ and $Y^{(n)}(s)$. We obtain

$$Y^{(f)}(s) = \frac{u(t)}{5} - \frac{1}{5} e^{-2t} u(t)$$

$$Y^{(n)}(s) = 2 \cdot e^{-2t} u(t)$$

(b) $\frac{d^2}{dt^2} y(t) + 5\frac{d}{dt} y(t) + 6y(t) = -4x(t) - 3\frac{d}{dt} x(t),$

$$y(0^+) = -1, \quad \frac{d}{dt} y(t)\Big|_{t=o^+} = 5$$

$$x(t) = e^{-t} u(t)$$

Solution: Taking the Laplace transform of both sides of the differential equation

We obtain,

$$(s^2 + 5s + 6)\, Y(s) - sy(0^+) - \frac{d}{dt} y(t)\Big|_{t=o^+} - 5\, y(0^+)$$

$$= -\left[(3s+4)\, X(s) - (3x(0^+))\right] = -(3s+4)\, X(s)$$

Solving for $Y(s)$, we get

$$Y(s) = \frac{-(3s+4)\, X(s) + 3x(0^+)}{(s^2+5s+6)} + \frac{sy(0^+) + \frac{d}{dt} y(t)\big|_{t=0^+} + 5y(0^+)}{(s^2+5s+6)}$$

Substituting for $X(s)$, $x(0^+)$, and initial conditions, we obtain

$$Y^{(f)}(s) = \frac{-(s+4)}{(s^2+5s+6)(s+1)} = \frac{-(s+4)}{(s+1)(s+2)(s+3)}$$

and $Y^{(f)}(s) = \frac{-s+5-5}{s^2+5s+6} = \frac{-s}{(s^2+5s+6)} = \frac{-s}{(s+2)(s+3)}$

Expanding both of these in a partial fraction expansion yeilds.

$$Y^{(f)}(s) = \frac{-2}{(s+1)} + \frac{-4}{(s+2)} - \frac{2}{(s+3)}$$

and $\quad Y^{(n)}(s) = \dfrac{2}{(s+2)} + \dfrac{-3}{(s+3)}$

Next, taking the inverse unilateral Laplace transforms of $Y^{(f)}(s)$ and $Y^{(n)}(s)$, we obtain

$$y^{(f)}(s) = -2e^{-t} - 4e^{-2t} - 2e^{-3t}$$

and $\quad y^{(n)}(s) = 2e^{-2t} - 3e^{-3t}$

(c) $\dfrac{d^2}{dt^2} y(t) + 4y(t) = 8x(t),\ y(0^+) = 1,$

$\dfrac{d}{dt} y(t)\Big|_{t=0^+} = 2,\ x(t) = u(t)$

Taking the Laplace transform of both sides of the differential equation, we obtain,

$$(s^2 + 4)\, Y(s) - sy(0^+) - \frac{d}{dt} y(t)\Big|_{t=0^+} = 8\, X(s)$$

Solving for $Y(s)$, we get

$$Y(s) = \frac{8\, X(s)}{(s^2+4)} + \frac{sy(0^+) + \dfrac{d}{dt} y(t)\Big|_{t=0^+}}{s^2+4}$$

Substituting for $X(s)$ and the initial conditions
We obtain,

$$Y^{(f)}(s) = \frac{8}{s(s^2+4)}$$

and $\quad Y^{(n)}(s) = \dfrac{s+2}{s^2+4}$

or $\quad Y^{(f)}(s) = \dfrac{-2s}{s^2+4} + \dfrac{2}{s}$

$$Y^{(n)}(s) = \frac{s}{s^2+4} + \frac{2}{s^2+4}$$

Now, taking the inverse Laplace transforms of $Y^{(n)}(s)$ and $Y^{(n)}(s)$, we get

$$y^{(f)}(t) = -2\cos(2t)\, u(t) + 2u(t)$$
$$y^{(n)}(t) = \cos(2t)\, u(t) + \sin(2t)\, u(t)$$

(*d*) $\frac{d^2}{dt^2} y(t) + 2\frac{d}{dt} y(t) + 5y(t) = \frac{d}{dt} x(t),$

$$y(0^+) = 2, \frac{d}{dt} y(t)|_{t=0^+} = 0, x(t) = e^{-t} u(t)$$

$$\frac{d^2}{dt^2} y(t) + 2\frac{d}{dt} y(t) + 5y(t) = \frac{d}{dt} x(t)$$

Taking the unilateral Laplace transform of both sides of the differential equation, we obtain

$$(s^2 + 2s + 5)\, Y(s) - sy(0^+) - \frac{d}{dt} y(t)\Big|_{t=0^+} - 2y(0^+)$$
$$= sX(s) - x(0^+)$$

Solving for $Y(s)$, we get

$$Y(s) = \frac{sX(s) - x(0^+)}{(s^2 + 2s + 5)} + \frac{sy(0^+) + \frac{d}{dt} y(t)\big|_{t=0^+} + 2y(0^+)}{(-s^2 + 2s + 6)}$$

The first term is associated with the forced response of the system. $Y^{(f)}(s)$. The second term corresponds to the natural response, $Y^{(n)}(s)$. Substituting for $X(s)$, $x(0^+)$, and the initial conditions, we obtain.

$$Y^{(f)}(s) = \frac{\frac{s}{(s+1)} - 1}{(s^2 + 2s + 5)} = \frac{-1}{(s+1)(s^2 + 2s + 5)}$$

$$= \frac{-1}{(s+1)\left[(s+1)^2 + 2^2\right]}$$

and $$Y^{(n)}(s) = \frac{2s + 4}{(s^2 + 2s + 5)} = \frac{2(s+2)}{(s+1)^2 + 2}$$

$$= \frac{2(s+1)}{(s+1)^2 + 2^2} + \frac{2}{(s+1)^2 + 2^2}$$

Next, taking the inverse unilateral Laplace transforms of $Y^{(f)}(s)$ and $Y^{(n)}(s)$, we obtain

$$Y^{(f)}(s) = \frac{1}{4}\left[\frac{s+1}{s^2+2s+5} - \frac{1}{(s+1)}\right]$$

$$y^{(f)}(y) = \frac{1}{4}\left[e^{-t} \cdot \cos(2t)\, u(t) - e^{-t} u(t)\right]$$

$$= \frac{e^{-t}}{4}\ [\cos(2t)\ u(t) - u(t)]$$

$$y^{(n)}(t) = 2 \cdot e^{-t} \cos(2t)\ u(t) + e^{-t} \sin(2t)\ u(t)$$

$$= e^{-t}\,[2 \cos(2t)\ u(t) + \sin(2t)\ u(t)]$$

Problem 6.12: Use Laplace transform circuit models to determine the current $y(t)$ in the circuit shown below in response to an applied voltage $x(t) = e^{-t}u(t)\ V$, assuming normalized value $R = 1\ \Omega$ and $L = \frac{1}{2}$ H. The current through the inductor at time $t = 0^+$ is 2A.

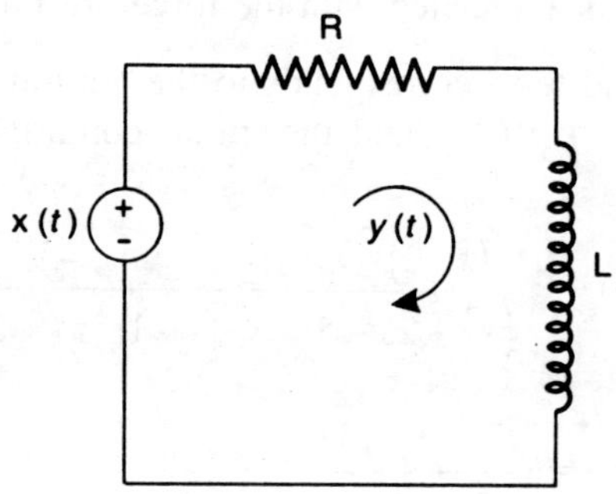

Fig. P. 6.12 (a)

Solution:

Applying KVL

$$x(t) = 1\,\dot{y}(t) + \frac{1}{2}\frac{d\dot{y}(t)}{dt}$$

We have $x(t) = e^{-t}\, u(t)\ V$

$$e^{-t}u(t)\ V = y(t) + \frac{1}{2}\frac{dy(t)}{dt}$$

$$\frac{1}{(s+1)} = Y(s) + \frac{1}{2}sY(s) - \frac{1}{2}y(0^+)$$

$$Y(s)\left[1+\frac{s}{2}\right] = \frac{1}{s+1} + 1$$

$$Y(s)\left[\frac{(s+2)}{2}\right] = \frac{(s+2)}{(s+1)}$$

$$Y(s) = \frac{2}{(s+1)}$$

$$y(t) = 2\,e^{-t}A$$

> **Problem 6.13:** A stable system has input $x(t)$ and output $y(t)$ as given below. Use Laplace transforms to determine the transfer function and impulse response of the system.
>
> (*a*) $x(t) = u(t),\ y(t) = e^{-t}\cos(2t)\,u(t)$

Solution: $X(s) = \dfrac{1}{s}$

$$Y(s) = \frac{s+1}{(s+1)^2 + 2^2}$$

$$\text{Transfer function} = H(s) = \frac{Y(s)}{X(s)} = \frac{(s+1)\,s}{\left[(s+1)^2 + 2^2\right]}$$

$$\text{Impulse response} = L^{-1}(H(s))$$

$$H(s) = 1 + \frac{-s-5}{\left[(s+1)^2 + 2^2\right]}$$

$$= 1 - \left[\frac{s+1}{\left[(s+1)^2 + 2^2\right]}\right] - \left[\frac{4}{\left[(s+1)^2 + 2^2\right]}\right]$$

$$L^{-1}(H(s)) = t - e^{-t}\cos 2t - 2e^{-t}\sin 2t$$

(*b*) $x(t) = e^{-2t}u(t),\ y(t) = -2e^{-t}u(t) + 2e^{-3t}u(t)$

Solution:

$$Y(s) = \frac{1}{(s+2)}$$

$$Y(s) = \frac{-2}{(s+1)} + \frac{2}{(s+3)} = \frac{-4}{(s+1)(s+3)}$$

Transfer function

$$H(s) = \frac{Y(s)}{H(s)} = \frac{-4(s+2)}{(s+1)(s+3)}$$

Thus impulse response of the system is

L^4 $[H(s)]$

$$H(s) = \frac{-4(s+2)}{(s+1)(s+3)} = \frac{-2}{(s+1)} + \frac{-2}{(s+3)} = -2e^{-t} - 2e^{-3t}$$

Problem 6.14: The relationship between the input $x(t)$ and output $y(t)$ of a causal system is described by each of the differential equations given below. Use Laplace transform to determine the transfer function and impulse response of each system.

(*a*) $5\frac{d}{dt}y(t) + 10y(t) = 2x(t)$

(*b*) $\frac{d^2}{dt^2}y(t) + 5\frac{d}{dt}y(t) + 6y(t) = x(t) + \frac{d}{dt}x(t)$

(*c*) $\frac{d^2}{dt^2}y(t) - 2\frac{d}{dt}y(t) + 10y(t) = x(t) + 2\frac{d}{dt}x(t)$

Solution:

(*a*) Taking the Laplace transform of both sides of the given differential equation

$(5s + 10)Y(s) = 2X(s)$

Hence the transfer function of the system is

$$H(s) = \frac{Y(s)}{X(s)} = \frac{2}{(5s+10)} = \frac{2}{5(s+2)}$$

Thus the impulse response is

$$L^{-1}[H(s)] = \frac{2}{5} e^{-2t}u(t)$$

(*b*) Taking the Laplace transform of both sides of the given differential equation.

$(s^2 + 5s + 6)\, Y(s) = (s + 1)\, X(s)$

Hence the transfer function of the system is

$$H(s) = \frac{Y(s)}{X(s)} = \frac{s+1}{s^2+5s+6} = \frac{(s+1)}{(s+2)(s+3)}$$

Impulse response of the system is

$L^{-1}[H(s)]$

$$H(s) = \frac{(s+1)}{s^2+5s+6} = \frac{-1}{(s+2)} + \frac{2}{(s+3)}$$

$$\therefore \quad L^{-1}[H(s)] = -e^{-2t}u(t) + 2e^{-2t}u(t)$$

(*c*) Taking the Laplace transform of both sides of the given differential equations.

$(s^2 - 2s + 10)\, Y(s) = (2s + 1)\, X(s)$

Hence the transfer function of the system is

$$H(s) = \frac{Y(s)}{X(s)} = \frac{(2s+1)}{s^2-2s+10}$$

Impulse response of the system is

$L^{-1}[H(s)]$

$$H(s) = \frac{(2s+1)}{(s-1)^2+3^2} = \frac{2s}{(s-1)^2+3^2} + \frac{1}{(s-1)^2+3^2} = \frac{2s}{(s-1)^2+3^2}$$

$$\therefore \quad L^{-1}[H(s)] = 2\, e^{t} \cos(3t)\, u(t) + e^{t} \sin(3t)\, u(t)$$

Problem 6.15: Determine the differential equation description for a system with each of the following transfer functions:

(*a*) $H(s) = \dfrac{2s+1}{s(s+2)}$

(*b*) $H(s) = \dfrac{3s}{s^2+2s+10}$

Solution: (*a*) $H(s) = \dfrac{Y(s)}{X(s)} = \dfrac{2s+1}{s(s+2)}$

or $$[s(s+2)]\,Y(s) = X(s)\,(2s+1)$$

or, $$\frac{d^2}{dt^2}y(t) + 2\frac{d}{dt}y(t) = 2\frac{d}{dt}x(t) + x(t)$$

(*b*) $$H(s) = \frac{Y(s)}{X(s)} = \frac{3s}{s^2+2s+10}$$

$\therefore$ $$= (s^2 + 2s + 10)\,Y(s) = 3s\,X(s)$$

or, $$\frac{d^2}{dt^2}y(t) + 2\frac{d}{dt}y(t) + 10y(t) = 3\frac{d}{dt}x(t).$$

Problem 6.16: Determine the Laplace transform and ROC for the signal

$$y(t) = -e^{at}u(-t)$$

Solution: The Laplace transform for any arbitrary signal

$$X(s) = \int_{-\infty}^{\infty} -x(t)\,e^{-st}\,dt \tag{1}$$

Substitute for $x(t)$ in Eq. (1), obtaining

$$Y(s) = \int_{-\infty}^{\infty} -e^{at}u(-t)\,dt = -\int_{-\infty}^{\infty} e^{-(s-a)t}\,dt = \frac{1}{s-a}e^{-(s-a)t}\Big|_{-\infty}^{0}$$

$$= \frac{1}{s-a}\quad Re(s) < a$$

The ROC and pole and zero locations are depicted in Fig. 6.16 (*a*).

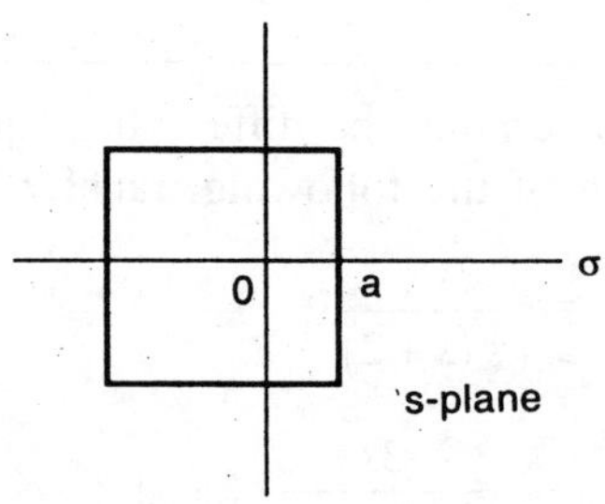

Fig. P. 6.16 (a)

Problem 6.17: Find the unilateral Laplace transform of

$$x(t) = (-e^{3t}u(t)) * (tu(t)))$$

Solution: First we note that

$$(-e^{3t}u(t)) \overset{L}{\longleftrightarrow} \frac{-1}{s-3}$$

and

$$u(t) \overset{L}{\longleftrightarrow} \frac{1}{s}$$

Apply the s-domain differentiation property to obtain

$$tu(t) \overset{L}{\longleftrightarrow} \frac{1}{s^2}$$

Now use the convolution property to obtain

$$x(t) = (-e^{3t}u(t)) * [tu(t)] \overset{L}{\longleftrightarrow} X(s) = \frac{-1}{s^2(s-3)}$$

Problem 6.18: Verify the differentiation property of Eq. $\frac{d}{dt}x(t) \overset{L}{\longleftrightarrow} sX(s) - x(0^+)$ for the signal $x(t) = e^{at}u(t)$.

Solution: The derivative of $x(t)$ is

$$\frac{d}{dt}e^{at} = ae^{at},\ t > 0$$

The unilateral Laplace transform of $ae^{at}u(t)$ is a times the unilateral Laplace transform of $e^{at}u(t)$; that is,

$$\frac{d}{dt}x(t) = ae^{at}u(t) \overset{L}{\longleftrightarrow} \frac{a}{s-a}$$

Next, let us rederive this result using the differentiation property. We have

$$\frac{d}{dt}x(t) \overset{L}{\longleftrightarrow} sX(s) - x(0^+) = s\frac{1}{s-a} - 1 = \frac{s-(s-a)}{s-a} = \frac{a}{s-a}$$

Problem 6.19: Determine the initial and final values of a signal $x(t)$ whose unilateral Laplace transform is

$$X(s) = \frac{7s+10}{s(s+2)}$$

Solution: We may apply the initial value theorem to obtain

$$x\ (0^+) = \lim_{s\to\infty} s\frac{7s+10}{s(s+2)} = \lim_{s\to\infty} \frac{7s+10}{s+2} = 7$$

The final value theorem is applicable since $X(s)$ has only a single pole at $s = 0$ and the remaining poles are in the left half of the s–plane. We have

$$x\ (\infty) = \lim_{s\to\infty} s\frac{7s+10}{s(s+2)} = \lim_{s\to\infty} \frac{7s+10}{s+2} = 5$$

Problem 6.20: Find the inverse Laplace transform of

$$X(s) = \frac{3s+1}{(s+1)(s+2)^2}$$

Solution: Use a partial fraction expansion of $X(s)$ to write

$$X(s) = \frac{A_1}{s+1}+\frac{A_2}{s+2}+\frac{A_3}{(s+2)^2}$$

Solving for A_1, A_2 and A_3 using the method of residues, we obtain

$$X(s) = \frac{1}{s+1}-\frac{1}{s+2}+\frac{2}{(s+2)^2}$$

We may construct $x(t)$ from the inverse Laplace transform of each term in the partial fraction expansion:

The pole of the first term is at $s = -1$

$$e^{-t}u(t) \xleftrightarrow{L} \frac{1}{s+1}$$

The second term has a pole at $S = -2$

$$-e^{-2t}u(t) \xleftrightarrow{L} -\frac{1}{s+2}$$

The double pole in the last term is at $s = -2$

$$2te^{-2t}u(t) \xleftrightarrow{L} \frac{2}{(s+2)^2}$$

Combining these three terms, we obtain

$$x(t) = e^{-t}u(t) - e^{-2t}u(t) + 2te^{-2t}u(t)$$

Problem 6.21: Find the inverse Laplace transform of

$$X(s) = \frac{4s^2 + 6}{s^3 + s^2 - 2}$$

Solution: There are three poles in $X(s)$. By trial and error we find that $s = 1$ is a pole. We factor $s - 1$ out of $s^3 + s^2 - 2$ to obtain $s^2 + 2s + 2 = 0$ as the equation defining the remaining two poles. Rooting this quadratic equation gives the complex–conjugate poles $s = -1 \pm j$.

We may write the quadratic equation $s^2 + 2s + 2$ in terms of the perfect square $(s^2 + 2s + 1) + 1 = (s + 1)^2 + 1$, and so the partial fraction expansion for $X(s)$ takes the form

$$X(s) = \frac{A}{s-1} + \frac{B_1 s + B_2}{(s+1)^2 + 1} \qquad (1)$$

The expansion coefficient A is easily obtained by the method of residues. That is, multiply both sides of Eq. (1) by $(s - 1)$ and evaluate at $s = 1$ to obtain.

$$A = X(s)\,(s-1)\,\Big|_{s=1} = \left.\frac{4s^2 + 6}{(s+1)^2 + 1}\right|_{s=1} = 2$$

The remaining expansion coefficients B_1 and B_2 are obtained by placing both terms in Eq. (1) over a common denominator and equating the numerator to the numerator of $X(s)$. We may thus write

$$\begin{aligned} 4s^2 + 6 &= 2[(s + 1)^2 + 1] + (B_1 s + B_2)\,(s - 1) \\ &= (2 + B_1)s^2 + (4 - B_1 + B_2)s + (4 - B_2) \end{aligned}$$

Equating coefficients on s^2 gives $B_1 = 2$, and equating the coefficients of s^0 gives $B_2 = -2$,

Thus

$$X(s) = \frac{2}{s-1} + \frac{2s-2}{(s+1)^2 + 1} = \frac{2}{s-1} + 2\frac{s+1}{(s+1)^2 + 1} - 4\frac{1}{(s+1)^2 + 1}$$

In arriving at the second equation we have factored $2s - 2$ as $2(s + 1) - 4$. Now take the inverse Laplace transform of each term. Putting these results together, we obtain

$$x(t) = 2e^t u(t) + 2e^{-t} \cos(t)\, u(t) - 4e^{-t} \sin(t)\, u(t)$$

Problem 6.22: Use the Laplace transform to find the output of the system described by the differential equation

$$\frac{d}{dt}y(t) + 5y(t) = x(t)$$

in response to the input $x(t) = 3e^{-2t}\, u(t)$ and initial condition $y(0^+) = -2$.

Solution: Take the unilateral Laplace transform of each side of the differential equation and apply the differentiation property to obtain

$$sY(s) - y(0^+) + 5Y(s) = X(s)$$

Solving for $Y(s)$, we get

$$Y(s) = \frac{1}{s+5}\left[X(s) + y(0^+)\right]$$

Now substitute $X(s) = \dfrac{3}{(s+2)}$ and the initial condition $y(0^+) = -2$, obtaining

$$Y(s) = \frac{3}{(s+2)(s+5)} + \frac{-2}{s+5}$$

Expanding $Y(s)$ in partial fractions,

$$Y(s) = \frac{1}{s+2} + \frac{-1}{s+5} + \frac{-2}{s+5}$$

and taking the inverse unilateral Laplace transform, we get

$$y(t) = e^{-2t}u(t) - 3e^{-5t}u(t)$$

Problem 6.23: Consider a system H described by the differential equation

$$\frac{d}{dt}y(t) + 2y(t) = \frac{d^2}{dt^2}x(t) + 2\frac{d}{dt}x(t) - 3x(t)$$

Find the transfer function of the inverse system for H. Does a stable and causal inverse system exist?

Solution: First find the system transfer function $H(s)$ by taking the Laplace transform of both sides of the given differential equation, obtaining

$$Y(s)(s+2) = X(s)(s^2 + 2s - 3)$$

Hence the transfer function of the system is

$$H(s) = \frac{Y(s)}{X(s)} = \frac{s^2+2s-3}{s+2}$$

and the inverse system has the transfer function

$$H^{-1}(s) = \frac{1}{H(s)} = \frac{s+2}{s^2+2s-3} = \frac{s+2}{(s-1)(s+3)}$$

The inverse system has poles at $s = 1$ and $s = -2$. The pole at $s = 1$ is in the right half of the s–plane. Therefore the inverse system cannot be both stable and causal.

Problem 6.24: A system has the transfer function

$$H(s) = \frac{2}{s+3} + \frac{1}{s-2}$$

Find the impulse response assuming
(*a*) the system is stable and (*b*)
the system is causal. Can this system be both stable and causal?

Solution: This system has a pole at $s = -3$ and at $s = 2$. If the system is stable, then the pole at $s = -3$ contributes a right-sided term to the impulse response, while the pole at $s = 2$ contributes a left-sided term. We have

$$h(t) = 2\,e^{-3t}u(t) - e^{2t}u(-t)$$

If the system is causal, then both poles must contribute right-sided terms to the impulse response and we have

$$h(t) = 2\,e^{-3t}u(t) + e^{2t}u(t)$$

Note that this system is not stable since the term $e^{2t}u(t)$ is not absolutely integrable. In fact, thes system cannot be both stable and causal since the pole at $s = 2$ is in the right half of the s-plane.

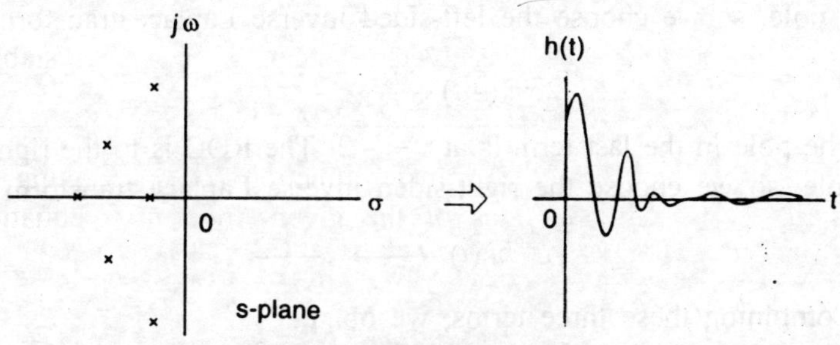

Fig. P. 6.24 (a)

Problem 6.25: Find the inverse Laplace transform of

$$X(s) = \frac{-5s-7}{(s+1)(s-1)(s+2)} \text{ with ROC } -1 < \text{Re}(s) < 1$$

Solution: Use a partial fraction expansion

$$X(s) = \frac{A_1}{s+1} + \frac{A_2}{s-1} + \frac{A_3}{s+2}$$

Solving for A_1, A_2 and A_3 gives

$$X(s) = \frac{1}{s+1} - \frac{2}{s-1} + \frac{1}{s+2}$$

The ROC and pole locations are depicted in Fig. 6.25 (*a*). We find the inverse Laplace transform of each term using the relationship between the pole location and ROC.

The pole of the first term is at $s = -1$. The ROC lies to the right of this pole, so we choose the right-sided inverse Laplace transform

$$e^{-t}u(t) \overset{L}{\longleftrightarrow} \frac{1}{s+1}$$

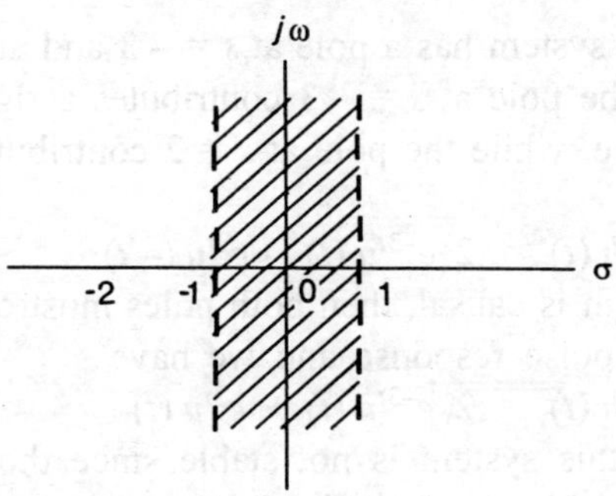

Fig. P. 6.25 (a)

The second term has a pole at $s = 1$. Here the ROC is to the left of the pole, so we choose the left-sided inverse Laplace transform

$$2e^{t}u(-t) \overset{L}{\longleftrightarrow} -\frac{2}{s-1}$$

The pole in the last term is at $s = -2$. The ROC is to the right of this pole, so we choose the right-sided inverse Laplace transform

$$e^{-2t}u(t) \overset{L}{\longleftrightarrow} \frac{1}{s+2}$$

Combining these three terms, we obtain

$$x(t) = e^{-t}u(t) + 2e^{t}u(-t) + e^{-2t}u(t)$$

Problem 6.26: Find the Laplace transform of

$$x(t) = \frac{d^2}{dt^2}(e^{-3(t-2)}u(t-2))$$

Solution: We know that

$$e^{-3t}u(t) \xleftrightarrow{L} \frac{1}{s+3} \text{ with ROC } Re(s) > -3$$

The time–shift property implies

$$e^{-3(t-2)}u(t-2) \xleftrightarrow{L} \frac{1}{s+3}e^{-2s} \text{ with ROC } Re(s) > -3$$

Now apply the time differentiation property twice, as shown by

$$x(t) = \frac{d^2}{dt^2}(e^{-3(t-2)}u(t-2)) \xleftrightarrow{L} X(s)$$

$$= \frac{s^2}{s+3}e^{-2s} \text{ with ROC } Re(s) > -3$$

Problem 6.27: Consider the two signals

$$x_1(t) = e^{-2t}u(t) + e^{-t}u(-t)$$

and

$$x_2(t) = e^{-t}u(t) + e^{-2t}u(-t)$$

Identify the ROC for each signal.

Solution: Check the absolute integrability of $|x_1(t)|\, e^{-\sigma t}$ by writing

$$I_1(\sigma) = \int_{-\infty}^{\infty} |x_1(t)|\, e^{-\sigma t}\, dt = \int_{-\infty}^{0} e^{-(1+\sigma)t}\, dt + \int_{0}^{\infty} e^{-(2+\sigma)t}\, dt$$

$$= \frac{-1}{1+\sigma} e^{-(1+\sigma)t}\Big|_{-\infty}^{0} + \frac{-1}{2+\sigma} e^{-(2+\sigma)t}\Big|_{0}^{\infty}$$

The first term converges for $\sigma < -1$, while the second term converges for $\sigma > -2$. Hence both terms converge for $-2 < \sigma < -1$. This is the intersection of the ROC for each term. The ROC for each term is depicted in Fig. 6.27 (a). The intersection of ROCs is shown as the doubly shaded region. The reader may verify that the Laplace transform of $x_1(t)$ is

$$X_1(s) = \frac{-1}{s+1} + \frac{1}{s+2} = \frac{-1}{(s+1)(s+2)}$$

which has poles at $s = -1$ and $s = -2$. We see that the ROC is the strip of the s-plane located between the poles.

For the second signal $x_2(t)$, we have

$$I_1(\sigma) = \int_{-\infty}^{\infty} |x_2(t)|\, e^{-\sigma t}\, dt = \int_{-\infty}^{0} e^{-(2+\sigma)t}\, dt + \int_{0}^{\infty} e^{-(1+\sigma)t}\, dt$$

$$= \frac{-1}{2+\sigma} e^{-(2+\sigma)t}\Big|_{-\infty}^{0} + \frac{-1}{1+\sigma} e^{-(1+\sigma)t}\Big|_{0}^{\infty}$$

The first term converges for $\sigma < -2$ and the second term for $\sigma > -1$. Here there is no value of σ for which both terms converge, and so the intersection is empty. Hence there are no values of s for which $X_2(s)$ converges. This situation is illustrated in Fig. 6.27 (b).

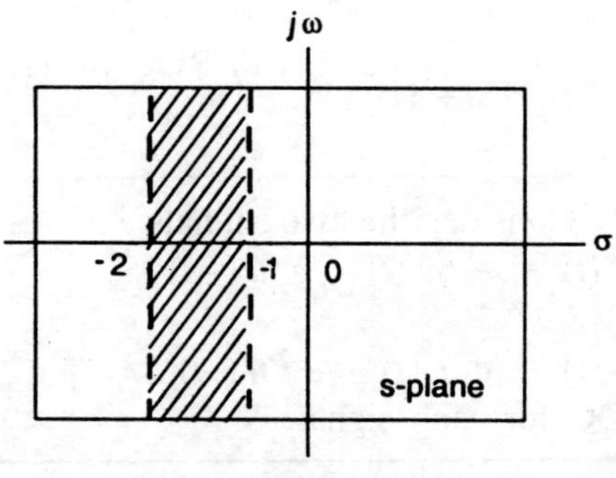

Fig. P. 6.27 (a)

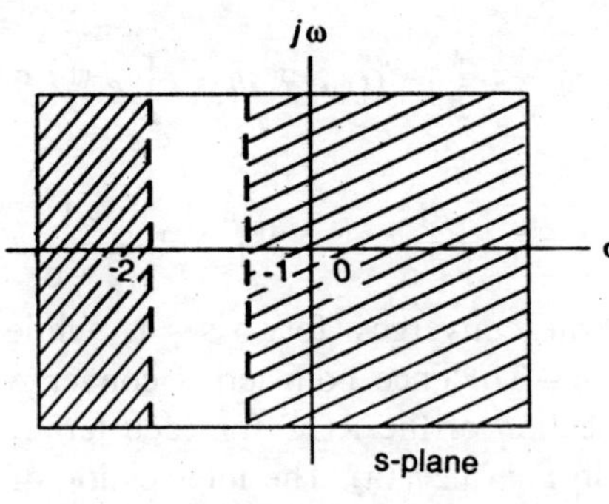

Fig. P. 6.27 (b)

Problem 6.28: Find the inverse transform of

$$F(s) = \frac{e^{-s}}{4s(s^2+1)}$$

Solution:

$$F(s) = \frac{-e^{-s}}{4s\left(s^2+1\right)}$$

Using after factorisation, we have,

$$F(s) = \frac{e^{-s}\cdot\left(s^2+1-s^2\right)}{4s\left(s^2+1\right)} = \frac{e^{-s}}{4s} - \frac{e^{-s}\cdot s}{4\left(s^2+1\right)}$$

Now, taking inverse laplace transformation both sides, we have

$$f(t) = L^{-1}[F(s)] = L^{-1}\left[\frac{e^{-s}}{4s} - \frac{e^{-s}\cdot s}{4\left(s^2+1\right)}\right]$$

Using shifting theorem,

We get $$f(t) = L^{-1}\left[\frac{e^{-s}}{4s}\right] - L^{-1}\left[\frac{e^{-s}\cdot s}{4\left(s^2+1\right)}\right]$$

$$f(t) = \frac{1}{4}U(t-1) - \frac{1}{4}\cos(t-1)\cdot U(t-1)$$

or, $$f(t) = \frac{1}{4}[1-\cos(t-1)]\cdot U(t-1)$$

Problem 6.29: Find the Laplace transform of

$$F(s) = \frac{(s+3)e^{-s/2}}{s^2+4s+9}$$

Solution:

$$F(s) = \frac{(s+3)\,e^{-\frac{s}{2}}}{s^2+4s+9}$$

Rearranging $F(s)$ as,

$$\text{or} \qquad F(s) = \frac{(s+3)\,e^{-\frac{s}{2}}}{(s+2)^2+\left(\sqrt{5}\right)^2} + \frac{\sqrt{5}\cdot e^{-\frac{s}{2}}}{\sqrt{5}\left[(s+2)^2+\left(\sqrt{5}\right)^2\right]}$$

Taking inverse Laplace transformation, we have,

Now, $\qquad f(t) = L^{-1} F(s)$

$$= L^{-1} \frac{(s+2).e^{-\frac{s}{2}}}{(s+2)^2+\left(\sqrt{5}\right)^2} + L^{-1} \frac{\sqrt{5}e^{-\frac{s}{2}}}{\sqrt{5\left[(s+2)^2+\left(\sqrt{5}\right)^2\right]}}$$

Using shifting theorem and

$$L^{-1}\left(\frac{\omega}{s^2+\omega^2}\right) = \sin \omega t,\; L^{-1}\left(\frac{s}{s^2+\omega^2}\right) = \cos \omega t,$$

Hence, the inverse Laplace transform,

$$f(t) = \left\{\cos\left[\sqrt{5}\left(t-\frac{1}{2}\right)\right] + \sin\left[\sqrt{5}\left(t-\frac{1}{2}\right)\right]\right\}\cdot e^{-2\left(t-\frac{1}{2}\right)} \cdot U\left(t-\frac{1}{2}\right)$$

Problem 6.30: Find the inverse transform of

$$F(s) = \frac{s^2+5}{s^3+2s^2+4s}$$

Solution:

$$F(s) = \frac{s^2+5}{s\left(s^2+2s+4\right)} = \frac{s^2+5}{s\left[(s+1)^2+\left(\sqrt{3}\right)^2\right]}$$

$$= \frac{A}{s} + \frac{Bs+C}{(s+1)^2+\left(\sqrt{3}\right)^2} \qquad \ldots(1)$$

Now, $s^2 + 2As + 4A + Bs^2 + Cs = s^2 + 5$

Equating co-efficients of s^2, s^1, s^0, we get

$$A + B = 1 \quad \text{...(2)}$$
$$4A = 5 \quad \text{...(3)}$$
$$2A + C = 0 \quad \text{...(4)}$$

Solving eqn (2), (3) and (4), we have

$$A = \frac{5}{4},\ B = \frac{1}{4},\ C = -\frac{5}{2}$$

Taking inverse Laplace of equation (1), we have

$$f(t) = \frac{5}{4}L^{-1}\left(\frac{1}{s}\right) + L^{-1}\frac{s+10}{4\left[(s+1)^2\left(\sqrt{3}\right)^2\right]}$$

$$= \frac{5}{4}L^{-1}\ \frac{1}{s} - \frac{1}{4}L^{-1}\left(\frac{s+1}{(s+1)^2+\left(\sqrt{3}\right)^2}\right)$$

$$+\frac{9}{\sqrt{3}\cdot 4}L^{-1}\frac{\sqrt{3}}{(s+1)^2+\left(\sqrt{3}\right)^2}$$

$$= \frac{5}{4} - \frac{1}{4}\left(\cos\sqrt{3t} + \frac{9}{\sqrt{3}}\sin\sqrt{3t}\right)\ e^{-t}$$

Problem 6.31: Find the inverse Laplace transform of

$$F(s) = \frac{s}{s^4+5s^2+4}$$

Solution:

$$F(s) = \frac{s}{\left(s^2+1\right)\left(s^2+4\right)}$$

After factorisation, $F(s)$ can be written as,

$$= \frac{1}{3}\cdot\left[\frac{s}{\left(s^2+1\right)} - \frac{s}{\left(s^2+4\right)}\right]$$

Taking inverse Laplace transformation both sides, we have

$$f(t) = L^{-1}\,\frac{1}{3}\left(\frac{s}{s^2+1}\right) - L^{-1}\frac{s}{3(s^2+4)}$$

$$f(t) = \frac{1}{3}\,(\cos t - \cos 2\,t)$$

Problem 6.32: Find the inverse Laplace transform of

$$F(s) = \frac{3s+1}{5s^3\,(s-2)^2}$$

Solution:

By using Partial fraction expansion gives,

$$= \frac{a_{11}}{s^3} + \frac{a_{12}}{s^2} + \frac{a_{13}}{s} + \frac{b_{11}}{(s-2)} + \frac{b_{12}}{(s-2)^2}$$

The coefficients are then calculated as,

$$a_{11} = F(s)\big|_{s=0} = \frac{1}{20}$$

$$a_{12} = \frac{d}{ds}[F(s)\cdot s^3]\Big|_{s=0} = \left.\frac{-3s-8}{(s-2)^3}\right|_{s=0} = \frac{1}{5}$$

$$a_{13} = \frac{d^2}{dt^2}[F(s)\cdot s^3]\Big|_{s=0} = \left.\frac{6s+30}{(s-2)^4}\right|_{s=0} = \frac{3}{8}$$

$$b_{11} = \left.\frac{3s+1}{5s^3-(s-2)^2}\cdot(s-2)^2\right|_{s=0} = \frac{7}{40}$$

$$b_{12} = \left.\frac{dF(s)}{ds}\cdot(s-2)^2\right|_{s=0} = \frac{1}{5}\left(\frac{-6}{s^3} - \frac{3}{s^4}\right) = -\frac{3}{8}$$

Now, $F(s)$ can be rewritten as,

$$\frac{3s+1}{5s^3(s-2)^2} = \frac{1}{20s^3} + \frac{1}{5s^2} + \frac{3}{8s} + \frac{7}{40(s-2)} - \frac{3}{8(s-2)^2}$$

Hence, the inverse Laplace transform can be written as,

$$L^{-1}\frac{3s+1}{5s^3(s-2)^2} = L^{-1}\frac{1}{20s^3}+L^{-1}\frac{1}{5s^2}+L^{-1}\frac{3}{8s}+$$

$$L^{-1}\frac{7}{40(s-2)}-L^{-1}\frac{3}{8(s-2)^2}$$

$$= \frac{1}{40}\left[\left(t^2+8t+\frac{15}{2}\right)+\left(7t-\frac{15}{2}\right)e^{2t}\right]$$

> **Problem 6.33:** Find the inverse Laplace transform of
>
> $$F(s) = \frac{1-e^{-4s}}{3s^3+2s^2}$$

Solution:

$$F(s) = \frac{1-e^{-4s}}{3s^3+2s^2} = \frac{1-e^{-4s}}{s^2(3s+2)} = \frac{1}{s^2(3s+2)} - \frac{e^{-4s}}{s^2(3s+2)}$$

Partial Fraction of

$$F'(s) = \frac{1}{s^2(3s+2)} \quad \text{can be written as,}$$

$$F'(s) = \frac{1}{3s^2\left(s+\frac{2}{3}\right)} = \frac{A}{s^2}+\frac{B}{s}+\frac{C}{s+\frac{2}{3}}$$

The coefficients are then calculated as,

or $$A = F'(s)\cdot s^2\Big|_{s=0} = \frac{1}{2}$$

$$B = \left.\frac{d\,[F'(s)\cdot s^2]}{ds}\right|_{s=0} = -\left.\frac{3}{(3s+2)^2}\right|_{s=0} = -\frac{3}{4}$$

$$C = F'(s)\cdot\left(s+\frac{2}{3}\right)\Big|_{s=-\frac{2}{3}} = \frac{3}{4}$$

Then, $$F'(s) = \frac{1}{2s^2} - \frac{3}{4s} + \frac{3}{4}\frac{1}{\left(s+\frac{2}{3}\right)}$$

Hence, the inverse Laplace transform can be expressed as,

$$L^{-1}F(s) = L^{-1}\left[\frac{1}{2s^2} - \frac{3}{4s} + \frac{3}{4\left(s+\frac{2}{3}\right)}\right] -$$

$$L^{-1}\left[\frac{1}{2s^2} - \frac{3}{4s} + \frac{3}{4\left(s+\frac{2}{3}\right)}\right] e^{-4s}$$

$$= \frac{1}{4}\left[2t - 3\left(1 - e^{-\frac{2t}{3}}\right)\right] -$$

$$\frac{1}{4}\left\{2(t-4) - 3\left[1 - e^{-\frac{2(t-4)}{3}}\right]\right\} \cdot U(t-4)$$

> **Problem 6.34:** Find the inverse Laplace transform of
>
> $$F(s) = \ln\left(\frac{s}{s^2+9}\right)$$

Solution:

$$F(s) = \ln\left(\frac{s}{s^2+9}\right) = \ln s - \ln(s+3j) - \ln(s-3j)$$

$$\left[\text{Using } \ln\left(\frac{a}{b}\right) = \log a - \log b\right]$$

Differentiating with respect to s, we have

$$-\frac{dF(s)}{ds} = -\frac{1}{s} + \frac{1}{s+3j} + \frac{1}{s-3j}$$

Using $-\dfrac{dF(s)}{ds} = L[t \cdot f(t)]$, we get

The inverse Laplace transform of $F(s)$,

$$f(t) = -\frac{1}{t} L^{-1}\left[\frac{dF(s)}{ds}\right]$$

$$= -\frac{1}{t} L^{-1}\left[-\frac{1}{s}+\frac{1}{s+3j}+\frac{1}{s-3j}\right]$$

$$= -\frac{1}{t}\left[-1 + e^{-3j} + e^{3j}\right]$$

$$f(t) = -\frac{1}{t}\left[2\cos 3t - 1\right] \textbf{ Ans}$$

Problem 6.35: Find the inverse Laplace transforms of the following functions by using the convolution theorem:

(*a*) $\frac{1}{s^3}$ (*b*) $\left(\frac{s}{s+1}\right)^2$ (*c*) $\frac{s+1}{s(s^2+4)}$

Solution : (*a*) Let $F_1(s) = \frac{1}{s^2}$ and $F_2(s) = \frac{1}{s}$

then, $f_1(t) = tU(t)$ and $f_2(t) = U(t)$

Using convolutional theorem, we get

$$f(t) = L^{-1}[F(s)]$$

$$= L^{-1}\left(\frac{1}{s^3}\right) = \int_0^t f_1(\tau)\cdot f_2(t-\tau)\cdot d\tau$$

$$= \int_0^t \tau U(\tau)\cdot U(t-\tau)\, d\tau$$

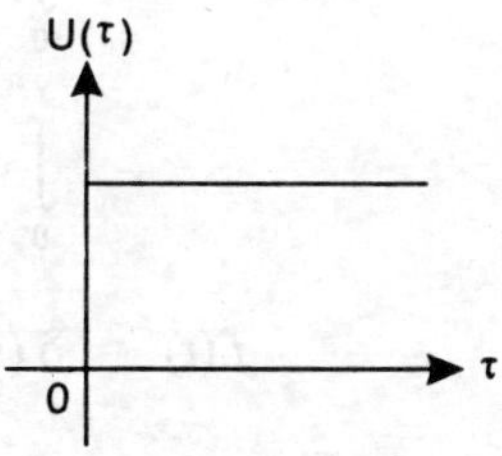

Fig. P. 6.35 (a)

From Fig. (c), we have,

$$U(\tau)\,.\,U(t-\tau) = 1$$

in interval $0 < \tau < t$

$$\text{Thus, } f(t) = \int_0^t \tau\, d\tau = \left.\frac{\tau^2}{2}\right|_0^t = \frac{t^2}{2}$$

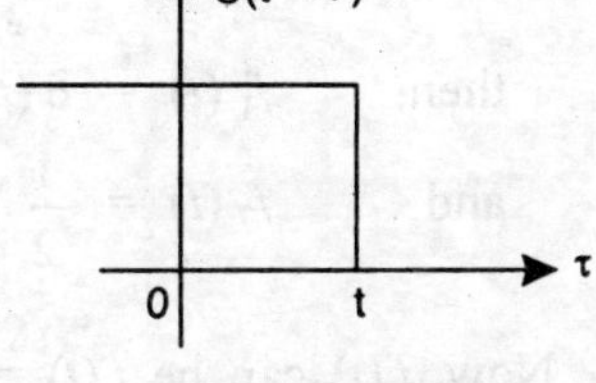

Fig. P. 6.35 (b)

(*b*) Let, $F_1(s) = F_2(s) = \frac{s}{s+1}$

Then, $f_1(t) = f_2(t) = L^{-1}\left(\frac{s}{s+1}\right) = L^{-1}\left(\frac{s+1-1}{s+1}\right) = \delta(t) - e^{-t}$

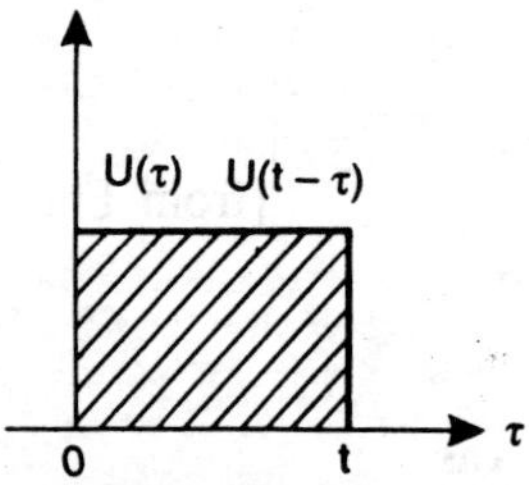

Fig. P. 6.35 (c)

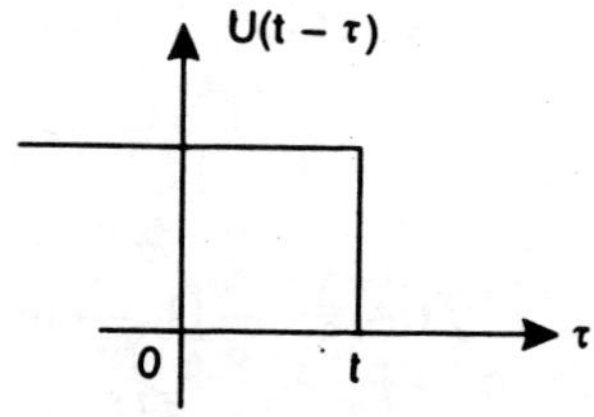

Fig. P. 6.35 (d)

Now, $f(t) = L^{-1}\left(\frac{s}{s+1}\right)^2 = L^{-1}[F_1(s) \cdot F_2(s)]$

$$= \int_0^t f_1(\tau) \cdot f_2(t-\tau)\, d\tau$$

$$= \int_0^t [\delta(\tau) - e^{-\tau}] \cdot [\delta(t-\tau) - e^{-(t-\tau)}]\, d\tau$$

$$= \int_0^t \delta(\tau) \cdot \delta(t-\tau)\, d\tau - \int_0^t e^{-\tau}(t-\tau)\, d\tau$$

$$\int_0^t e^{-(t-\tau)} \cdot \delta(t)\, d\tau + e^{-t}\int_0^t d\tau$$

$$f(t) = \delta(t) - e^{-t} - e^{-t} + te^{-t} = \frac{\delta(t) - 2e^{-t} + te^{-t}}{t}$$

$$\left[\text{Using } \int_0^t \delta(\tau) \cdot f(t-\tau)\, d\tau = f(t)\right]$$

(*c*) Here, $F_1(s) = \frac{s+1}{s} = 1 + \frac{1}{s}$ $\qquad F_2(s) = \frac{1}{2} \cdot \frac{2}{s^2+4}$

then, $f_1(t) = \delta(t) + U(t)$,

and $f_2(t) = \frac{1}{2} \cdot \sin 2t$

Now, $f(t)$ can be $f(t) = \int_0^t f_2(\tau) \cdot f_1(t-\tau)\, d\tau$

$$= \frac{1}{2}\int_0^t \sin(2\tau)\cdot[\delta(t-\tau)+U(t-\tau)]\,d\tau$$

$$= \frac{1}{2}\int_0^t \sin(2\tau)\ \delta(t-\tau)\,d\tau + \frac{1}{2}\int_0^t \sin(2\tau)\cdot U(t-\tau)\,d\tau$$

$$= \frac{1}{2}\cdot\sin 2t + \frac{1}{2}\int_0^t \sin 2t\,dt \qquad \text{[from Fig. (a)]}$$

$$f(t) = \frac{1}{4}[1-\cos 2t + 2\sin 2t] \ \textbf{Ans}$$

Problem 6.36: The convolution theorem extended to three functions can be written as follows:

$$L[f_1(t) * f_3(t)] = F_1(s)\ F_2(s)\ F_3(s)$$

Find the inverse Laplace transform of

$$F(s) = \frac{s}{(s+1)(s+2)(s+3)} = \left(\frac{1}{s+1}\right)\left(\frac{1}{s+2}\right)\left(\frac{s}{s-3}\right)$$

by making use of the above relationship. Check your answer by Heaviside's expansion theorem.

Solution:

Let, $F_1(s) = \dfrac{1}{s+1}$, $F_2(s) = \dfrac{1}{s+2}$, and $F_3(s) = \dfrac{1}{s+3}$,

$f_1(t) * f_2(t) = f_4(t)$, we have

$$f_4(t) = \int_0^t e^{-\tau}\cdot e^{-2(t-\tau)}\,d\tau$$

[Here, $f_1(t) = L^{-1}[F_1(s)] = L^{-1}\left[\dfrac{1}{s+1}\right] = e^{-t}$

and, $f_2(t) = L^{-1}\dfrac{1}{s+2} = \delta(t) - 3e^{-3t}] = e^{-2t}\displaystyle\int_0^t e^{\tau}\,d\tau$

or $f_4(t) = e^{-2t}(e^t - 1) = e^{-t} - e^{-2t}$

Now, $f_3(t) = L^{-1}\left(\dfrac{s}{s+3}\right) = L^{-1}\left[1-\dfrac{3}{s+3}\right] = \delta(t) - 3e^{-3t}$

Then, $f(t) = L^{-1}F(s) = L^{-1}\cdot[F_4(s)\cdot F_3(s)]$

$$= \int_0^t f_3(\tau) \cdot f_4(t-\tau) \cdot d\tau$$

$$= \int_0^t [\delta(\tau) - 3e^{-3\tau}] \cdot e^{-2(t-\tau)} [e^{(t-\tau)} - 1] \cdot d\tau$$

$$= \int_0^t \delta(\tau) \cdot e^{-2(t-\tau)} [e^{(t-\tau)} - 1] \cdot d\tau - 3\int_0^t e^{-2t/\tau} [e^{(t-\tau)} - 1] \cdot d\tau$$

$$= e^{-2t}(e^t - 1) + 3e^{-2t}\int_0^t e^{-\tau} \cdot d\tau - 3\int_0^t e^{-2t/\tau} [e^{(t-\tau)} - 1] \cdot d\tau$$

$$= e^{-2t}(e^t - 1) + 3e^{-2t}(1 - e^t) + \frac{3}{2}e^{-t}(e^{-2t} - 1) \cdot d\tau$$

$$= -\frac{e^{-t}}{2}(1 - 4 \cdot e^{-2t} + 3e^{-2t})$$

Problem 6.37: Solve the following integral equation for $f(x)$ by the method of Laplace transforms:

$$2f(x) + \beta\int_0^x \sin\beta(x-u) \cdot f(u)\, du = \cos\beta x.$$

Solution:

Taking Laplace transform on both sides to the given expression, we get

$$2L[f(x)] + \beta L\left[\int_0^x \sin\beta(x-u) \cdot f(u)du\right] = L[\cos\beta x]$$

$$\text{or} \quad 2F(s) + \beta \cdot \frac{\beta}{s^2+\beta^2} \cdot F(s) = \frac{s}{s^2+\beta^2}$$

[Using convolution theorem]

$$\text{or } F(s)\left[\frac{2s^2+3\beta^2}{s^2+\beta^2}\right] = \frac{s}{s^2+\beta^2}$$

$$\text{or} \quad F(s) = \frac{s}{2s^2+3\beta^2} = \frac{1}{2} \cdot \frac{s}{s^2 + \left(\sqrt{\frac{3}{2}} \cdot \beta\right)^2}$$

Now, takin inverse Laplace transform of both sides, we have

$$L^{-1}F(s) = f(x) = \frac{1}{2}L^{-1}\frac{s}{s^2+\left(\sqrt{\frac{3}{2}}\cdot\beta\right)^2}$$

$$\left[\text{Using } L[\cos(\omega x)] = \frac{s}{s^2+\omega^2}\right]$$

Hence, $f(x)$ is derived as,

$$f(x) = \frac{1}{2}\cos\sqrt{\frac{3}{2}}\,\beta x.$$

Problem 6.38: The step response of a certain initially relaxed device is $w_u(t) = \left(1-\frac{1}{2}e^{-t/3}\right)U(t)$

Determine the impulse response of a system consisting of two such devices in cascade, assuming that the loading effect due to cascading is negligible.

Solution:

Laplace transform for the step response gives,

$$w_u(s) = \left[\frac{1}{s}-\frac{1}{2}\cdot\frac{1}{s+\frac{1}{3}}\right] = \left(\frac{1}{s}-\frac{3}{6s+2}\right) = \frac{3s+2}{2s(3s+1)}$$

then, the transfer function can be given as,

$$H(s) = \frac{w_u(s)}{E(s)}\left[\text{Here } E(s)=\frac{1}{s}\right] = \frac{3s+2}{2(3s+1)}$$

Since, device consists of such two system, their transfer function together, is expressed as,

$$H'(s) = [H(s)]^2 = \frac{(3s+2)^2}{4.(3s+1)^2}$$

then, impulse response,

$$h(t) = L^{-1}\,\frac{(3s+2)^2}{4.(3s+1)^2} = L^{-1}\left[\frac{1}{4}+\frac{6s+3}{4.(3s+1)^2}\right]$$

$$= L^{-1}\left[\frac{1}{4}+\frac{1}{36}\cdot\frac{1}{\left(s+\frac{1}{3}\right)^2}+\frac{2}{12\left(s+\frac{1}{3}\right)}\right]$$

$$= L^{-1}\,\frac{1}{4}\left[1+\frac{1}{9}\cdot\frac{1}{\left(s+\frac{1}{3}\right)^2}+\frac{2}{3}\cdot\frac{1}{\left(s+\frac{1}{3}\right)}\right]$$

$$= \frac{1}{4}\left[\delta(t)+\frac{2}{3}\left(1+\frac{t}{6}\right)e^{-\frac{t}{3}}\cdot U(t)\right]$$

Problem 6.39: It is found that when a voltage $v_1(t) = tU(t)$ is applied at the input of an initially relaxed one-stage linear amplifier, the output voltage $v_2(t)$ is given by $v_2(t) = (e^{-2t} - 1)\,U(t)$. Suppose that two such identical stages are connected in cascade and a uhit step voltage is applied at the input. Find the voltage at the output of the second stage.

Solution:

Input voltage,

$$v_1(t) = tU(t)$$

Then its Laplace transform,

$$V_1(s) = \frac{1}{s^2}\quad\left[\text{As, } L[t^n]\cdot U(t) = \frac{Ln}{s^{n+1}}\right]$$

and, Laplace transform of output voltage,

$$V_2(s) = \left(\frac{1}{s+2}-\frac{1}{s}\right) = \frac{-2}{s(s+2)}$$

Then, $$\frac{V_2(s)}{V_1(s)} = \frac{-2s}{s(s+2)}$$

Hence, transfer function of the cascade arrangement,

$$H(s) = \frac{4s^2}{(s+2)^2}$$

and output voltage at the second stage,

$$w_u(s) = H(s) \cdot E(s)$$

$$w_u(s) = \frac{4s^2}{(s+2)^2} \cdot \frac{1}{s} = \frac{4s}{(s+2)^2} \qquad \left[\because E(s) = \frac{1}{s}\right]$$

then, $\quad w_u(t) = L^{-1}[w_u(s)] = L^{-1}\left[\frac{4s}{(s+2)^2}\right]$

Now, using convolution theorem, we have

$$f_1(t) = L^{-1}\left(\frac{s}{s+2}\right) = \delta(t) - 2e^{-2t}$$

and, $\quad f_2(t) = L^{-1}\left(\frac{1}{s+2}\right) = e^{-2t}$

Now, the output voltage of the second stage,

$$w_u(t) = \int_0^t f_1(\tau) \cdot f_2(t-\tau) \cdot d\tau$$

$$= 4\int_0^t [\delta(\tau) - 2 \cdot e^{-2\tau}] \cdot e^{-2(t-\tau)} \cdot d\tau$$

$$= 4\int_0^t \delta(\tau) \cdot e^{-2(t-\tau)} \cdot d\tau - (\delta \cdot e^{-2t})\int_0^t d\tau$$

$$= 4e^{-2t}\, U(t) - \delta \cdot e^{-2t} \cdot t \cdot U(t)$$

$$w_u(t) = 4(1 - e^{-2t})\, e^{-2t} \cdot U(t)$$

is required voltage at the output.

Problem 6.40: Find the inverse Laplace transforms of the following functions:

$(a)\ s^{-5/12}$	$(b)\ \dfrac{e^{-2s}}{(s+1)\sqrt{(s+5)}}$	$(e)\ \dfrac{1}{\sqrt{s^2+4s+13}}$
$(c)\ \dfrac{1}{\sqrt{s}\,(\sqrt{s}+\sqrt{a})}$		$(d)\ \dfrac{1}{\sqrt{s+1}}$

Solution:

(*a*)

$$F(s) = s^{-5/12}$$

$$f(t) = \mathrm{L}^{-1} F(s) = \mathrm{L}^{-1}[s^{-1/2} \cdot s^{-2}]$$

here, $\mathrm{L}^{-1} \cdot [s^{-1/2}] = \dfrac{1}{\sqrt{\pi t}} * t = \dfrac{1}{\sqrt{\pi}} \displaystyle\int_0^t \dfrac{1}{\sqrt{\tau}} (t - \tau) \cdot d\tau$

$$= \frac{1}{\sqrt{\pi}} \int_0^t \frac{1}{\sqrt{\tau}}\, d\tau \quad \frac{1}{\sqrt{\pi}} \int_0^t \tau^{1/2} \cdot d\tau$$

$$= \frac{2t}{\sqrt{\pi}} t^{1/2} - \frac{2}{\sqrt{\pi}} \cdot \frac{t^{3/2}}{3} = \frac{4t^{3/2}}{3\sqrt{\pi}}$$

(*b*)

$$F'(s) = \frac{1}{\sqrt{s+2}} \cdot \frac{1}{s+1}$$

$$f(t) = \mathrm{L}^{-1} F'(s) = \mathrm{L}^{-1} \frac{1}{\left(\sqrt{s+5}\right)(s+1)} = e^{-5t}\, \mathrm{L}^{-1}\left[\frac{1}{\sqrt{5}}, \frac{1}{(s-4)}\right]$$

Now, $\mathrm{L}^{-1} \dfrac{1}{\sqrt{5}} = \dfrac{1}{\sqrt{\pi t}}$ and $\mathrm{L}^{-1}\left[\dfrac{1}{s-4}\right] = e^{4t}$.

then, $f'(t) = \left[\left(e^{4t}\right) * \dfrac{1}{\sqrt{\pi t}}\right] . e^{-5t} = \left[\displaystyle\int_0^t e^{4(t-\tau)} \cdot \dfrac{1}{\sqrt{\pi t}} \cdot d\tau\right] e^{-5t}$

$$= \left[\frac{e^{4t}}{\sqrt{\pi}} \cdot \int_0^t e^{4\tau} \cdot \frac{1}{\sqrt{\tau}} \cdot d\tau\right] e^{-5t}$$

Let, $\sqrt{\tau} = x,\ x^2 = \tau,$

then $$f'(t) = \frac{e^{-t}}{2}\cdot\frac{2}{\sqrt{\pi}}\int_0^{\sqrt{4t}} e^{-(2x)^2}\cdot d\,(2x) = \frac{e^{-t}}{2}\,erf\left(\sqrt{4t}\right)$$

$$= \frac{e^{-t}}{2}\,erf\left(2\sqrt{t}\right)\cdot\left[\because\ erf(x) = \frac{2}{\pi}\int_0^x e^{-u^2}.du\right]$$

$$= \frac{e^{t}}{\sqrt{\pi}}\int_0^t \frac{e^{-\tau}}{\sqrt{\tau}}\cdot d\tau$$

Let $\sqrt{\tau} = x,\ x^2 = \tau,\ d\tau = 2x\cdot dx$

then, $$f'(t) = \frac{2e^{t}}{\sqrt{\pi}}\int_0^{\sqrt{t}} \frac{e^{-x^2}}{x}\,x\,dx = e^{t}\cdot\frac{2}{\sqrt{\pi}}\int_0^{\sqrt{t}} e^{-x^2}\cdot dx = e^{t}\cdot erf\left(\sqrt{t}\right)$$

Now, $F(s) = e^{-2s}\cdot F'(s)$

then, $$f(t) = \frac{1}{2}\left[e^{-(t-2)}erf\ 2\sqrt{t-2}\right]\cdot U(t-2)$$

(*c*) $$F(s) = \frac{\sqrt{s}-\sqrt{a}}{\sqrt{s}\left(\sqrt{s}+\sqrt{a}\right)\left(\sqrt{s}+\sqrt{a}\right)} = \frac{1}{s-a} - \frac{\sqrt{a}}{\sqrt{s}(s-a)}$$

then, $$F(t) = L^{-1}F(s) = L^{-1}\left[\frac{1}{s-a}\right] - \sqrt{a}\cdot L^{-1}\frac{1}{\sqrt{s}(s-a)}$$

$$= e^{at} - \sqrt{a}\int_0^t \frac{1}{\sqrt{\pi\tau}}e^{a(t-\tau)}\cdot d\tau$$

$$= e^{at} - \frac{\sqrt{a}.e^{at}}{\sqrt{\pi}}\int_0^t \frac{1}{\sqrt{\tau}}e^{-at}\cdot d\tau$$

Let $\sqrt{\tau} = x,\quad x^2 = \tau,\quad d\tau = 2x\ dx$

then, $$f(t) = e^{at}\ \frac{\sqrt{a}}{\sqrt{\pi}}\cdot e^{at}\cdot 2\int_0^{\sqrt{t}} e^{-\left(\sqrt{a}\cdot x\right)^2}\cdot dx$$

$$= e^{at}\left[1-\frac{2}{\pi}\int_0^{\sqrt{at}} e^{-(\sqrt{a}.x)^2}\cdot d(\sqrt{a}x)\right]$$

$$= e^{at}\left[1-erf\sqrt{at}\right] = e^{at}\cdot erfc\ \sqrt{at}$$

(*d*) $$F(s) = \frac{1}{\sqrt{s}+1}$$

$F(s)$ can be rewritten as, $F(s) = \dfrac{1}{\sqrt{s}}+\dfrac{1}{\sqrt{s}(s-1)}-\dfrac{1}{s-1}$ ...(1)

Now, $$F'(s) = \frac{1}{\sqrt{s}(s-1)}$$

then, $$f'(t) = \mathrm{L}^{-1}F'(s) = \frac{1}{\sqrt{\pi t}} * e^{(t)} = \frac{1}{\sqrt{\pi}}\int_0^t \frac{1}{\sqrt{t}}e^{(t-\tau)}\cdot d\tau$$

Hence, from equation (1), we get

$$f(t) = \mathrm{L}^{-1}\frac{1}{\sqrt{s}}+L^{-1}\frac{1}{\sqrt{s}(s-1)}-L^{-1}\frac{1}{(s-1)}$$

$$= \frac{1}{\sqrt{\pi t}}+e^{t}erf\sqrt{t}-e^{t} = \frac{1}{\sqrt{\pi t}}e^{t}\left(1-erf\sqrt{t}\right)$$

(*f*) $$F(s) = \frac{1}{s^2+4s+13} = \frac{1}{\sqrt{(s+2)^2+9}} = \mathrm{L}^{-1}F(s)$$

$$= \mathrm{L}^{-1}\frac{1}{\sqrt{(s+2)^2+9}} = e^{-2t}\mathrm{L}^{-1}\frac{1}{\sqrt{s^2+9}}$$

$$= e^{-2t}\cdot \mathrm{L}^{-1}\frac{1}{S}\left[1+\left(\frac{3}{S}\right)^2\right]^{-1/2}$$

$$= e^{-2t}\,\mathrm{L}^{-1}\left[\frac{1}{s}-\frac{9}{2s^3}+\frac{3.3^4}{\underline{|2.2^2s^5}}-\frac{3.53^6}{\underline{|3.2^3s^7}}\cdots\right]$$

$$= e^{-2t}\left[1-\frac{(3t)^2}{2^2}+\frac{(3t)^4}{2^24^2}-\frac{(3t)^6}{2^24^26^2}+\ldots\right]$$

$$= e^{-2t}\sum_{k=0}^{\infty}\frac{(-1)^k}{(\mathrm{L}k)^2}\left(\frac{3t}{2}\right)^{2k}$$

Problem 6.41: Find the Laplace transform of a pair of rectangular pulses, each of duration τ, as shown in the figure below.

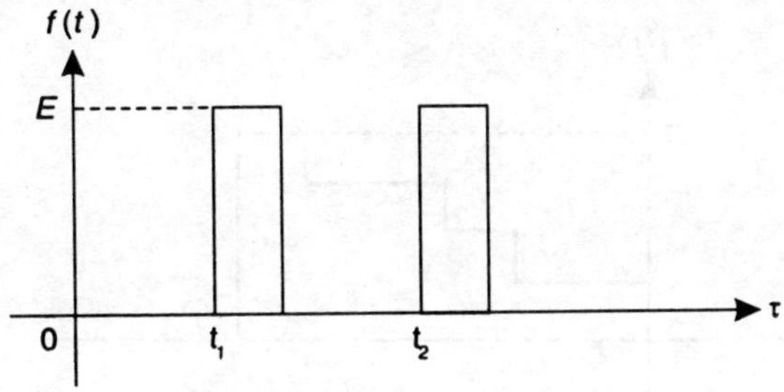

Fig. P. 6.41 (a)

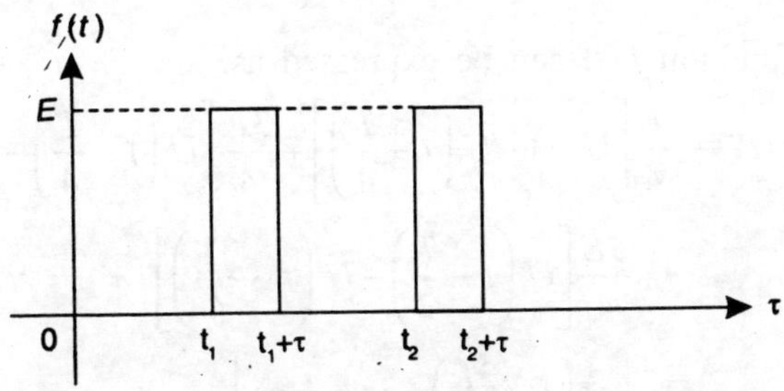

Fig. P. 6.41 (b)

Solution:

From the figure, $f(t)$ can be expressed as,

$$f(t) = E,\ t_1 < t < (t_1 + \tau)$$

and $$t_2 < t < t < t_2 + \tau$$

$$= 0, \text{ otherwise}$$

Now, using Gate function,

$f(t)$ can be written as,

$$f(t) = E\,[U(t - t_1) - U\{t - (t_1 + \tau)\}] + E\,[U(t - t_2) - U(t - t_2 - \tau)]$$

then $$F(s) = \mathrm{L}\,[f(t)]$$

$$= \frac{E}{s}\left[e^{-t_1 s} - e^{-(t_1+\tau)s}\right] + \frac{E}{s}\left[e^{-t_2 s} - e^{-(t_1+\tau)s}\right]$$

$$= \frac{E}{s}\left(e^{-t_1 s} + e^{-t_2 s}\right)\left(1 - e^{-\tau s}\right)$$

Problem 6.42: Find the Laplace transform of the staircase function shown in the figure below.

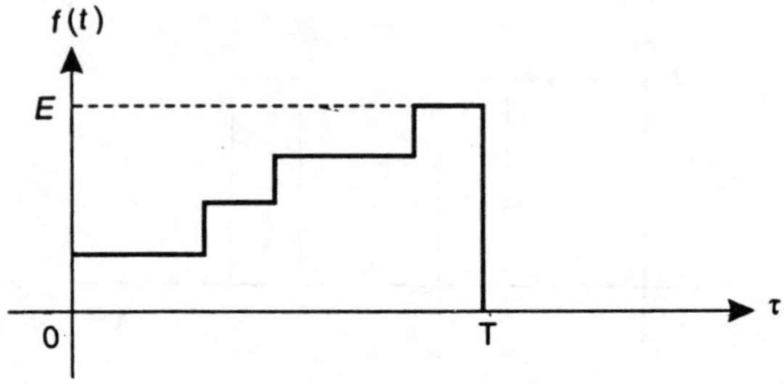

Fig. P. 6.42 (a)

Solution:

The function $f(t)$ can be expressed as,

$$f(t) = \frac{E}{4}\left[U(t) - U\left(t - \frac{T}{4}\right)\right] + \frac{3E}{4}\left[U\left(t - \frac{T}{4}\right) - U\left(t - \frac{T}{2}\right)\right]$$

$$+ \frac{3E}{4}\left[U\left(t - \frac{T}{2}\right) - U\left(t - \frac{3T}{4}\right)\right] +$$

$$E\left[U\left(t - \frac{3T}{4}\right) - U(t - T)\right]$$

or $$f(t) = \frac{E}{4}U(t) + \frac{E}{4}U\left(t - \frac{T}{4}\right) +$$

$$\frac{E}{4}U\left(t - \frac{T}{2}\right) + \frac{E}{4}\cdot U\left(t - \frac{3T}{4}\right) - EU(t - T)$$

Using Laplace transformation, we get,

$$F(s) = \text{L}\,[F(t)]$$

$$= \frac{E}{4s} + \frac{E}{4s}e^{-\frac{T}{4}.s} + \frac{E}{4s}e^{-T/2.s} + \frac{E}{4s}e^{-3T/4.s} - \frac{E}{s}\cdot e^{-Ts}$$

$$= \frac{E}{4s}[(1 + e^{-T/2.s})(1 + e^{-T/4.s}) - 4\,e^{-Ts}]$$

$$= \frac{E}{4s}\left[\frac{\left(1 - e^{-Ts}\right)}{\left(1 - e^{-T/4s}\right)} - 4e^{-Ts}\right]$$

Chapter 7

Z-transform

Problem 7.1: Determine the *z*-transform and ROC for the following time signals. Sketch the ROC, poles, and zeros in the z-plane.

(*a*) $x[n] = \delta[n]$ (*b*) $x[n] = \delta[n-k],\ k > 0$

(*c*) $x[n] = \delta[n+k],\ k > 0$ (*d*) $x[n] = u[n]$

(*e*) $x[n] = \left(\frac{1}{2}\right)^n (u[n] - u[n-10])$ (*f*) $x[n] = \left(\frac{1}{2}\right)^n u[-n]$

(*g*) $x[n] = 2^n u[-n-1]$ (*h*) $x\{n\} = \left(\frac{1}{2}\right)^{|n|}$

Solution:

(*a*) $x[n] = \delta[n]$

Its Z-transform is given by

$$\delta[n] \xleftrightarrow{Z} \sum_{n=-\infty}^{+\infty} \delta[n]\, z^{-n} = 1$$

ROC is the entire z-plane, including $z = 0$ and $z = \infty$

(*b*) $x[n] = \delta[n-k],\ k > 0$

Its *z*-transform is given by

$$\delta[n-k] \xleftrightarrow{Z} \sum_{n=-\infty}^{+\infty} \delta[n-1]\, z^{-n} = z^{-n}$$

This *z*-transform is well defined except at $z = 0$. Where there is a pole. thus, the ROC consists of the entire *z* plane, including $z = \infty$ but excluding $z = 0$.

(*c*) $x[n] = \delta[n+k],\ k > 0$

Its z-transform is given by

$$\delta[n+k] \overset{Z}{\longleftrightarrow} \sum_{n=-\infty}^{+\infty} \delta[n+k]\, z^{-n} = z^{k}$$

Which is well defined for all finite values of z, thus, the ROC consists of the entire finite z plane (including $z = 0$)

(*d*) $x[n] = u[n]$

Its z-transform is given by

$$x[n] \overset{Z}{\longleftrightarrow} \sum_{n=-\infty}^{+\infty} u[n]\, z^{-n} = \sum_{n=0}^{\infty} z^{-n} = \sum_{n=0}^{\infty} (z^{-1})^{n}$$

This is a geometric series of infinite length in the ratio $\frac{1}{z}$; the sum converges provided

$$\left|\frac{1}{z}\right| < 1 \text{ or } |z| > |1|$$

Hence $$X[z] = \frac{1}{1-z^{-1}},$$

$$|z| > |1| = \frac{z}{z-1}$$

There is a pole at $z = 1$ and a zero at $z = 0$ as shown in the fig. The ROC is depicted as the shaded region of the z-plane

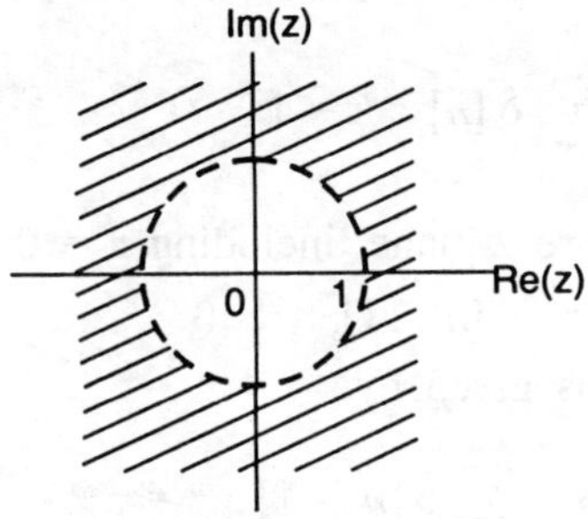

Fig. P. 7.1 (a)

(*e*) $$x[n] = \left(\frac{1}{2}\right)^{n} (u[n] - u[n-10])$$

Its z-transform is given by

$$x[n] \overset{Z}{\longleftrightarrow} \sum_{n=-\infty}^{\infty} \left[\left(\frac{1}{2}\right)^n (u[n]-u[n-10])\right] z^{-n}$$

$$= \sum_{n=-\infty}^{\infty} \left(\frac{1}{2z}\right)^n - \sum_{10}^{\infty}\left(\frac{1}{2z}\right)^n = \sum_{n=0}^{9}\left(\frac{1}{2z}\right)^n$$

$$= \frac{1-\left(\frac{1}{2z}\right)^{10}}{1-\left(\frac{1}{2z}\right)} = \left[\frac{(2z)^{10}-1}{2z-1}\right]\cdot\frac{1}{(2z)^9}$$

$$= \left[\frac{z^{10}-\left(\frac{1}{2}\right)^{10}}{z-\left(\frac{1}{2}\right)}\right]\cdot\frac{1}{z^9}$$

The ROC consists of all values of z except $z = 0$.

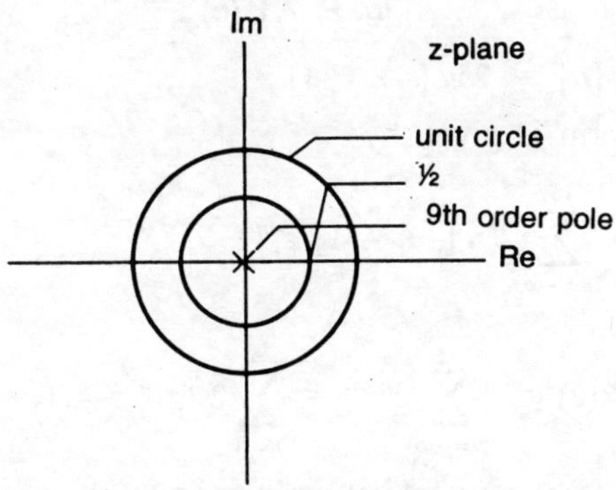

Fig. P. 7.1 (b)

(*f*) $$x[n] = \left(\frac{1}{2}\right)^n u[-n]$$

Its z transform is given by

$$x[n] \overset{z}{\longleftrightarrow} \sum_{n=-\infty}^{+\infty} \left(\frac{1}{2}\right)^n u[-n]\, z^{-n}$$

$$X[z] = \sum_{n=0}^{\infty}\left(\frac{1}{2}\right)^{-n} z^n = \sum_{n=0}^{\infty}(2z)^n$$

If $(2z) < 1$, the sum converges and $X[z] = \dfrac{1}{1-2z}$

$$2z < 1 \text{ or } |z| < \left|\frac{1}{2}\right|$$

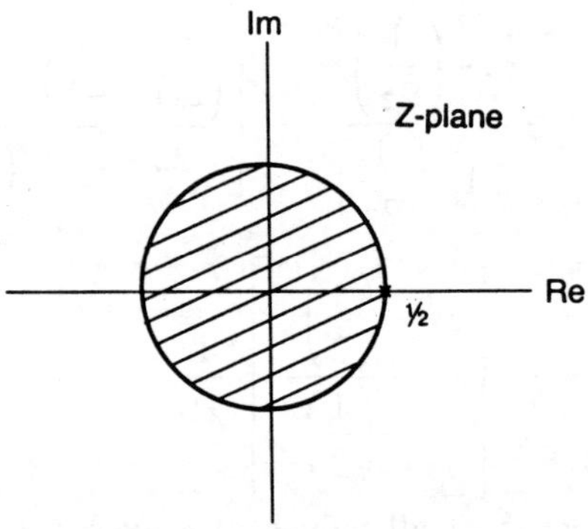

Fig. P. 7.1 (c)

(*g*) $x[n] = 2^n u[-n-1]$

Its *z*-transform is given by

$$x[n] \xleftrightarrow{z} \sum_{n=-\infty}^{+\infty} 2^n u[-n-1]\, z^{-n}$$

$$X(z) = \sum_{n=-\infty}^{-1} 2^n z^{-n} = \sum_{n=0}^{\infty} 2^{-n} z^{-n} = -1 + \sum_{n=0}^{\infty} (1^{-1}z)^n$$

If $|2^{-1}z| < 1$, or equivalently, $|z| < |2|$,
the of above eqn converges and

$$X(z) = \frac{1}{1-\frac{z}{2}} - 1 = \frac{2}{2-z} - 1 = \frac{z}{2-z} \; |z| < |2|$$

(*h*) $x[n] = \left(\dfrac{1}{2}\right)^{|n|}$

The *z*-transform for the sequence can be obtained by expressing it as the sum of a right sided and a left-sided sequence

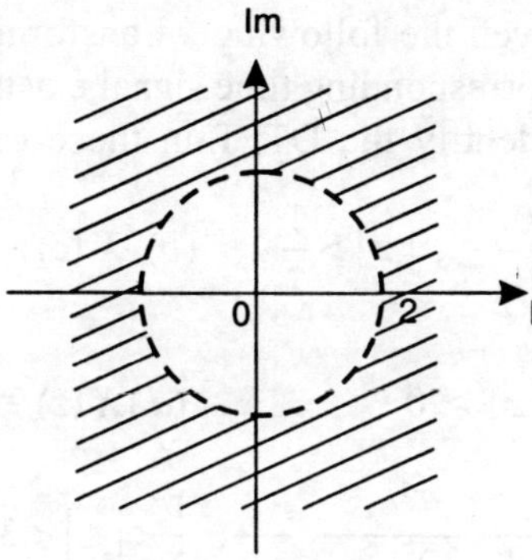

Fig. P. 7.1 (d)

$$x[n] = \left(\frac{1}{2}\right)^n u[n] + \left(\frac{1}{2}\right)^{-n} u[-n-1] \xleftrightarrow{z} \left(\frac{1}{2}\right)^n u(n)$$

$$\xleftrightarrow{z} \frac{1}{1-\frac{1}{2}z^{-1}}, \; |z| > \frac{1}{2}$$

$$\text{also} \left(\frac{1}{2}\right)^n u[-n-1] \xleftrightarrow{z} \frac{-1}{1-\left(\frac{z}{2}\right)^{-1}} |z| < 2$$

$$X(z) = \frac{1}{1-\left(\frac{1}{2}\right)z^{-1}} - \frac{1}{1-2z^{-1}} \cdot \frac{1}{2} < |z| < 2$$

$$= -\frac{3}{2} \cdot \frac{z}{\left(z-\frac{1}{2}\right)(z-2)}, \frac{1}{2} < |z| < 2$$

Pole-zero and ROC

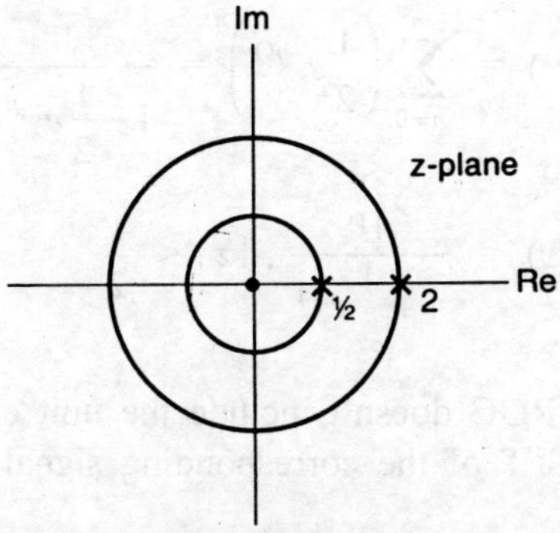

Fig. P. 7.1 (e)

Problem 7.2: Given the following z-transforms, determine whether the DTFT of the corresponding time signal exists without determining the time signal. Identify the DTFT in those cases where it exists.

(*a*) $X(z) = \dfrac{10}{1+\frac{1}{2}z^{-1}}, \; |z| > \dfrac{1}{2}$ (*b*) $X(z) = \dfrac{10}{1+\frac{1}{2}z^{-1}}, \; |z| < \dfrac{1}{2}$

(*c*) $X(z) = z^{-5}, \; |z| > 0$ (*d*) $X(z) = z^{5}, \; |z| < \infty$

(*e*) $X(z) = \dfrac{z^{-1}}{\left(1-\frac{1}{3}z^{-1}\right)\left(1+3z^{-1}\right)}, \; \dfrac{1}{3} < |z| < 3$

Solution:

(*a*)
$$X(z) = \frac{10}{1+\frac{1}{2}z^{-1}}, \; |z| > \frac{1}{2}$$

Here the ROC includes the unit circle in the z plane, thus the DTFT of the corresponding time signal exists.

Also, since the pole is inside the unit circle thus the right–sided inverse z-transform is chosen.

$$\therefore \quad x[n] = \left(\frac{1}{2}\right)^n u[n]$$

$$\text{Thus,} \quad X(e^{j\Omega}) = \sum_{n=-\infty}^{\infty} \left(\frac{1}{2}\right)^n u[n]\, e^{-j\Omega n} = \sum_{n=-\infty}^{\infty} \left(\frac{1}{2}\right)^n e^{-j\Omega n}$$

$$X(e^{j\Omega}) = \sum_{n=0}^{\infty} \left(\frac{1}{2}e^{-j\Omega}\right)^n = \frac{1}{1-\frac{1}{2}e^{-j\Omega}}$$

(*b*)
$$X(z) = \frac{10}{1+\frac{1}{2}z^{-1}}, \; |z| < \frac{1}{2}$$

In this case ROC doesn't include the unit circle in the z-plane Thus the DTFT of the corresponding signal doesn't exist.

(*c*) $X(z) = z^{-5}, |z| > 0$

This z-transform is well defined except at $z = 0$. Thus the ROC consists of the entire z-plane including $z = \infty$ but excluding $z = 0$.

Since in this case ROC include the unit circle in the z plane, thus, the DTFT of corresponding signal exist.

$$x[n] = \delta[n-5]$$

$$\therefore \quad X(e^{j\Omega}) = \sum_{n=-\infty}^{\infty} \delta[n-5]\, e^{-j\Omega n} = 1 \cdot e^{-5j\Omega}$$

(*d*) $X(z) = z^{5}, |z| < \infty$

This well defined for all finite values of z, thus the ROC consists of the entire finite z-plane (including $Z = 0$).

Since ROC includes the unit circle in the Z plane. Thus the DTFT of the corresponding signal exist.

$$x[n] = \delta[n+5] \xleftrightarrow{Z} z^5$$

$$\therefore \quad X(e^{+j\Omega}) = \sum_{n=-\infty}^{\infty} \delta[n+5]\, e^{-j\Omega n} = 1 \cdot e^{j5\Omega}$$

(*e*) $X(z) = \dfrac{z^{-1}}{\left(1-\frac{1}{3}z^{-1}\right)\left(1+3z^{-1}\right)}, \dfrac{1}{3} < |Z| < 3 = \dfrac{1}{\left(z-\frac{1}{3}\right)(z+3)}$

Problem 7.3: Use the tables of z-transforms and z-transform properties to determine the z-transforms of the following signals:

(*a*) $x[n] = \left(\frac{1}{2}\right)^n u[n] * \left(\frac{1}{3}\right)^n u[n]$

(*b*) $x[n] = n\left(\left(\frac{1}{2}\right)^n u[n] * \left(\frac{1}{2}\right)^n u[n]\right)$

(*c*) $x[n] = u[-n]$

(*d*) $x[n] = \sin\left(\frac{\pi}{8}n - \frac{\pi}{4}\right) u[n-2]$

Solution:

(*a*) $$x[n] = \left(\frac{1}{2}\right)^n u[n] * \left(\frac{1}{3}\right)^n u[n]$$

Let $$y_1[n] = \left(\frac{1}{2}\right)^n u[n], \quad y_2[n] = \left(\frac{1}{3}\right)^n$$

then $$Y_1(z) = \frac{1}{1-\left(\frac{1}{2}\right)z^{-1}}$$

$$Y_2(z) = \frac{1}{1-\left(\frac{1}{3}\right)z^{-1}}$$

We know $x[n] * y[n] \longleftrightarrow X(z)\, Y(z)$

Thus $$X(z) = \frac{1}{\left(1-\left(\frac{1}{2}\right)z^{-1}\right)\left(1-\left(\frac{1}{3}\right)z^{-1}\right)}$$

(*b*) $$x[n] = n\left[\left(\frac{1}{2}\right)^n u[n] * \left(\frac{1}{2}\right)^n u[n]\right]$$

Let $$y[n] = \left(\frac{1}{2}\right)^n u[n] * \left(\frac{1}{2}\right)^n u[n]$$

$$\therefore\ y[n] \longleftrightarrow \frac{1}{\left[1-\left(\frac{1}{2}\right)z^{-1}\right]^2}$$

$$x[n] = ny[n]$$

$$X(z) = -z\frac{d}{dz}\left(\frac{1}{\left[1-\left(\frac{1}{2}\right)z^{-1}\right]^2}\right) = -z \cdot \frac{d}{dz}\left(\frac{4z^2}{(2z-1)^2}\right)$$

$$= -z \cdot \left[\frac{(2z-1)^2(8z) - 4z^2(2\cdot(2z-1).2)}{(2z-1)^4}\right]$$

$$= \frac{-z\left(-48z^2+8z\right)}{(2z-1)^4} = \frac{8z^2(6z-1)}{(2z-1)^4}$$

(*c*) $x[n] = u[-n]$

From time reversal property $x[-n] \longleftrightarrow X(z^{-1})$

Thus $u[-n] \longleftrightarrow \frac{1}{1-\frac{1}{z^{-1}}} = \frac{1}{1-z}$

(*d*) $x[n] = \sin\left(\frac{\pi}{8}n - \frac{\pi}{4}\right) u[n-2]$

$$x[n] = \sin\frac{\pi}{8}(n-2)\, u[n-2]$$

From time shifting property we have

$$x[n-n_o] \longleftrightarrow Z^{n_o}X(z)$$

And we have

$$[\sin w_o n]\, u[n] \longleftrightarrow \frac{(\sin w_o)\, z^{-1}}{1-[2\cos w_o]\, z^{-1} + z^{-2}}$$

$$\therefore\ \sin\frac{\pi}{8}(n-2)\, u[n-2] \longleftrightarrow z^{-2}\left[\frac{\left(\sin\frac{\pi}{8}\right) z^{-1}}{1-\left(2\cos\frac{\pi}{8}\right) z^{-1} + z^{-2}}\right]$$

$$X(z) = \frac{z^{-3}\cdot\sin\frac{\pi}{8}}{1-\left(2\cos\frac{\pi}{8}\right) z^{-1} + z^{-2}}$$

Problem 7.4: Given the *z*-transform pair $x\{n\} \xleftrightarrow{z} z/(z^2+4)$ with ROC $|z| < 2$, use the *z*-transform properties to determine the *z*-transform of the follwing signals:

(*a*) $y[n] = x[n-4]$ (*b*) $y[n] = 2^n x[n]$

(*c*) $y[n] = x[-n]$ (*d*) $y[n] = nx[n]$

(*e*) $y[n] = x[n+1] + x[n-1]$ (*f*) $y[n] = \underbrace{x[n] * x[n] * \ldots * x[n]}_{m \text{ times}}$

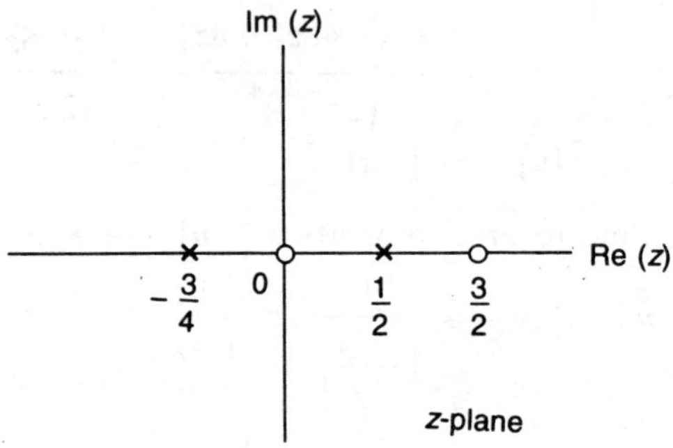

Fig. P. 7.4 (a)

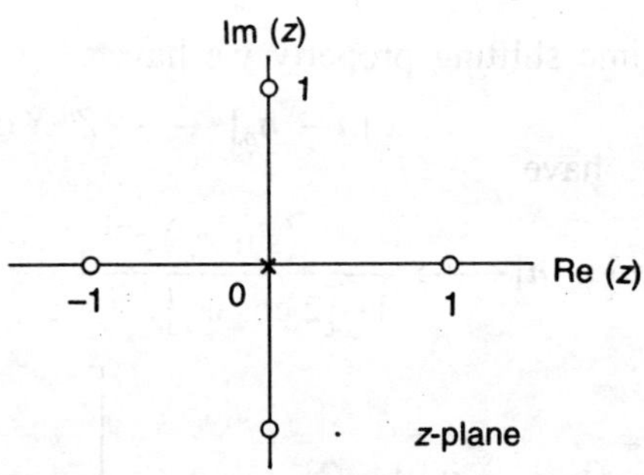

Fig. P. 7.4 (b)

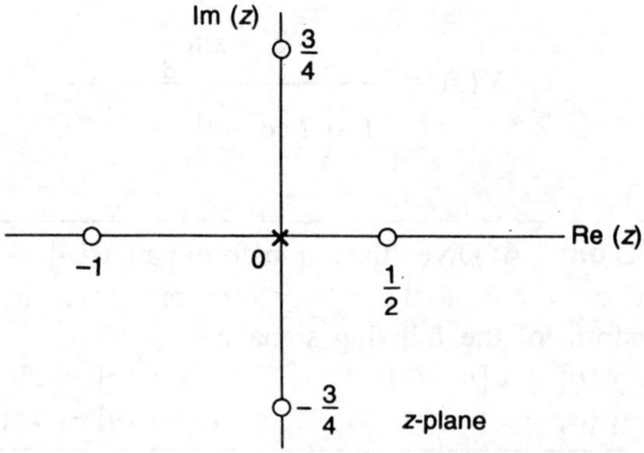

Fig. P. 7.4 (c)

Solution:

(*a*) $y[n] = x[n-4]$

given $x[n] \overset{z}{\longleftrightarrow} \dfrac{z}{(z^2+4)}$

Using time shifting property

$$x[n-4] \longleftrightarrow \frac{z^{-4} \cdot z}{(z^4+4)} = \frac{z^{-3}}{(z^2+4)}$$

(*b*) $y[n] = 2^n x[n]$

We have from scaling property

$$a^n x[n] \longleftrightarrow X(a^{-1}z)$$

Also we have,

$$x[n] \overset{z}{\longleftrightarrow} \frac{z}{z^2+4}, \text{ thus } Y(z) = \frac{\frac{z}{2}}{\left(\frac{z}{2}\right)^2+4} = \frac{2z}{z^2+16}$$

(*c*) $y[n] = x[-n]$

We have from Time reversal property

$$x[-n] \overset{z}{\longleftrightarrow} X(z^{-1})$$

Thus, $$Y(z) = \frac{z^{-1}}{z^{-2}+4} \left[\begin{array}{c}\text{because} \\ x[n] \overset{z}{\longleftrightarrow} \dfrac{z}{(z^2+4)} = \dfrac{z}{1+4z^2}\end{array}\right]$$

(*d*) $y[n] = nx[n]$

From the property of differentiation in the z–domain

$$nx[n] \overset{z}{\longleftrightarrow} -z\frac{dX(z)}{dz}$$

and we have,

$$x[n] \overset{z}{\longleftrightarrow} \frac{z}{(z^2+4)}$$

Thus, $$Y(z) = -z \cdot \frac{d}{dz}\left(\frac{z}{z^2+4}\right) = -z\left[\frac{z^2+4-z(2z)}{(z^2+4)^2}\right]$$

$$= -z\left[\frac{-z^2+4}{\left(z^2+4\right)^2}\right] = \frac{z(z^2-4)}{(z^2+4)^2}$$

(*e*) $y[n] = x[n+1] + x[n-1]$

Let $y_1[n] = x[n+1],\ y_2[n] = x[n-1]$

Then from time shifting property

$$Y_1(z) = z^1 \cdot X(z)$$

We have

$$X(z) = \frac{z}{\left(z^2+4\right)} \qquad Y_1(z) = \frac{z^2}{\left(z^2+4\right)} \qquad Y_2(z) = \frac{z^{-1}\cdot z}{z^2+4}$$

Thus from Linearity property

$$Y(z) = \frac{z^2}{z^2+4} + \frac{1}{z^2+4} = \frac{z^2+1}{z^2+4}$$

(*f*) $y[n] = \underbrace{x[n]*x[n]*x[n]*\ldots*x[n]}_{m \text{ times}}$

From the convolution property

$$X(z) = \underbrace{X(z)\cdot X(z)\ldots X(z)}_{m \text{ times}}$$

given $X(z) = \dfrac{z}{\left(z^2+4\right)}$

$$\therefore \qquad Y(z) = \left[\frac{z}{\left(z^2+4\right)}\right]^m$$

(*g*) $y[n] = (n-3)\,x[n-2]$

$= nx[n-2] - 3\cdot x[n-2]$

Let $y_1[n] = x[n-2]$

Using time shifting property

$$Y_1(z) = \frac{z^{-2}\cdot z}{\left(z^2+4\right)} = \frac{z^{-1}}{\left(z^2+4\right)}$$

$$\therefore \quad ny_1[n] = -z\frac{dY_1(z)}{dz} = -z\left[\frac{-1(3z^2+4)}{\left[z(z^2+4)\right]^2}\right]$$

$$= \frac{3z^3-4z}{z^2(z^2+4)^2} = \frac{3z^2-4}{z^2(z^2+4)^2}$$

and, $3x[n-2] \longleftrightarrow \dfrac{3\cdot z^{-1}}{(z^2+4)^2}$

$\therefore$ Using Linearity property

$$Y(z) = \frac{3z^2-4}{z(z^2+4)^2} - \frac{3}{z(z^2+4)} = \frac{2z^2-4-3z^2-12}{z(z^2+4)^2}$$

$$Y(z) = \frac{-16}{z(z^2+4)^2}$$

Problem 7.5: Given the z-transform pair $3^n u[n] \overset{z}{\longleftrightarrow} X(z)$, use the z-transform properties to determine the time-domain signals corresponding to the following z-transforms:

(*a*) $Y(z) = X(3z)$ (*b*) $Y(z) = X(z^{-1})$

(*c*) $Y(z) = \dfrac{d}{dz}X(z)$ (*d*) $Y(z) = \dfrac{z+z^{-1}}{2}X(z)$

(*e*) $Y(z) = X(z)\,X(9z)$

Solution:

(*a*) $Y(z) = X(3z)$

we have $3^n u[n] \overset{z}{\longleftrightarrow} X(z)$

$x[n] = 3^n u[n]$

we have,

$$a^n x[n] \overset{z}{\longleftrightarrow} X(a^{-1}z)$$

Thus $y[n] = \left(\dfrac{1}{3}\right)^n x[n] = \left(\dfrac{1}{3}\right)^n \cdot 3^n u[n] = u[n]$

(*b*) $$Y(z) = X(z^{-1})$$

From the time reversal property

$$x[-n] \overset{z}{\longleftrightarrow} X(z^{-1})$$

$$\therefore \quad y[n] = x[-n]$$

and we have

$$x[n] = 3^n u[n]$$

$$\therefore \quad y[n] = 3^n u[-n]$$

(*c*) $$Y(z) = \frac{d}{dz} X(z)$$

or, $$Y(z) = -\left[\frac{-z}{z}\frac{d}{dz} X(z)\right]$$

Then by using time shifting property and differentiation in the z domain

$$y[n] = -nx[n-1]$$

we have

$$x[n] = 3^n u[n] = -n \cdot 3^{(n-1)} u[n-1]$$

or $$y[n] = -\frac{n}{3} \cdot 3^n u[n-1]$$

(*d*) $$Y(z) = \frac{z+z^{-1}}{2} X(z) = \frac{z}{2} X(z) + \frac{z^{-1}}{2} X(z)$$

Thus from time shifting property

$$\frac{x[n+1]}{2} \longleftrightarrow \frac{z}{2} X(z)$$

$$\frac{x[n-1]}{2} \longleftrightarrow \frac{z^{-1}}{z} X(z)$$

By using Linearity property

$$y[n] = \frac{1}{2}[x[n+1] + x[n-1]]$$

We have,

$$x[n] = 3^n u[n]$$

$$y[n] = \frac{1}{2}[3^{n+1} u[n+1] + 3^{n-1} u[n-1]]$$

$$= \frac{3^n}{2}\left[3\,u[n+1] + \frac{1}{3}u[n-1]\right]$$

(*e*) $$Y(z) = X(z)\,X(9z)$$

we have,

$$3^n u[n] \longleftrightarrow X(z)$$

From scaling property

$$\left(\frac{1}{9}\right)^n 3^n u[n] \longleftrightarrow X(9z)$$

$$\left(\frac{3}{9}\right)^n u[n] \longleftrightarrow X(9z)$$

$$\left(\frac{1}{3}\right)^n u[n] \longleftrightarrow X(9z)$$

Now, from property of convolution

$$y[n] = 3^n u[n] * \left(\frac{1}{3}\right)^n u[n]$$

Problem 7.6: Use the method of partial fractions to obtain the time-domain signals corresponding to the following *z*-transforms:

(*a*) $$X(z) = \frac{\frac{1}{4}z^{-1}}{\left(1-\frac{1}{2}z^{-1}\right)\left(1-\frac{1}{4}z^{-1}\right)},\quad |z| > \frac{1}{2}$$

(*b*) $$X(z) = \frac{\frac{1}{4}z^{-1}}{\left(1-\frac{1}{2}z^{-1}\right)\left(1-\frac{1}{4}z^{-1}\right)},\quad |z| > \frac{1}{4}$$

(c) $$X(z) = \frac{\frac{1}{4}z^{-1}}{\left(1-\frac{1}{2}z^{-1}\right)\left(1-\frac{1}{4}z^{-1}\right)},\quad \frac{1}{4} < |z| < \frac{1}{2}$$

(d) $X(z) = \dfrac{z^2 - 3z}{z^2 + \frac{3}{2}z - 1}, \quad \frac{1}{2} < |z| < 2$

(e) $X(z) = \dfrac{12(11z^2 - 3z)}{12z^2 - 7z + 1}, \quad |z| > \frac{1}{3}$

(f) $X(z) = \dfrac{8z^2 + 4z}{4z^2 - 4z + 1}, \quad |z| > \frac{1}{2}$

(g) $X(z) = \dfrac{z^3 + z^2 + \frac{3}{2}z + \frac{1}{2}}{z^3 + \frac{3}{2}z^2 + \frac{1}{2}z}, \quad |z| < \frac{1}{2}$

(h) $X(z) = \dfrac{z^3 + z^2 + \frac{3}{2}z + \frac{1}{2}}{z^2 + \frac{3}{2}z + \frac{1}{2}}, \quad |z| > 1$

(i) $X(z) = \dfrac{2z^4 - 2z^3 - 2z^2}{z^2 - 1}, \quad |z| > 1$

Solution:

(a)
$$X(z) = \frac{\frac{1}{4}z^{-1}}{\left(1 - \frac{1}{2}z^{-1}\right)\left(1 - \frac{1}{4}z^{-1}\right)}, \quad |z| > \frac{1}{2}$$

Use a partial fraction expansion to write

$$X(z) = \frac{A}{\left(1 - \frac{1}{2}z^{-1}\right)} + \frac{B}{\left(1 - \frac{1}{4}z^{-1}\right)}$$

Solving for A and B gives

$$X(z) = \frac{1}{\left(1 - \frac{1}{2}z^{-1}\right)} - \frac{1}{\left(1 - \frac{1}{4}z^{-1}\right)}$$

Now, we find the inverse z-transform of each term using the relationship between the pole location and the ROC of $X(z)$.

The ROC has a radius greater than the pole at $z = \frac{1}{2}$, so this term has the right-sided inverse transform

$$\left(\frac{1}{2}\right)^n u[n] \xleftrightarrow{z} \frac{1}{1-\frac{1}{2}z^{-1}}$$

The ROC has a radius greater than the pole at $z = \frac{1}{4}$, so this term has the right-sided inverse z transform

$$-\left(\frac{1}{4}\right)^n u[n] \longleftrightarrow \frac{1}{1-\frac{1}{4}z^{-1}}$$

Combining the individual terms gives

$$x[n] = \left(\frac{1}{2}\right)^n u[n] - \left(\frac{1}{4}\right)^n u[n]$$

(*b*)
$$X(z) = \frac{\frac{1}{4}z^{-1}}{\left(1-\frac{1}{2}z^{-1}\right)\left(1-\frac{1}{4}z^{-1}\right)}, \ |z| < \frac{1}{4}$$

Use a partial fraction expansion to write

$$X(z) = \frac{A}{\left(1-\frac{1}{2}z^{-1}\right)} + \frac{B}{\left(1-\frac{1}{4}z^{-1}\right)}$$

solving for A and B gives

$$X(z) = \frac{1}{\left(1-\frac{1}{2}z^{-1}\right)} - \frac{1}{\left(1-\frac{1}{4}z^{-1}\right)}$$

The ROC has a radius less than the pole at $z = \frac{1}{2}$, so this term has the left-sided inverse transform

$$-\left(\frac{1}{2}\right)^n u[-n-1] \longleftrightarrow \frac{1}{1-\frac{1}{2}z^{-1}}$$

The ROC has a radius less than the pole at $z = \frac{1}{4}$, so this term has the left-sided inverse transform

$$-\left[-\left(\frac{1}{4}\right)^n u[-n-1]\right] \longleftrightarrow -\frac{1}{1-\frac{1}{4}z^{-1}}$$

Combining the individual terms gives

$$x[n] = \left(\frac{1}{4}\right)^n u[-n-1] - \left(\frac{1}{2}\right)^n u[-n-1]$$

(c) $$X(z) = \frac{\frac{1}{4}z^{-1}}{\left(1-\frac{1}{2}z^{-1}\right)\left(1-\frac{1}{4}z^{-1}\right)}, \quad \frac{1}{4} < |z| < \frac{1}{2}$$

use a partial fraction expansion to write

$$X(z) = \frac{A}{\left(1-\frac{1}{2}z^{-1}\right)} + \frac{B}{\left(1-\frac{1}{4}z^{-1}\right)}$$

solving for A and B gives

$$X(z) = \frac{1}{1-\frac{1}{2}z^{-1}} - \frac{1}{1-\frac{1}{4}z^{-1}}$$

The ROC has a radius less than the pole at $z = \frac{1}{2}$, so this term has the left sided inverse transform

$$-\left(\frac{1}{2}\right)^n u[-n-1] \longleftrightarrow -\frac{1}{1-\frac{1}{4}z^{-1}}$$

The ROC has a radius greater than the pole at $z = \frac{1}{4}$, so this term has the right-sided inverse z transform.

$$-\left(\frac{1}{4}\right)^n u[n] \longleftrightarrow -\frac{1}{1-\frac{1}{4}z^{-1}}$$

combining the individual term gives

$$x[n] = -\left(\frac{1}{2}\right)^n u[-n-1] - \left(\frac{1}{4}\right)^n u[n]$$

(*d*) $$X(z) = \frac{z^2 - 3z}{z^2 + \frac{3}{2}z - 1}, \frac{1}{2} < |z| < 2$$

or, $$X(z) = \frac{1-3z^{-1}}{1+\frac{3}{2}z^{-1} - z^{-2}}$$

Now, using a partial fraction expansion,

$$\frac{1-3z^{-1}}{1+\frac{3}{2}z^{-1} - z^{-2}} = \frac{-2}{\left(z^{-1}-2\right)} - \frac{1}{\left(z^{-1}+\frac{1}{2}\right)} = \frac{1}{1-\frac{1}{2}z^{-1}} - \frac{2}{\left(1+2z^{-1}\right)}$$

The ROC has a radius less than one pole and greater than the other pole.

Thus, $$\left(\frac{1}{2}\right)^n u[n] \longleftrightarrow \frac{1}{1-\frac{1}{4}z^{-1}}$$

also $$-2 \cdot (-2^n)\, u[-n-1] \longleftrightarrow \frac{2}{\left(1+2z^{-1}\right)}$$

combining the individual terms

$$x[n] = \left(\frac{1}{2}\right)^n u[n] + 2 \cdot (-2)^n\, u[-n-1]$$

(*e*) $$X(z) = \frac{12\left(11z^2 - 3z\right)}{12z^2 - 7z + 1}, \quad |z| > \frac{1}{3}$$

$$X(z) = \frac{12\left(11-3z^{-1}\right)}{12-7z^{-1}+z^{-2}} = \frac{12\left(11-3z^{-1}\right)}{\left(z^{-1}-4\right)\left(z^{-1}-3\right)}$$

using a partial fraction expansion,

$$\frac{12\left(11-3z^{-1}\right)}{\left(z^{-1}-4\right)\left(z^{-1}-3\right)} = \frac{12}{\left(4-z^{-1}\right)}+\frac{24}{\left(3-z^{-1}\right)}$$

$$= \frac{3}{\left(1-\frac{1}{4}z^{-1}\right)}+\frac{8}{\left(1-\frac{1}{3}z^{-1}\right)}$$

The ROC has a greater radius than either pole and so the inverse z-transform of $X(z)$ is

$$x[n] = 3\left(\frac{1}{4}\right)^n u[n] + 8\left(\frac{1}{3}\right)^n u[n]$$

(*f*) $$X(z) = \frac{8z^2+4z}{4z^2-4z+1}, \; |z| > \frac{1}{2}$$

or, $$X(z) = \frac{8+4z^{-1}}{4-4z^{-1}+z^{-2}} = \frac{8+4z^{-1}}{\left(2-z^{-1}\right)^2}$$

$$= \frac{8}{\left(2-z^{-1}\right)^2}+\frac{4z^{-1}}{\left(2-z^{-1}\right)^2} = \frac{4}{\left(1-\frac{1}{2}z^{-1}\right)^2}+\frac{z^{-1}}{\left(1-\frac{1}{2}z^{-1}\right)^2}$$

$$= \frac{z\cdot 8\cdot\frac{1}{2}z^{-1}}{\left(1-\frac{1}{2}z^{-1}\right)^2}+\frac{2\cdot\frac{1}{2}z^{-1}}{\left(1-\frac{1}{2}z^{-1}\right)^2}$$

Let $$w(z) = \frac{\frac{1}{2}z^{-1}}{\left(1-\frac{1}{2}z^{-1}\right)^2}$$

we have,

$$na^n u[n] \xleftrightarrow{z} -z\frac{d}{dz}\left(\frac{1}{1-az^{-1}}\right) = \frac{az^{-1}}{\left(1-az^{-1}\right)^2}, \ |z| > |a|$$

$$\therefore \qquad w[n] = n\left(\frac{1}{2}\right)^n u[n]$$

Also, $\quad X(z) = 8zw(z) + 2w(z)$

The time shift property indicates that

$$x[n] = 8w[n+1] + 2w[n]$$

Thus

$$x[n] = 8(n+1)\left(\frac{1}{2}\right)^{n+1} u[n+1] + 2n\left(\frac{1}{2}\right)^n u[n]$$

(g) $$X(z) = \frac{z^3 + z^2 + \frac{3}{2}z + \frac{1}{2}}{z^3 + \frac{3}{2}z^2 + \frac{1}{2}z}$$

or, $$X(z) = \frac{1 + z^{-1} + \frac{3}{2}z^{-2} + \frac{1}{2}z^{-3}}{1 + \frac{3}{2}z^{-1} + \frac{1}{2}z^{-2}}$$

Using long division to reduce the order of the numerator polynomial, we have

$$\begin{array}{r|l} & z^{-1} \\ \hline \frac{1}{2}z^{-2} + \frac{3}{2}z^{-1} + 1 & \frac{1}{2}z^{-3} + \frac{3}{2}z^{-2} + z^{-1} + 1 \\ & \frac{1}{2}z^{-3} + \frac{3}{2}z^{-2} + z^{-1} \\ \hline & \qquad\qquad\qquad\qquad 1 \end{array}$$

Thus we may write

$$\frac{1 + z^{-1} + \frac{3}{2}z^{-2} + \frac{1}{2}z^{-3}}{1 + \frac{3}{2}z^{-1} + \frac{1}{2}z^{-2}} = z^{-1} + \frac{1}{\left(1+z^{-1}\right)\left(\frac{1}{2}z^{-1}+1\right)}$$

Next using a partial fraction expansion, we write

$$\frac{1}{\left(1+z^{-1}\right)\left(1+\frac{1}{2}z^{-1}\right)} = \frac{2}{\left(1+z^{-1}\right)} - \frac{1}{\left(\frac{1}{2}z^{-1}+1\right)}$$

and thus,

$$X(z) = z^{-1} + \frac{2}{\left(1+z^{-1}\right)} - \frac{1}{\left(1+\frac{1}{2}z^{-1}\right)}$$

The ROC has smaller radius than either pole and so the inverse z-transform of $X(z)$ is

$$x[n] = \delta[n-1] - 2(-1)^n u[-n-1] + \left(\frac{1}{2}\right)^n u[-n-1]$$

(*h*) $$X(z) = \frac{z^3 + z^2 + \frac{3}{2}z + \frac{1}{2}}{z^2 + \frac{3}{2}z + \frac{1}{2}}, \ |z| > 1$$

convert $X(z)$ into a ratio of polynomials in z^{-1}.

We do this by factoring z^3 from the numerator and z^2 from the denominator, yielding

$$X(z) = z\left[\frac{1+z^{-1}+\frac{3}{2}z^{-2}+\frac{1}{2}z^{-3}}{1+\frac{3}{2}z^{-1}+\frac{1}{2}z^{-2}}\right]$$

Using long division to reduce the order of the numerator polynomial, we have

$$\begin{array}{r|l} & z^{-1} \\ \hline \frac{1}{2}z^{-2}+\frac{3}{2}z^{-1}+1 & \frac{1}{2}z^{-3}+\frac{3}{2}z^{-2}+z^{-1}+1 \\ & \underline{\frac{1}{2}z^{-3}+\frac{3}{2}z^{-2}+z^{-1}} \\ & 1 \end{array}$$

Thus we may write

$$\left(\frac{1+z^{-1}+\frac{3}{2}z^{-2}+\frac{1}{2}z^{-3}}{1+\frac{3}{2}z^{-1}+\frac{1}{2}z^{-2}}\right) = z^{-1} + \frac{1}{1+\frac{3}{2}z^{-1}+\frac{1}{2}z^{-2}}$$

Next, using a partial fraction expansion,

$$\frac{1}{1+\frac{3}{2}z^{-1}+\frac{1}{2}z^{-2}} = \frac{1}{\left(1+z^{-1}\right)\left(\frac{1}{2}z^{-1}+1\right)}$$

$$= \frac{2}{\left(1+z^{-1}\right)} - \frac{1}{\left(1+\frac{1}{2}z^{-1}\right)}$$

and thus define

$$X(z) = zw(z)$$

where $w(z) = z^{-1} + \dfrac{2}{\left(1+z^{-1}\right)} - \dfrac{1}{\left(1+\frac{1}{2}z^{-1}\right)}$

The ROC has a greater radius than either pole and so the inverse z–transform of $w(z)$ is

$$w[n] = \delta[n-1] + 2(-1)^n u[n] - 1 \cdot \left(\frac{-1}{2}\right)^n u[n]$$

The time-shift property indicates that

$$x[n] = w[n+1]$$

and thus

$$x[n] = \delta[n] + 2 \cdot (-1)^{n+1} u[n+1] - \left(-\frac{1}{2}\right)^{n+1} u[n+1]$$

(*i*) $$X(z) = \frac{2z^4 - 2z^3 - 2z^2}{z^2 - 1}, \ |z| > 1$$

Convert $X(z)$ into a ratio of polynomials in z^{-1}. We do this by factoring $2z^4$ from the numerator and z^2 from the denominator,

$$X(z) = 2z^2\left(\frac{1 - z^{-1} - z^{-2}}{1 - z^{-1}}\right)$$

Using long division to reduce the order of the numerator polynomial

$$\begin{array}{r} z^{-1} + 2 \\ -z^{-1} + 1 \overline{\big) -z^{-2} - z^{-1} + 1} \\ \underline{-z^{-2} + z^{-1} } \\ -2z^{-1} + 1 \\ \underline{-2z^{-1} + 2} \\ -1 \end{array}$$

Thus, we may write

$$\frac{1-z^{-1}-z^{-2}}{1-z^{-1}} = z^{-1}+2-\frac{1}{1-z^{-1}}$$

and thus define

$$X(z) = 2z^2 w(z)$$

where,

$$w(z) = z^{-1} + 2 - \frac{1}{1-z^{-1}}, \text{ with } |z| > 1$$

The ROC has a greater radius than pole, so, the inverse z-transform of $w(z)$ is

$$w[n] = \delta[n-1] + 2\delta[n] - (1)^n u[n]$$

The time shift property indicates that

$$x[n] = 2w[n+2]$$

and thus

$$x[n] = 2\delta[n+1] + 4\delta[n+2] - 2(1)^{n+2} u[n+2]$$

Problem 7.7: Determine the time-domain signals corresponding to the following z-transforms:
(*a*) $X(z) = 1 + 2z^{-2} + 4z^{-4}, |z| > 0$
(*b*) $X(z) = \sum_{k=5}^{10}(1/k)z^{-k}, |z| > 0$
(*c*) $X(z) = (1 + z^{-1})^4, |z| > 0$

Solution:

(*a*) $$X(z) = 1 + 2z^{-2} + 4z^{-4}, |z| > 0$$

we have,

$$\delta[n-j] \xleftrightarrow{z} z^{-j}$$

$$\therefore \quad x[n] = \delta[n] + 2\delta[n-2] + 4\delta[n-4]$$

(*b*) $$X(z) = \sum_{k=5}^{10}\left(\frac{1}{k}\right) z^{-k}$$

$$= \frac{1}{5}z^{-5} + \frac{1}{6}z^{-6} + \frac{1}{7}z^{-7} + \frac{1}{8}z^{-8} + \frac{1}{9}z^{-9} + \frac{1}{10}z^{-10}$$

$$= \frac{\delta[n-5]}{5} + \frac{\delta[n-6]}{6} + \frac{\delta[n-7]}{7} + \frac{\delta[n-8]}{8} + \frac{\delta[n-9]}{9}$$

$$+ \frac{\delta[n-10]}{10}$$

(*c*) $X(z) = z^4 + 2z^2 + 3 + 2z^{-2} + z^{-4},\ 0 < |z| < \infty$

$$= \delta[n+4] + 2\delta[n+2] + 3\delta[n] + 2[n-2] + \delta[n-4]$$

Problem 7.8: Determine the impulse response corresponding tot he following transfer functions if

(*i*) the system is stable; (*ii*) the system is causal:

(*a*) $H(z) = \dfrac{3z^{-1}}{\left(1-2z^{-1}\right)^2}$ (*b*) $H(z) = \dfrac{4z}{z^2 - \frac{1}{2}z + \frac{1}{16}}$

Solution:

(*a*) $$H(z) = \frac{3z^{-1}}{\left(1-2z^{-1}\right)^2}$$

This system has two poles at $z = 2$ if the system is stable, then the ROC includes the unit circle, thus the pole outside the unit circle contributes a left sided term. Hence,

$$H(z) = \frac{3}{2} \cdot \frac{2z^{-1}}{\left(1-2z^{-1}\right)^2} = -\frac{3}{2} \cdot n \cdot (2)^n\, u[-n-1]$$

If the system is causal, then all the poles contribute right-sided terms to the impulse response

$$h[n] = \frac{3}{2} n\,(2)^n\, u[n]$$

* This system cannot be both stable and causal since there is a pole outside the unit circle.

(*b*) $$H(z) = \frac{4z}{z^2 - \frac{z}{2} + \frac{1}{16}} = \frac{z \cdot 4}{1 - \frac{z^{-1}}{2} + \frac{z^{-2}}{16}} = z \cdot \frac{4}{\left(1 - \frac{1}{4}z^{-1}\right)^2}$$

This system has poles at

$$z = \frac{1}{4}, \frac{1}{4}$$

If the system is stable, then ROC includes the unit circle. The two poles inside the unit circle contribute right sided terms to the impulse response.

$$H(z) = z^2 \cdot 4 \cdot 4 \left[\frac{\frac{1}{4}z^{-1}}{\left(1 - \frac{1}{4}z^{-1}\right)^2} \right]$$

Let $$X(z) = \frac{\frac{1}{4}z^{-1}}{\left(1 - \frac{1}{4}z^{-1}\right)^2}$$

$$H(z) = 16z^2 X(z)$$

$$h[n] = 16 \cdot [n + 2]$$

$$\therefore \quad h[n] = 16 \cdot (n + 2) \left(\frac{1}{4}\right)^{n+2} u[n + 2]$$

Problem 7.9: use a power series expansion to determine the time-domain signal corresponding to the following *z*-transforms:

(*a*) $X(z) = \dfrac{1}{1 - z^{-2}}, \ |z| > 1$

(*b*) $X(z) = \dfrac{1}{1 - z^{-2}}, \ |z| < 1$

(*c*) $X(z) = \cos(2z), \ |z| < \infty$

(*d*) $X(z) = \ln(1 + z^{-1}), \ |z| > 0$

Solution:

This expression can be expanded in a power series by long division

$$\begin{array}{r} 1+z^{-2}+z^{-4}+- \\ 1-z^{-2}\;\overline{)1\qquad\qquad} \\ \underline{1-z^{-2}} \\ z^{-2} \\ \underline{z^{-2}-z^{-4}} \\ z^{-4} \\ \underline{z^{-4}-z^{-6}} \\ z^{-6} \end{array}$$

$$\frac{1}{1-z^{-2}} = 1 + z^{-2} + z^{-4} + -$$

Thus, $$x[n] = \begin{cases} 0, n < 0 \text{ and odd} \\ u[n], \text{otherwise} \end{cases}$$

(*b*) $$X(z) = \frac{1}{1-z^{-2}}, |z| < 1$$

The ROC of $X(z)$ is specified as $|z| < 1$

We can obtain a convergent power series by long division again.

$$\begin{array}{r} -z^{2}-z^{+4} \\ -z^{-2}+1\;\overline{)1\qquad} \\ \underline{1-z^{+2}} \\ z^{2} \\ \underline{z^{2}-z^{-4}} \\ +z^{-4} \end{array}$$

or, $$\frac{1}{1-z^{-2}} = -z^{2} - z^{4} \ldots$$

In this case

$$[n] = 0;\ n \geq 0$$

then $$x[n] = \begin{cases} 0, n \geq 0 \text{ and odd} \\ -(1)^{n} u[-n-1], \text{othewise} \end{cases}$$

(*c*) $$X(z) = \cos(2z),\ |z| < \infty$$

Using Taylor's series expansion

$$\cos\theta = 1 - \frac{1}{2!}\theta^{2} + \frac{1}{4!}\theta^{4} \ldots$$

also, $$\cos\theta = \frac{1}{2}(e^{j\theta} + e^{-j\theta})$$

Thus

$$\cos(2z) = \frac{1}{2}\left[e^{j2z} + e^{-j2z}\right]$$

$$= \frac{1}{2}\left[\sum_{k=0}^{\infty}\frac{1}{k!}(j2z)^k + \sum_{k=0}^{\infty}\frac{1}{k!}(-j2z)^k\right]$$

$$X(z) = \frac{1}{2}\left[\sum_{k=0}^{\infty}\frac{(j2)^k}{k!}\cdot z^k + \sum_{k=0}^{\infty}\frac{(-2j)^k}{k!}z^k\right]$$

Thus,

$$x[n] = \begin{cases} 0, & n > 0 \\ \frac{1}{2}\left[\frac{(2j)^n}{n!} + \frac{(-2j)^n}{n!}\right] \end{cases}$$

or,

$$x[n] = \begin{cases} 0, & n > 0 \\ \frac{1}{2n!}\left[(2j)^n + (-2j)^n\right], & \text{otherwise} \end{cases}$$

(*d*) $X(z) = \ln(1 + z^{-1}),\ |z| > 0$

using the Taylor's series expansion

$$\log(1 + x) = \sum_{n=1}^{\infty}\frac{(-1)^{n+1}x^n}{n}$$

Thus, $$X(z) = \sum_{n=0}^{\infty}\frac{(-1)^{n+1}z^{-n}}{n}$$

Therefore $$x[n] = \begin{cases} \frac{(-1)^{nH}}{n}, & n \geq 1 \\ 0, & n \leq 0 \end{cases}$$

Problem 7.10: A causal system has input $x[n]$ and output $y[n]$ use the transfer function to determine the impulse response of this system

(*a*) $$x[n] = \delta[n] + \frac{1}{4}\delta[n-1] - \frac{1}{8}\delta[n-2]$$

$$y[n] = \delta[n] - \frac{3}{4}\delta[n-1]$$

Solution:

$$Y(z) = 1 - \frac{3}{4}z^{-1}$$

$$X(z) = 1 + \frac{1}{4}z^{-1} - \frac{1}{8}z^{-2}$$

$$H(z) = \frac{1 - \frac{3}{4}z^{-1}}{1 + \frac{1}{4}z^{-1} - \frac{1}{8}z^{-2}} = \frac{\left(1 - \frac{3}{4}z^{-1}\right)}{\left(1 - \frac{1}{4}z^{-1}\right)\left(1 + \frac{1}{2}z^{-1}\right)}$$

using partial fraction,

$$H(z) = \frac{-2}{3\left(1 - \frac{1}{4}z^{-1}\right)} + \frac{5}{3\left(1 + \frac{1}{2}z^{-1}\right)}$$

Thus, impulse response

$$h[n] = -\frac{2}{3}\left(\frac{1}{4}\right)^n u[n] + \frac{5}{3}\left(-\frac{1}{2}\right)^n u[n]$$

(*b*) $$x[n] = \left(-\frac{1}{3}\right)^n u[n]$$

$$y[n] = 3(-1)^n u[n] + \left(\frac{1}{3}\right)^n u[n]$$

Solution:

Thus, $$X(z) = \frac{1}{1 + \frac{1}{3}z^{-1}}$$

$$X(z) = \frac{3}{1 + z^{-1}} + \frac{1}{3\left(1 - \frac{1}{3}z^{-1}\right)}$$

$$\therefore\ Y(z) = \frac{9 - 3z^{-1} + 1 + z^{-1}}{3\left(1 + z^{-1}\right)\left(1 - \frac{1}{3}z^{-1}\right)} = \frac{10 - 2z^{-1}}{3\left(1 + z^{-1}\right)\left(1 - \frac{1}{3}z^{-1}\right)}$$

Thus, $$H(z) = \frac{Y(z)}{X(z)} = \frac{(10 - 2z^{-1})}{3(1 + z^{-1})\left(1 - \frac{1}{3}z^{-1}\right)\left(1 + \frac{1}{3}z^{-1}\right)}$$

using partial fraction

$$H(z) = \frac{1}{3}\left[\frac{A}{(1 + z^{-1})} + \frac{B}{\left(1 - \frac{1}{3}z^{-1}\right)} + \frac{C}{\left(1 + \frac{1}{3}z^{-1}\right)}\right]$$

Thus, solving for A, B *and* C

$$H(z) = \frac{1}{3}\left[\frac{9}{2(1 + z^{-1})} + \frac{1}{2\left(1 - \frac{1}{3}z^{-1}\right)} - \frac{4}{\left(1 + \frac{1}{3}z^{-1}\right)}\right]$$

Thus impulse response,

$$h[n] = \frac{3}{2}(-1)^n u[n] + \frac{1}{6}\left(-\frac{1}{3}\right)^n u[n] - \frac{4}{3}\left(-\frac{1}{3}\right)^n u[n]$$

(*c*) $$x[n] = (-3)^n u[n]$$

$$y[n] = 4 \cdot (2)^n u[n] - \left(\frac{1}{2}\right)^n u[n]$$

$$X(z) = \frac{1}{1 + 3z^{-1}}$$

$$Y(z) = \frac{4}{1 - 2z^{-1}} - \frac{1}{\left(1 - \frac{1}{2}z^{-1}\right)}$$

$$= \frac{4 - 2z^{-1} - 1 + 2z^{-1}}{(1 - 2z^{-1})\left(1 - \frac{1}{2}z^{-1}\right)} = \frac{3}{(1 - 2z^{-1})\left(1 - \frac{1}{2}z^{-1}\right)}$$

Hence transfer function

$$H(z) = \frac{Y(z)}{X(z)} = \frac{3}{(1 + 3z^{-1})(1 - 2z^{-1})\left(1 - \frac{1}{2}z^{-1}\right)}$$

Using partial fraction,

$$H(z) = \frac{A}{\left(1+3z^{-1}\right)} + \frac{B}{5\left(1-2z^{-1}\right)} - \frac{1}{\left(1-\frac{1}{2}z^{-1}\right)}$$

solving for A, B and C, we get

$$H(z) = \frac{54}{35\left(1+3z^{-1}\right)} + \frac{8}{\left(1-2z^{-1}\right)} + \frac{1}{7\left(1-\frac{1}{2}z^{-1}\right)}$$

∴ Impulse response

$$h[n] = \frac{54}{35}(-3)^n u[n] + \frac{8}{5}(2)^n u[n] - \frac{1}{7}\left(\frac{1}{2}\right)^n u[n]$$

Problem 7.11: A system has impulse response $h[n] = \left(\frac{1}{2}\right)^n u[n]$.

Determine the input to the system if the output is given by

(*a*) $y[n] = \delta[n-2]$

Solution:

We have, $h[n] = \left(\frac{1}{2}\right)^n u[n]$; $H(z) = \dfrac{1}{1-\frac{1}{2}z^{-1}}$

also $Y(z) = z^{-2}$

$$H(z) = \frac{Y(z)}{X(z)}$$

$$\therefore \quad X(z) = \frac{z^{-2}}{\frac{1}{\left(1-\frac{1}{2}z^{-1}\right)}} = z^{-2} - \frac{1}{2}z^{-3}$$

$$x[n] = \delta[n-2] - \frac{1}{3}\delta[n-3]$$

(*b*) $$y[n] = \left(\frac{1}{2}\right)^n u[n] + \left(-\frac{1}{2}\right)^n u[n]$$

$$Y(z) = \frac{1}{1-\frac{1}{2}z^{-1}} + \frac{1}{1+\frac{1}{2}z^{-1}} = \frac{2}{\left(1-\frac{1}{2}z^{-1}\right)\left(1+\frac{1}{2}z^{-1}\right)}$$

$$X(z) = \frac{2\left(1-\frac{1}{2}z^{-1}\right)}{\left(1-\frac{1}{2}z^{-1}\right)\left(1+\frac{1}{2}z^{-1}\right)}$$

$$x[n] = 2\left(-\frac{1}{2}\right)^n u[n]$$

(*c*) $$y[n] = \frac{1}{3}u[n] + \frac{2}{3}\left(-\frac{1}{2}\right)^n u[n]$$

We, have

$$h[n] = \left(\frac{1}{2}\right)^n u[n]$$

$$H(z) = \frac{1}{1-\frac{1}{2}z^{-1}}$$

$$y[n] = \frac{1}{3}u[n] + \frac{2}{3}\left(-\frac{1}{2}\right)^n u[n]$$

$$\therefore \quad Y(z) = \frac{1}{3\left(1-z^{-1}\right)} + \frac{2}{3\left(1+\frac{1}{2}z^{-1}\right)} = \frac{1}{3}\left[\frac{1}{\left(1-z^{-1}\right)} + \frac{2}{\left(1+\frac{1}{2}z^{-1}\right)}\right]$$

$$H(z) = \frac{Y(z)}{X(z)}$$

$$\therefore X(z) = \frac{Y(z)}{X(z)} = \frac{1}{3}\left[\frac{\left(1+2z^{-1}+2-2z^{-1}\right)}{\left(1-z^{-1}\right)\left(1+\frac{1}{2}z^{-1}\right)}\right]\cdot\left(1-\frac{1}{2}z^{-1}\right)$$

$$= \frac{1-\frac{1}{2}z^{-1}}{\left(1-z^{-1}\right)\left(1+\frac{1}{2}z^{-1}\right)}$$

Using partial fraction

$$\frac{A}{\left(1-z^{-1}\right)}+\frac{B}{\left(1+\frac{1}{2}z^{-1}\right)}$$

Solving for A and B

$$X(z) = \frac{1}{3\left(1-z^{-1}\right)}+\frac{2}{3\left(1+\frac{1}{2}z^{-1}\right)}$$

$$\therefore \qquad x[n] = \frac{1}{3}u[n] + \frac{2}{3}\left(-\frac{1}{2}\right)^n u[n]$$

> **Problem 7.12:** Determine the z-transform of the signal
>
> $$x[n] = \begin{cases} 1, & n=-1 \\ 2, & n=0 \\ -1, & n=1 \\ 1, & n=2 \\ 0, & \text{otherwise} \end{cases}$$
>
> Use the z-transform to determine the DTFT of $x[n]$

Solution:

Substitute $x[n]$ into the z-transform of an arbitrary signal $x[n]$ is

$$X[z] = \sum_{n=-\infty}^{\infty} x[n]\, z^{-n}$$

to obtain $X(z) = z + 2 - z^{-1} + z^{-2}$

We obtain the DTFT from $X[n]$ by substituting $z = e^{j\Omega}$:

$$X(e^{j\Omega}) = e^{j\Omega} + 2 - e^{-j\Omega} + e^{-j2\Omega}$$

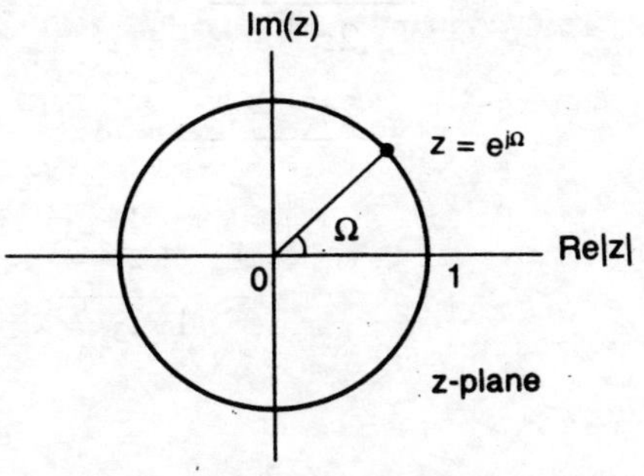

Fig. P. 7.12 (a)

> **Problem 7.13:** Find the inverse z-transform of
>
> $$X(z) = e^{z^2} \text{ with ROC all } z \text{ except } |z| = \infty$$

Solution:

Using the power series representation for e^a, $e^a = \sum_{k=0}^{\infty} \frac{a^k}{k!}$

we write $X(z) = \sum_{k=0}^{\infty} \frac{(z^2)^k}{k!} = \sum_{k=0}^{\infty} \frac{z^{2k}}{k!}$

Thus $$x[n] = \begin{cases} 0, & n > 0 \text{ and } n \text{ odd} \\ \dfrac{1}{\left(\dfrac{n}{2}\right)!}, & \text{otherwise} \end{cases}$$

> **Problem 7.14:** Find the inverse z-transform of
>
> $$X(z) = \frac{2 + z^{-1}}{1 - \frac{1}{2}z^{-1}}$$
>
> using a power series expansion.

Solution:

We use division to write $X(z)$ as a powe series in z^{-1}, since the ROC indicates that $x[n]$ is right sided. We have

$$\begin{array}{r|l} & 2 + 2z^{-1} + z^{-2} + \frac{1}{2}z^{-3} + \ldots \\ \hline 1 - \frac{1}{2}z^{-1} & 2 + z^{-1} \\ & \underline{2 - z^{-1}} \\ & 2z^{-1} \\ & \underline{2z^{-1} - z^{-2}} \\ & z^{-2} \\ & \underline{z^{-2} - \frac{1}{2}z^{-3}} \\ & \frac{1}{2}z^{-3} \end{array}$$

That is,

$$X(z) = 2 + 2z^{-1} + z^{-2} + \frac{1}{2}z^{-3} + \ldots$$

Thus we conclude that

$$\begin{aligned} x[n] &= 0, \qquad\qquad n < 0 \\ x[0] &= 2, \\ x[1] &= 2, \\ x[2] &= 1, \\ x[3] &= \frac{1}{2}, \end{aligned}$$

If the ROC is changed to $|z| < \dfrac{1}{2}$, then we expand $X(z)$ as a power series in z:

$$\begin{array}{r|l} & -2 - 8z - 16z^2 - 32z^3 + \ldots \\ \hline -\frac{1}{2}z^{-1} + 1 & z^{-1} + 2 \\ & \underline{z^{-1} - 2} \\ & \quad 4 \\ & \quad \underline{4 - 8z} \\ & \qquad 8z \\ & \qquad \underline{8z - 16z^2} \\ & \qquad\qquad 16z^2 \end{array}$$

That is,

$$X(z) = -2 - 8z - 16z^2 - 32z^3 + \ldots$$

Thus in this case we have

$$\begin{aligned} x[n] &= 0, \ n > 0 \\ x[0] &= -2, \\ x[-1] &= -8, \\ x[-2] &= -16, \\ x[-3] &= -32, \\ &\vdots \end{aligned}$$

Long division may be used to obtain the power series whenever $X(z)$ is a ratio of polynomials and is simple to perform. However, long division may not lead to a closed-form expression for $x[n]$.

Problem 7.15: Determine the z-transform of the signal

$$x[n] = \alpha^n u[n]$$

Depict the ROC and pole and zero locations of $X(z)$ in the z-plane.

Solution:

Substitute $x[n]$ into the z-transform of arbitrary signal $x[n]$ is

$$X[n] = \sum_{n=-\infty}^{\infty} x[n] z^{-n}$$

obtaining

$$X(z) = \sum_{n=-\infty}^{\infty} \alpha^n u[n]\, z^{-n} = \sum_{n=-\infty}^{\infty} \left(\frac{\alpha}{z}\right)^n$$

This is a geometric series of infinite length in the ratio $\frac{\alpha}{z}$; the sum converges provided $\left|\frac{\alpha}{z}\right| < 1$, or $|z| > |\alpha|$. Hence

$$X(z) = \frac{1}{1-\alpha z^{-1}},\ |z| > |\alpha| = \frac{z}{z-\alpha},\ |z| > |\alpha|$$

There is a pole at $z = \alpha$ and at $z = 0$ as illustrated in Fig. P. 7.15 (*a*). The ROC is depicted as the shaded region of the z-plane.

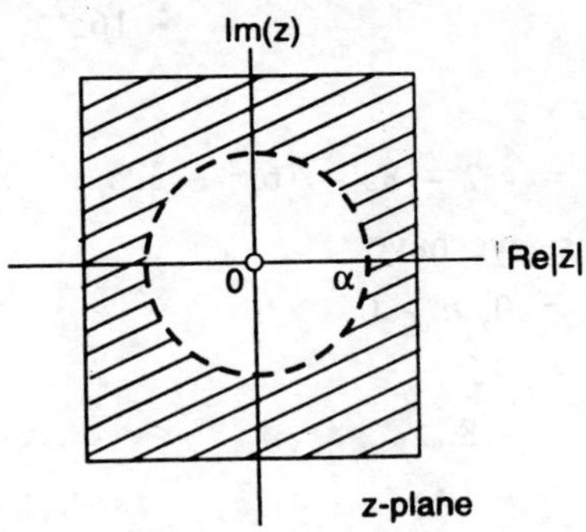

Fig. P. 7.15 (a)

Problem 7.16: Determine the z-transform of the signal

$$y[n] = -\alpha^n u[-n-1]$$

Depict the ROC and pole and zero locations in the z-plane.

Solution:

Substitute $y[n]$ into the z-transform of an arbitrary signal $x[n]$ is

$$X[n] = \sum_{n=-\infty}^{\infty} x[n]\, z^{-n}$$

and write

$$\begin{aligned} Y(z) &= \sum_{n=-\infty}^{\infty} -\alpha^n u[-n-1]\, z^{-n} \\ &= -\sum_{n=-\infty}^{-1}\left(\frac{\alpha}{z}\right)^n = -\sum_{n=1}^{\infty}\left(\frac{z}{\alpha}\right)^k = 1-\sum_{k=0}^{\infty}\left(\frac{z}{\alpha}\right)^k \end{aligned}$$

The sum converges provided $\left|\frac{z}{\alpha}\right| < 1$, or $|z| < |\alpha|$. Hence

$$Y(z) = 1 - \frac{1}{1-z\alpha^{-1}},\ |z| < |\alpha| = \frac{z}{z-\alpha},\ |z| < |\alpha|$$

Problem 7.17: Use the properties of linearity and multiplication by a complex exponential to find the z-transform of

$$x[n] = a^n \cos(\Omega_o n)\, u[n]$$

where a is real and positive.

Solution:

First note that $y[n] = a^n u[n]$ has the z-transform

$$Y(z) = \frac{1}{1-az^{-1}} \quad \text{with ROC } |z| > a$$

Now rewrite $x[n]$ as the sum

$$x[n] = \frac{1}{2}e^{j\Omega_o n}y[n] + \frac{1}{2}e^{-j\Omega_o n}y[n]$$

and apply the multiplication by a complex exponential property to each term, obtaining

$$\begin{aligned} X(z) &= \frac{1}{2}Y\left(e^{-j\Omega_o}z\right) + \frac{1}{2}Y\left(e^{j\Omega_o}z\right) \quad \text{with ROC } |z| > a \\ &= \frac{1}{2}\frac{1}{\left(1-ae^{j\Omega_o}z^{-1}\right)} + \frac{1}{2}\frac{1}{\left(1-ae^{-j\Omega_o}z^{-1}\right)} \end{aligned}$$

$$= \frac{1}{2}\left[\frac{\left(1 - ae^{-j\Omega_o}z^{-1} + 1 - ae^{j\Omega_o}z^{-1}\right)}{\left(1 - ae^{j\Omega_o}z^{-1}\right)\left(1 - ae^{-j\Omega_o}z^{-1}\right)}\right]$$

$$= \frac{1 - a\cos(\Omega_o)\, z^{-1}}{1 - 2a\cos(\Omega_o)\, z^{-1} + a^2 z^{-2}} \quad \text{with ROC } |z| > a$$

Problem 7.18: Find the z-transform of the signal

$$x[n] = \left(n\left(-\frac{1}{2}\right)^n u[n]\right) * \left(\frac{1}{4}\right)^{-4} u[-n]$$

Solution:

First, we find the z-transform of $w[n] = n\left(-\frac{1}{2}\right)^n u[n]$. We know that

$$\left(-\frac{1}{2}\right)^n u[n] \xleftrightarrow{z} \frac{z}{z + \frac{1}{2}} \quad \text{with ROC } |z| > \frac{1}{2}$$

Thus the z-domain differentiation property implies

$$w[n] = n\left(-\frac{1}{2}\right)^n u[n] \xleftrightarrow{z} W(z)$$

$$= -z\frac{d}{dz}\left(\frac{z}{z + \frac{1}{2}}\right) \quad \text{with ROC } |z| > \frac{1}{2}$$

$$= -z\left[\frac{z + \frac{1}{2} - z}{\left(z + \frac{1}{2}\right)^2}\right] = \frac{-\frac{1}{2}z}{\left(z + \frac{1}{2}\right)^2} \quad \text{with ROC } |z| > \frac{1}{2}$$

Next, we find the z-transform of $y[n] = \left(\frac{1}{4}\right)^{-n} u[-n]$. We do this by using the time reversal property. Note that

$$\left(\frac{1}{4}\right)^n u[n] \overset{z}{\longleftrightarrow} \frac{z}{z-\frac{1}{4}} \text{ with ROC } |z| > \frac{1}{4}$$

Using Time Reversal Property

$$y[n] \overset{z}{\longleftrightarrow} Y(z) = \frac{z^{-1}}{z^{-1} - \frac{1}{4}} \text{ with ROC } \frac{1}{|z|} > \frac{1}{4}$$

$$= \frac{-4z}{z-4} \text{ with ROC } |z| < 4$$

Lastly, we apply the convolution property to obtain $X(z)$ and thus write

$$\begin{aligned} x[n] &= w[n] * y[n] \overset{z}{\longleftrightarrow} X(z) \\ &= W(z)\, Y(z) \text{ with ROC } R_w \cap R_y \\ &= \frac{2x^2}{(z-4)\left(z+\frac{1}{2}\right)^2} \text{ with ROC } \frac{1}{2} < |z| < 4 \end{aligned}$$

Problem 7.19: Suppose

$$\left(\frac{1}{2}\right)^n u[n] - \left(\frac{3}{2}\right)^n u[-n-1] \overset{z}{\longleftrightarrow} X[z] = \frac{-z}{\left(z-\frac{1}{2}\right)\left(z-\frac{3}{2}\right)}$$

with ROC $\frac{1}{2} < |z| < \frac{3}{2}$

and

$$\begin{aligned} y[n] &= \left(\frac{1}{4}\right)^n u[n] - \left(\frac{1}{2}\right)^n u[n] \overset{z}{\longleftrightarrow} Y(z) \\ &= \frac{-\frac{1}{4}z}{\left(z-\frac{1}{4}\right)\left(z-\frac{1}{2}\right)} \text{ with ROC } |z| > \frac{1}{2} \end{aligned}$$

Evaluate the z-transform of $ax[n] + by[n]$.

Solution: The pole-zero plots and ROCs for $x[n]$ and $y[n]$ are illustrated in Fig. 7.19 (a) and (b), respectively. The linearity property indicates that

$$ax[n] + by[n] \overset{z}{\longleftrightarrow} a\frac{-z}{\left(a-\frac{1}{2}\right)\left(z-\frac{3}{2}\right)} + b\frac{-\frac{1}{4}z}{\left(z-\frac{1}{4}\right)\left(z-\frac{1}{2}\right)}$$

In general, the ROC is the intersection of individual ROCs, or $\frac{1}{2} < |z| < \frac{3}{2}$, which corresponds to the ROC depicted in Fig. 7.19 (a). Note, however, what happens when $a = b$. In this case, we have

$$ax[n] + ay[n] = a\left[-\left(\frac{3}{2}\right)^n\right]u[-n-1] + \left(\frac{1}{4}\right)^n u[n]$$

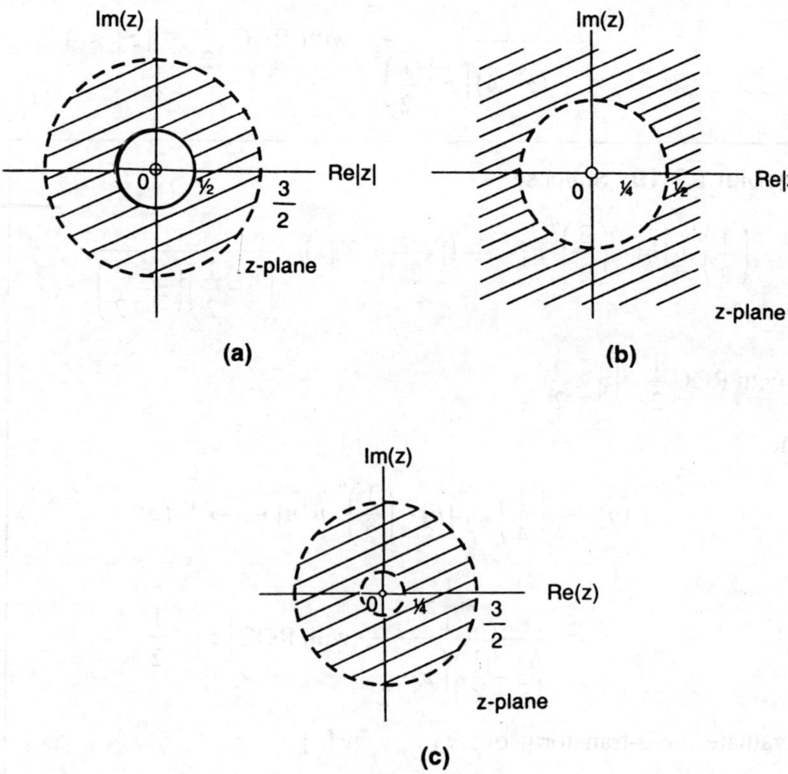

Fig. P. 7.19

and we see the term $\left(\frac{1}{2}\right)^n u[n]$ has been cancelled in the time-domain signal. The ROC is now easily verified to be $\frac{1}{4} < |z| < \frac{3}{2}$, as shown in Fig. 7.19 (c). This ROC is larger than the intersection of the individual ROCs, because the term $\left(\frac{1}{2}\right)^n u[n]$ is no longer present. Combining z-transforms and using the linearity property gives

$$aX(z) + aY(z) = a\left[\frac{-z}{\left(z-\frac{1}{2}\right)\left(z-\frac{3}{2}\right)} + \frac{-\frac{1}{4}z}{\left(z-\frac{1}{4}\right)\left(z-\frac{1}{2}\right)}\right]$$

$$= a\frac{-\frac{1}{4}z\left(z-\frac{3}{2}\right) - z\left(z-\frac{1}{4}\right)}{\left(z-\frac{1}{4}\right)\left(z-\frac{1}{2}\right)\left(z-\frac{3}{2}\right)}$$

$$= a\frac{-\frac{5}{4}z\left(z-\frac{1}{2}\right)}{\left(z-\frac{1}{4}\right)\left(z-\frac{1}{2}\right)\left(z-\frac{3}{2}\right)}$$

The zero at $z = \frac{1}{2}$ cancels the pole at $z = \frac{1}{2}$, and so we have

$$aX(z) + aY(z) = a\frac{-\frac{5}{4}z}{\left(z-\frac{1}{4}\right)\left(z-\frac{3}{2}\right)}$$

hence cancellation of the $\left(\frac{1}{2}\right)^n u[n]$ term in the time domain corresponds to cancellation of the pole at $z = \frac{1}{2}$ by a zero in the z-domain. This pole defined the ROC boundary, so the ROC enlarges when it is removed.

Chapter 8

Control Systems and Components

Problem 8.1: A transistor amplifier has a gain of 2500. Feedback is applied around the amplifier using a passive network that returns a fraction of the amplifier output to the input, namely, $\beta = 0.01$.

(*a*) Calculate the closed-loop gain of the feed-back amplifier.

(*b*) The gain of the transistor amplifier changes by 10% due to external factors. Calculate the corresponding change in the closed-loop gain of the feedback amplifier.

Solution:

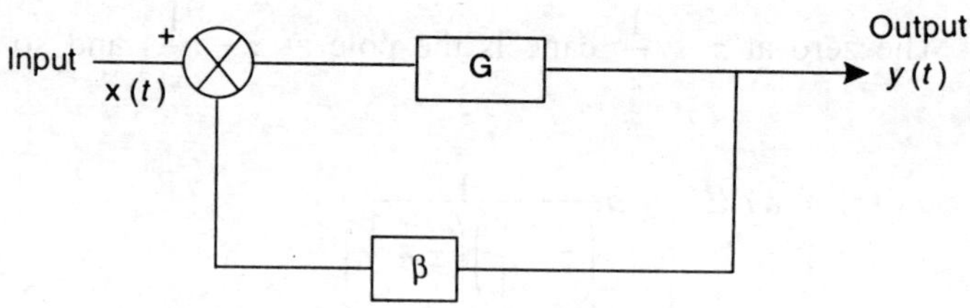

Fig. P. 8.1 (a)

(*a*) Closed-loop gain

$$G_c = \frac{Y(t)}{X(t)} = \frac{G}{1+G\beta} = \frac{2500}{1+2500\times 0.01} = \frac{2500}{26} = 96.15$$

(*b*) As,

$$G_c = \frac{G}{1+G\beta} \quad \text{...(1)}$$

Then, $$\frac{dG_c}{dG} = \frac{1+G\beta-G\beta}{(1+G\beta)^2} = \frac{1}{(1+G\beta)^2} \quad \ldots(2)$$

Dividing eqn (1) by eqn (2), we have

$$\frac{dG_c}{G_c} = \frac{dG}{G}\cdot\frac{1}{(1+G\beta)}$$

$$= (0.10)\frac{1}{1+2500\times 0.01} = \frac{0.10}{26} = 0.38\%$$

Problem 8.2: Figure 8.2 (*a*) shows the block diagram of a position-control system. The preamplifier has a gain G_a. The gain of the motor and load combination (i.e., plant) is G_p. The sensor in the feedback path returns a fraction H of the motor output to the input of the system.

(*a*) Determine the closed-loop gain T of the feedback system.

(*b*) Determine the sensitivity of T with respect to changes in G_p.

(*c*) Assuming that $H = 1$ and nominally $G_P = 1.5$, what is the value of G_a that would make the sensitivity of T with respect to changes in G_P equal to 1%?

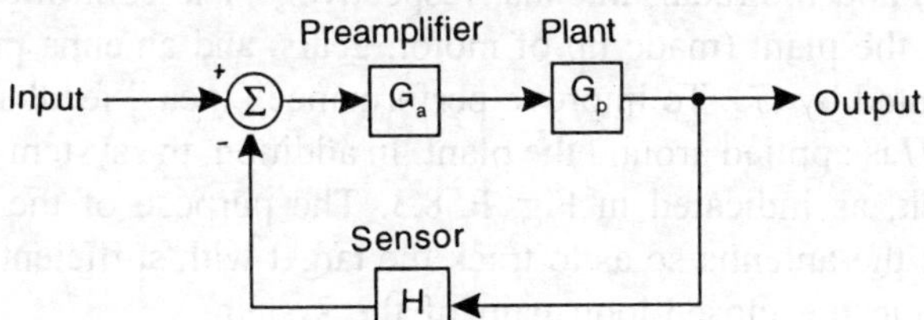

Fig. P. 8.2 (a)

Solution:

From the block diagram, we have, transfer function

(*a*) $$T = \frac{y(t)}{x(t)} = \frac{G_aG_p}{1+G_aG_pH}$$

(*b*) Sensitivity of T w.r.t. G_p,

$$S_{G_p} = \frac{\partial T}{T}\Big/\frac{\partial G_p}{G_p}$$

$$\text{or,} \quad S_{G_p} = \frac{\partial T}{\partial G_p} \cdot \frac{G_p}{T}$$

$$= \frac{G_p}{T}\left[\frac{(1+G_aG_pH)G_a - G_aG_p \cdot G_aH}{(1+G_aG_pH)^2}\right]$$

$$= \frac{G_p}{T} \cdot \frac{G_a}{(1+G_aG_pH)^2} = \frac{1}{1+G_aG_pH}$$

(c) $S_{G_p} = 0.01,\ H = 1,\ G_p = 1.5$

$$\text{then,} \quad 0.01 = \frac{1}{1+G_a \times 1 \times 1.5}$$

$$\text{or,} \quad G_a = 66$$

Problem 8.3: Figure 8.3 shows the simplified block diagram of a radar tracking system. The radar is represented by K_r, denoting some gain; θ_{in} and θ_{out} denote the angular positions of the target (being tracked) and the radar antenna, respectively. The controller has gain G_c, and the plant (made up of motor, gears, and antenna pedestal) is represented by G_P. To improve performance, :local" feedback via the sensor H is applied around the plant. In addition, the system uses unity feedback, as indicated in Fig. P. 8.3. The purpose of the system is to drive the antenna so as to track the target with sufficient accuracy. Determine the closed-loop gain of the system.

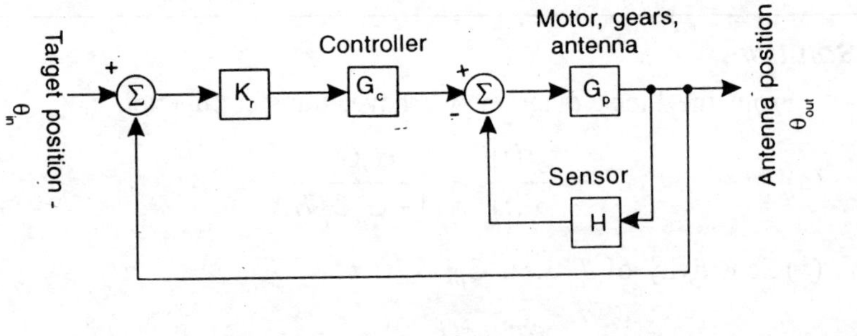

Fig. P. 8.3 (a)

Solution: Using Block-reduction technique, we have

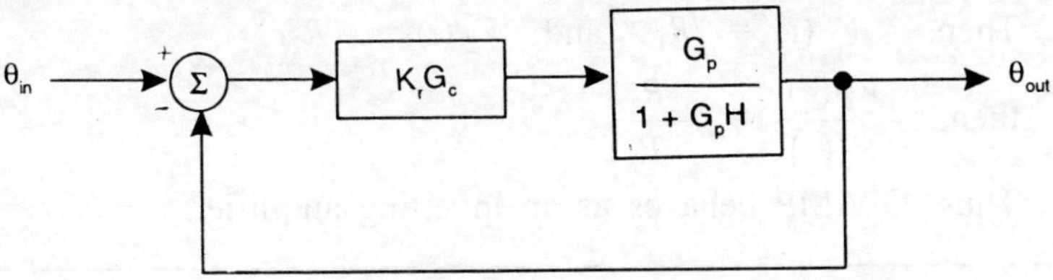

Fig. P. 8.3 (b)

Now, the closed-loop gain,

$$G = \frac{\dfrac{k_r G_c G_p}{1+G_p H}}{1+\dfrac{1+k_r G_c G_p}{1+G_p H}} = \frac{k_r G_c G_p}{1+G_p H+k_r G_c G_p}$$

Problem 8.4: Figure 8.4 shows the circuit diagram of an inverting op amp circuit. The op amp is modeled as having infinite input impedance, zero output impedance, and infinite voltage gain. Evaluate the transfer function of this circuit, that is, $V_2(s)/V_1(s)$.

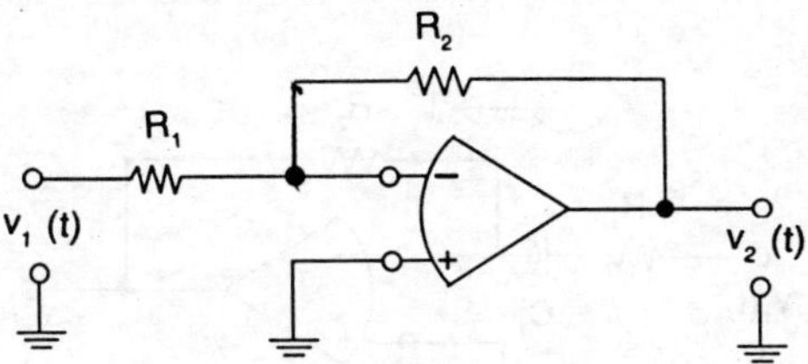

Fig. P. 8.4 (a)

Solution:

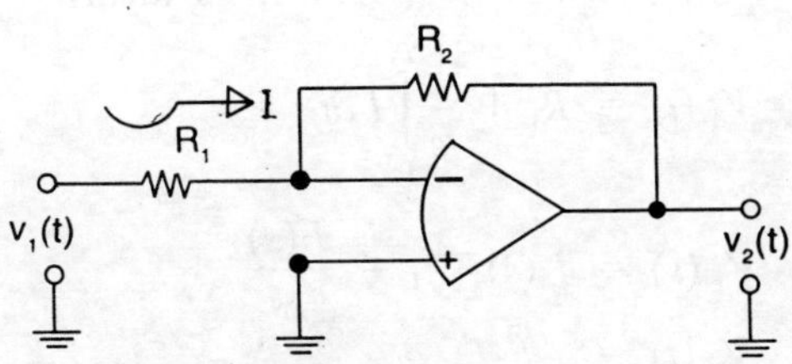

Fig. P. 8.4 (b)

Since, OPAMP is ideal,

$$V_- = V_+ \approx 0$$

Then, $V_1(t) = IR_1$ and $V_2(t) = -R_2 I$

then, $$\frac{V_2(s)}{V_1(s)} = -\frac{R_2}{R_1}$$

Thus, OPAMP behaves as an inverting amplifier.

Problem 8.5: Figure 8.5 shows a practical differentiator using an op amp. Assume that the op amp is ideal, having infinite input impedance, zero output impedance, and infinite gain. Determine the transfer function of this circuit.

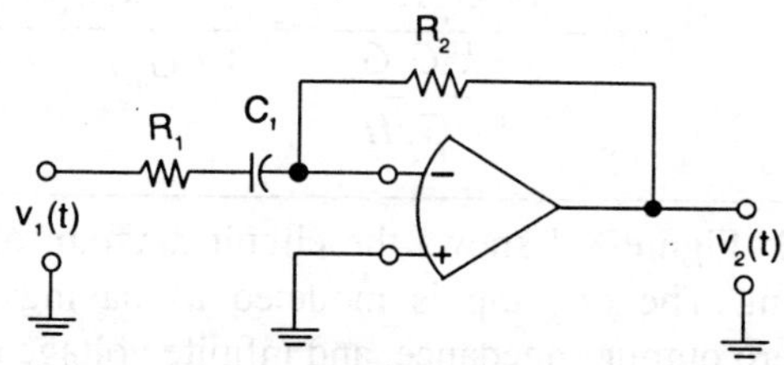

Fig. P. 8.5 (a)

Solution:

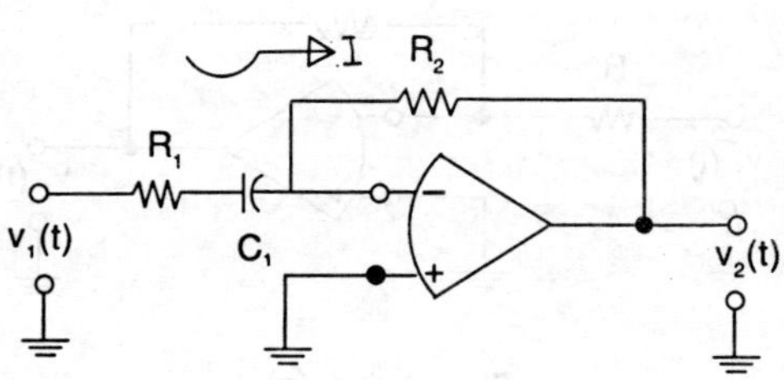

Fig. P. 8.5 (b)

Since $V_- = V_+ \approx 0$ (OPAMP is ideal)

then $$V_1(t) = R_1 + \frac{1}{c}\int_0^t I\,dt$$

or, $$V_1(s) = I(s) \cdot R_1 + \frac{I(s)}{cs} \quad \text{...(1)}$$

and, $V_2(t) = -R_2 I$

or, $V_2(s) = -R_2 I(s)$ ...(2)

Dividing (2) by (1), we get

$$\frac{V_2(s)}{V_1(s)} = \frac{-R_2 cs}{R_1 cs + 1}$$

Problem 8.6: Figure 8.6 shows a control system with unity feedback. Determine the steady-state error of the system for (*i*) unit-step input, (*ii*) unit-ramp input, and (*iii*) unit-parabolic input. Do these calculations for each of the following cases:

(*a*) $G(s) = \frac{15}{(s+1)(s+3)}$ (*b*) $G(s) = \frac{5}{s(s+1)(s+4)}$

(*c*) $G(s) = \frac{5(s+1)}{s^2(s+3)}$

You may assume that, in each case, the closed-loop transfer function of the system is stable.

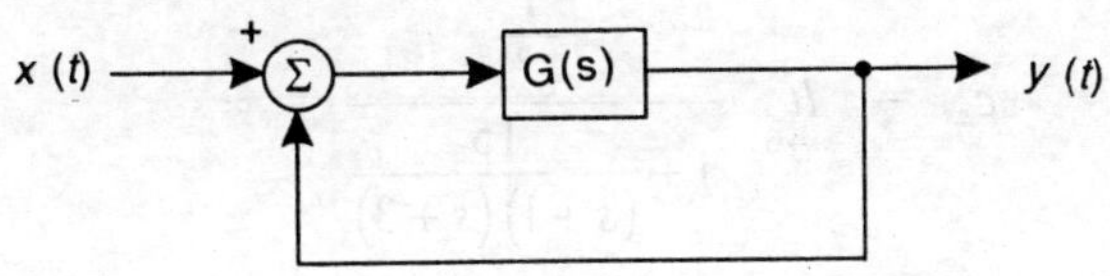

Fig. P. 8.6 (a)

Solution:

From the figure given,

$$\frac{y(t)}{x(t)} = \frac{G}{1+G}$$

or, $$\frac{x(t)-y(t)}{x(t)} = \frac{e(t)}{x(t)} = \frac{1}{1+G}$$

or, $$e(t) = x(t) \cdot \frac{1}{1+G}$$

or, e_{ss}. (steady-state emr) $= \underset{s\to 0}{Lt} \frac{sX(s)}{1+G(s)}$

(*a*) $$G(s) = \frac{15}{(s+1)(s+3)}$$

Here for unit step input $X(s) = \frac{1}{s}$

$$e_{ss} = \underset{s\to 0}{lt} \frac{1}{1+\frac{15}{(s+1)(s+3)}} = \frac{1}{6}$$

For unit ramp-input

$$X(s) = \frac{1}{s^2}$$

$$e_{ss} = \underset{s\to 0}{lt} \frac{\frac{1}{s}}{1+\frac{15}{(s+1)(s+3)}} = \infty$$

For unit–parabolic input

$$X(s) = \frac{1}{s^3}$$

$$e_{ss} = \underset{s\to 0}{lt} \frac{\frac{1}{s^2}}{1+\frac{15}{(s+1)(s+3)}} = \infty$$

(*b*) For, $X(s) = \frac{1}{s}$ (unit-setp)

$$e_{ss} = \underset{s\to 0}{lt} \frac{1}{1+\frac{5}{s(s+1)(s+4)}} = \underset{s\to 0}{lt} \frac{s(s+1)(s+4)}{s(s+1)(s+4)+5} = 0$$

For $X(s) = \frac{1}{s^2}$ (ramp-input)

$$e_{ss} = \underset{s\to 0}{lt} \frac{\frac{1}{s}}{1+\frac{5}{s(s+1)(s+4)}} = \underset{s\to 0}{lt} \frac{(s+1)(s+4)}{s(s+1)(s+4)+5} = \frac{4}{5}$$

For $X(s) = \frac{1}{s^3}$ (unit-parabolic)

$$e_{ss} = \underset{s\to 0}{lt} \frac{\frac{1}{s^2}}{1+\frac{5}{s(s+1)(s+4)}} = \underset{s\to 0}{lt} \frac{1}{s}\frac{(s+1)(s+4)}{[s(s+1)(s+4)+5]} = \infty$$

(*c*) For $X(s) = \frac{1}{s}$ (unit-step)

$$e_{ss} = \underset{s\to 0}{lt} \frac{1}{1+\frac{5(s+1)}{s^2(s+3)}} = \underset{s\to 0}{lt} \frac{s^2(s+3)}{s^2(s+3)+5(s+1)} = 0$$

For $X(s) = \frac{1}{s^2}$ (unit-ramp)

$$e_{ss} = \underset{s\to 0}{lt} \frac{\frac{1}{s}}{1+\frac{s(s+1)}{s^2(s+3)}} = \underset{s\to 0}{lt} \frac{s(s+3)}{s^2(s+3)+5(s+1)} = 0$$

For $X(s) = \frac{1}{s^3}$ (unit-parabolic)

$$e_{ss} = \underset{s\to 0}{lt} \frac{\frac{1}{s^2}}{1+\frac{5(s+1)}{s^2(s+3)}} = \underset{s\to 0}{lt} \frac{(s+3)}{s^2(s+3)+5(s+1)} = \frac{3}{5}$$

Problem 8.7: Use the Rough-Hurwitz criterion to determine the number of roots in the left half, on the imaginary axis, and in the right half of the *s*-plane for each of the following characteristic equation:
(*a*) $s^4 + 2s^2 + 1 = 0$
(*b*) $s^4 + s^3 + s + 0.5 = 0$
(*c*) $s^4 + 2s^3 + 3s^2 + 2s + 4 = 0$

Solution:

(*a*) Given $s^4 + 2s^2 + 1 = 0$

or, $(s^2 + 1)^2 = 0$

or, $s^2 + 1 = 0$

or, $s = \pm j$

Two roots lies on the *jw*-axis.

(b) $s^4 + s^3 + s + 0.5 = 0$

constructing routh–array, we have

$$\begin{array}{c|ccc} s^4 & 1 & \in & 0.5 \\ s^3 & 1 & 1 & 0 \\ s^2 & \in -1 & 0.5 & 0 \\ s^1 & \dfrac{\in -1.5}{\in -1} & 0 & \end{array}$$

As $\varepsilon \to 0$, $\in - 1 \to -\text{ve}$ and $\dfrac{\in -1.5}{\in -1} \to +\text{ve}$

and, since the number of sign changes of the first column of the routh array is two, therefore number of roots in RHS-plane is two and remaining two in LSH-plane.

(c) $s^4 + 2s^3 + 3s^2 + 2s + 4 = 0$

constructing Routh-array, we have

$$\begin{array}{c|ccc} s^4 & 1 & 3 & 4 \\ s^3 & 2 & 2 & 0 \\ s^2 & 2 & 4 & 0 \\ s^1 & -2 & 0 & \\ s^0 & 4 & & \end{array}$$

Since, the number of sign changes in the first column is two, therefore two roots lies in the RHS-plane and remaining two in the LHS-plane.

Problem 8.8: Using the Rough-Hurwitz criterion, find the range of parameter K for which the following characteristic equation represents a stable system:

$$s^3 + s^2 + s + K = 0$$

Solution:

Constructing Routh-array, we have,

$$\begin{array}{c|cc} s^3 & 1 & 1 \\ s^2 & 1 & k \\ s^1 & 1-k & 0 \\ s^0 & k & \end{array}$$

For stability, each term in the first column should be positive, then

$$k > 0$$

and $\quad 1 - k > 0$

Thus, range of parameter k for stability is $0 < k < 1$

Problem 8.9: (*a*) The characteristic equation of a third-order feedback system is defined, in general, by

$$A(s) = a_2 s^3 + a_2 s^2 + a_1 s + a_0 = 0$$

Using the Routh-Hurwitz criterion, determine the conditions which the coefficients a_0, a_1, a_2, and a_3 must satisfy for the system to be stable.

Solution:

(*a*) Constructing Routh-array

$$\begin{array}{c|cc} s^3 & a_3 & a_1 \\ s^2 & a_2 & a_o \\ s^1 & \dfrac{a_2 \cdot a_1 - a_o a_3}{a_2} & 0 \\ s^0 & a_o & \end{array}$$

For stability

$$a_3,\ a_2,\ a_o > 0$$

and

$$\frac{a_2 a_1 - a_0 a_3}{a_2} > 0$$

or,

$$a_2 a_1 > a_0 a_3$$

Problem 8.10: The loop transfer function of a feedback control system is defined by

$$L(s) = \frac{K}{s(s^2 + s + 2)},\ K > 0$$

The system uses unity feedback.

(*a*) Sketch the root locus of the system for varying K.

(*b*) What is the value of K for which the system is on the verge of instability?

Solution:

(*a*) The transfer function $L(s)$ has three poles one at $s = 0$ and other tow are at $s = \dfrac{-1 \pm \sqrt{7}j}{2} = -0.5 \pm 1.32j$

centroid $= -\frac{1}{3}$

The root locus is drawn next,

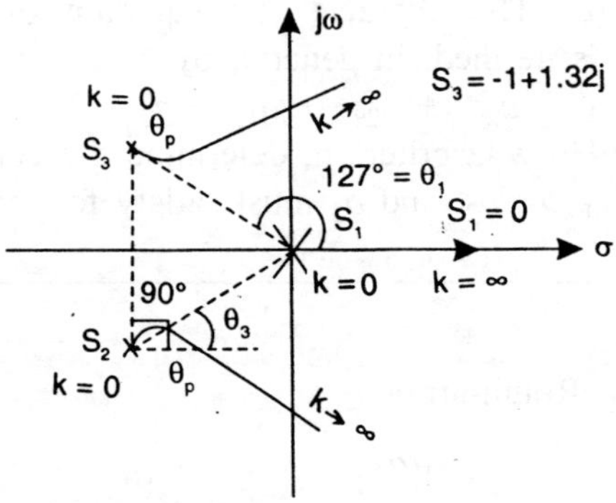

Fig. P. 8.10 (a)

Angle of departure at s_3,

$$\theta_{P3} = 180° - \theta_1 - \theta_2 = 180° - 127.1° - 90° = -37.1°$$

Angle of departure at s_2,

$$\theta_{P2} = 180° - \theta_3 - \theta_2 = 180° - 52.85° - 90° = 37.1°$$

(*b*) The characteristics equation,

$$1 + L(s) = 0$$

or, $$1 + \frac{k}{s\left(s^2 + s + 2\right)} = 0$$

or, $$s^3 + s^2 + 2s + k = 0$$

constructing routh-array, we have

s^3	1	1
s^2	1	k
s^1	$2 - k$	0
s^0	k	

For stability the elements in the first row should be +ve

then, $k > 0,\ 2 - k > 0$

or, $2 > k$

Thus if $k = 2$

then the system will be on the verge of stability.

Problem 8.11: The block diagram of a feedback control system using a *proportional* (*P*) *controller* is shown in Fig. 8.11. This form of compensation is used when satisfactory performance is attainable merely by setting the constant K_P. For the plant specified in Fig. 8.11, determine the value of K_P needed to realize a natural (undamped) frequency ω_n = 2. What are the corresponding values of
(*a*) the damping factor and
(*b*) the time constant of the system?

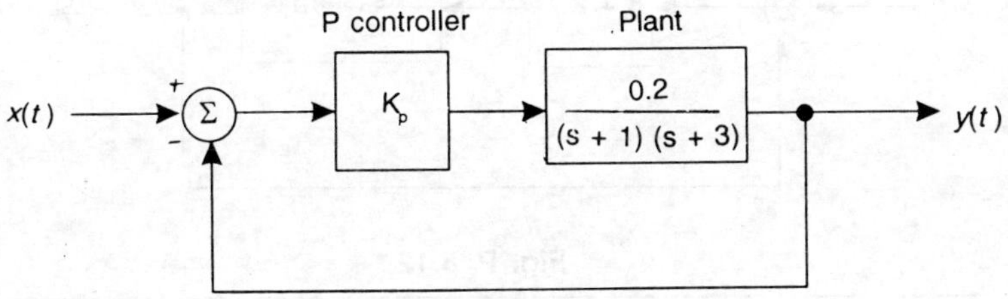

Fig. P. 8.11

Solution:

From the block diagram,

(*a*) $$\frac{y(t)}{x(t)} = \frac{0.2\,k_p}{s^2+4s+3+0.2\,k_p} = \frac{0.2k_p}{s^2+4s+(3+0.2k_p)}$$

From the transfer function,

$$\sqrt{3+0.2\,k_p} = \omega_n = 2$$

or $$3 + 0.2\,k_p = 4$$

or, $$k_p = 5$$

(*b*) Time constant

$$T = \frac{1}{r\omega_n}$$

since, $$2r\omega_n = 4$$

then, $$T = \frac{1}{2} = 0.5 \text{ sec.}$$

Problem 8.12: Figure 8.12 shows the block diagram of a feedback control system using a *proportional-plus-derivative* (*PD*) *controller*.

This form of compensation, characterized by the parameters K_P and K_D, is used to improve the transient response of the system. You are given that $K_P/K_D = 4$. Plot the root locus of the system for varying K_D. Hence determine the value of K_D needed to place a pair of poles of the closed-loop transfer function of the system at $s = -2 \pm j2$.

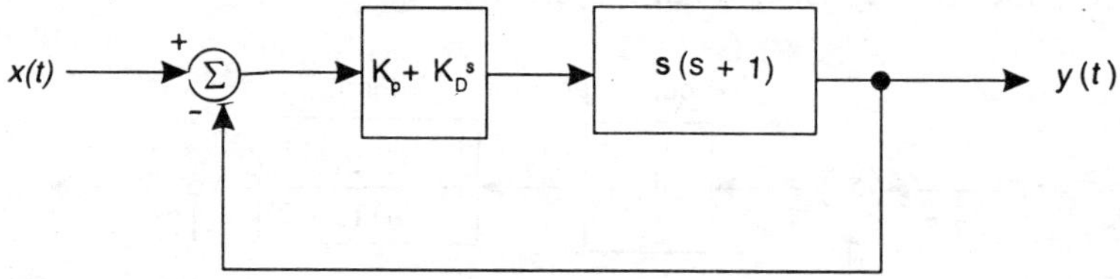

Fig. P. 8.12

Solution:

The characteristic equation,

$$1 + (k_p + k_D s)\frac{1}{s(s+1)} = 0$$

or, $$1 + \frac{(s+1)\,k_D}{s\,(s+1)} = 0 \quad \left(\text{using} \frac{k_p}{k_D} = 4\right)$$

the transfer function, $$T(s) = \frac{(s+4)\,k_D}{s\,(s+1)}$$

The root-locus is drawn next

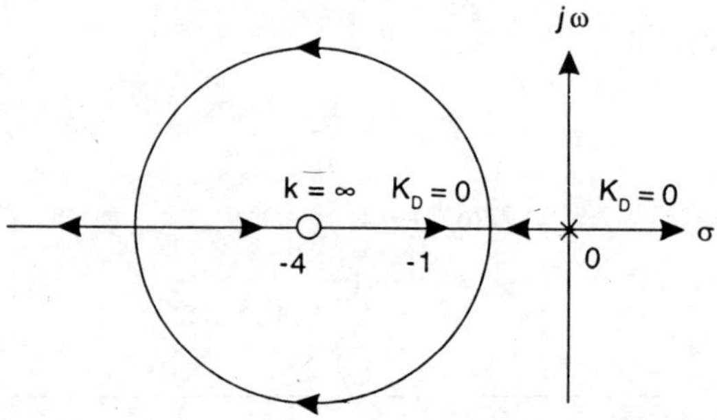

Fig. P. 8.12 (a)

The closed-loop transfer function,

$$T(s) = \frac{(k_p + k_D s)}{s^2 + (k_D + 1)s + k_p}$$

The poles are given by

$$s^2 + (k_D + 1\ s + k_p = 0$$

or, $$s = \frac{-(k_D + 1) \pm \sqrt{(k_D + 1)^2 - 4k_p}}{2})$$

Given, $$s = -2 \pm j2$$

we have, $$-(k_D + 1) = -2$$

or, $$k_D = 1$$

Problem 8.13: Figure 8.13 shows the block diagram of a feedback control system of type 1. Determine the damping ratio, natural frequency, damped frequency, and time constant of this system for $K = 20$.

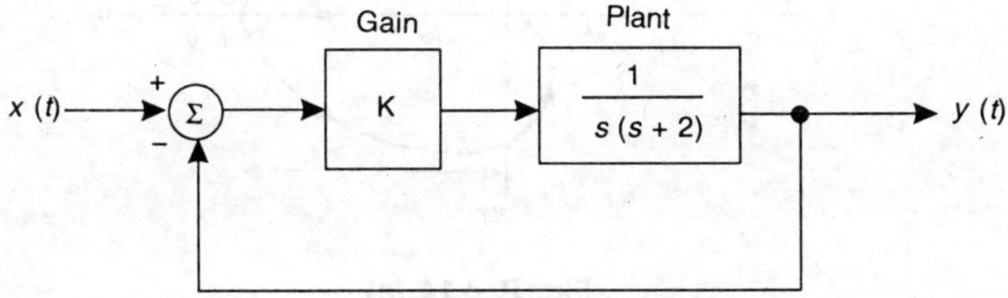

Fig. P. 8.13

Solution:

From the block-diagram,

$$\frac{y(t)}{x(t)} = \frac{k}{s^2 + 2s + k}$$

then, $2r\omega_n = 2$ and, $\omega_n = \sqrt{k}$

Given $k = 20,\ \omega_n = \sqrt{20} = 4.47$ rad/sec

and, $$r = \frac{1}{\sqrt{20}} = 0.22$$

Damped frequency

$$\omega_d = (\sqrt{1 - r^2})\ \omega_n = \sqrt{1 - (0.22)^2} \cdot 4.47 = 4.35 \text{ rad/sec}$$

Time constant

$$T = \frac{1}{r\omega_n} = \frac{1}{I} = 1 \text{ sec.}$$

Problem 8.14: Consider a two-stage feedback amplifier, for which the loop transfer function is

$$L(s) = \frac{K}{(s+1)^2}$$

Using the Nyquist criterion, show that this system is stable for all $K > 0$.

Solution:

The plot of $L(jw)$ is shown next

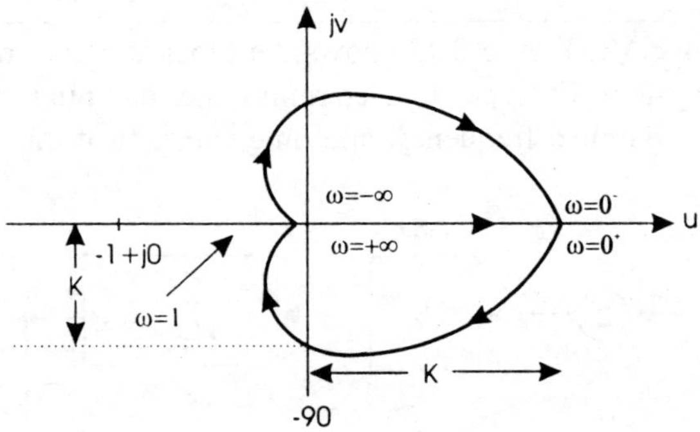

Fig. P. 8.14 (a)

The infinite semicircle arc of the Nyquist contour maps into origin. It is seen that the plot of $G(jw)\,H(jw)$ dows not encircle the point $(-1 + j0)$ for any positive value of k.

Hence, the system is stable for $k > 0$.

Problem 8.15: Consider a three-stage feedback amplifier, for which the loop transfer function is

$$L(s) = \frac{K}{(s+1)^2\,(s+5)}$$

(*a*) Using the Nyquist criterion, investigate the stability of this system for varying K.

(*b*) Find the value of the scaling factor K needed for a gain margin of 5 dB.

(*c*) For the value of K found in part (*b*), what is the phase margin?

Solution:

(*a*) The Locus for $L(j\omega)$ is drawn below

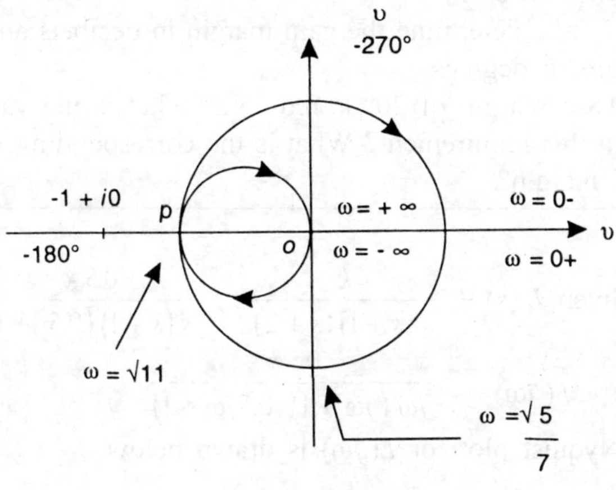

Fig. P. 8.15 (a)

$$OP = |L(j\omega)|_{\omega=\sqrt{11}} = \left|\frac{k}{(j\omega+1)^2(5+j\omega)}\right|_{\omega=\sqrt{11}} = \left(\frac{k}{72}\right)$$

For the system to be stable, Nyquist contour should not encircle $(-1+j0)$ point. For this $\frac{k}{72} < 1$ or, $k < 72$

(*b*) $$GM = 20 \log \frac{1}{OP} = 20 \log \frac{72}{k}$$

Given, $GM = 5dB$, then $20 \log \frac{72}{k} = 5$ or $k = 40.5$

(*c*) The phase margin,

$$PM = 180° + \angle L(j\omega)\big|_{\omega=\sqrt{11}} = 180° + \left(-2\tan^{-1}\omega - \tan^{-1}\frac{\omega}{5}\right)$$

$$= 180° - (146.44° - 33.55°) = 0.01°$$

Problem 8.16: The loop transfer function of a unity-feedback system of type 1 is defined by

$$L(s) = \frac{K}{(s+1)(s+2)}$$

(*a*) Using the Nyquist stability criterion, show that the system is stable for $0 < K < 6$. Also, verify your answer using the Rough-Hurwitz criterion.

(*b*) For $K = 2$, determine the gain margin in decibels and the phase margin in degrees.

(*c*) A phase margin of 20° is required. What is the value of K to attain this requirement? What is the corresponding value of the gain margin?

Solution:

(*a*) Given $L(s) = \dfrac{k}{s(s+1)(s+2)} = \dfrac{0.5k}{s(s+1)(0.5s+1)}$

or, $$L(j\omega) = \frac{0.5k}{j\omega(j\omega+1)(0.5j\omega+1)}$$

The Nyquist plot for $L(j\omega)$ is drawn below

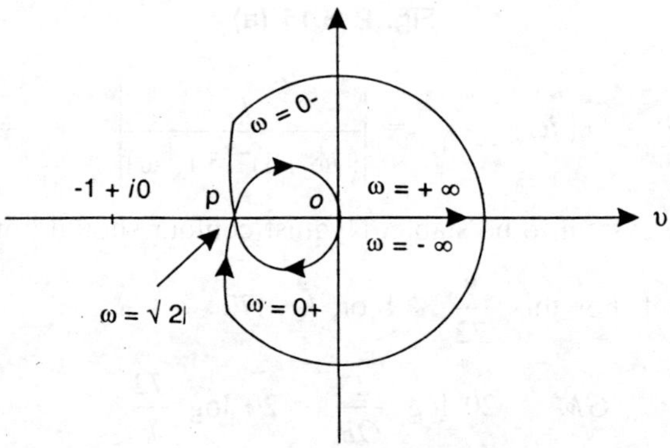

Fig. 8.16 P. (a)

At $$\omega = \sqrt{2}$$

$$OP = |L(j\omega)|_{\omega=\sqrt{2}} = \frac{0.5k}{\sqrt{2}\sqrt{3}\sqrt{1.5}}$$

For stability,

$$OP < 1 \text{ [zero encirclement occurs]}$$

or, $$\frac{0.5k}{\sqrt{9}} < 1$$

or, $$k < 6$$

Hence, the system will be stable for $0 < k < 6$

(*b*) Gain margin,

$$GM = \left|L(j\omega)\right|_{\omega=\sqrt{2}} = \frac{2}{\sqrt{2}\cdot\sqrt{3}\sqrt{6}} = \frac{2}{6} = \frac{1}{3}$$

In decibles,

$$GM = 20 \text{ logu}3 = 9.54 \text{ dB}$$

Phase margin,

$$PM = 180° + \angle L(j\omega)\big|_{\omega=\sqrt{2}}$$

$$= 180° - 90° - \tan^{-1}\sqrt{2} - \tan^{-1}\frac{1}{\sqrt{2}}$$

$$= 180° - 90° - 54.7° - 35.26° = 0.035°$$

(*c*) For $PM = 20°$

$$\angle L(j\omega) = 180° - PM = 180° - 20° = 160°$$

or, $$-90° - \tan^{-1}\omega - \tan^{-1}\frac{\omega}{2} = 160°$$

or, $$\tan^{-1}\frac{\frac{3\omega}{2}}{1-\frac{\omega^2}{2}} = -250$$

or, $$\frac{3\omega}{-2+\omega^2} = 2.74$$

or, $$\omega^2 - 1.09\omega - 2 = 0$$

$$\omega = \frac{1.09 \pm 0.03}{2}$$

$$= 2.06, \quad \text{[–ve value discarded]}$$

then, $$K\,L(j\omega)\big|_{\omega=2.06} = 1$$

or, $$k = 1\ (j\omega)\ (j\omega + 1)\ (j\omega + 2)\big|_{\omega=2.06}$$

$$= 2.06 \times 2.28 \times 2.87 = 13.48$$

Ans, the gain margin, is corresponding this is 1 or 0 dB

Problem 8.17: The loop transfer function of a unity-feedback control system of type 1 is defined by

$$L(s) = \frac{K}{s(s+1)}$$

> Using the Nyquist criterion, investigate the stability of this system for varying K. hence show that the system is stable for all $K > 0$.

Solution:

Given, $$L(s) = \frac{k}{s(s+1)}$$

$$L(jw) = \frac{k}{j\omega\left(j\omega + 1\right)}.$$

the Nyquist plot of $L(j\omega)$ is drawn next

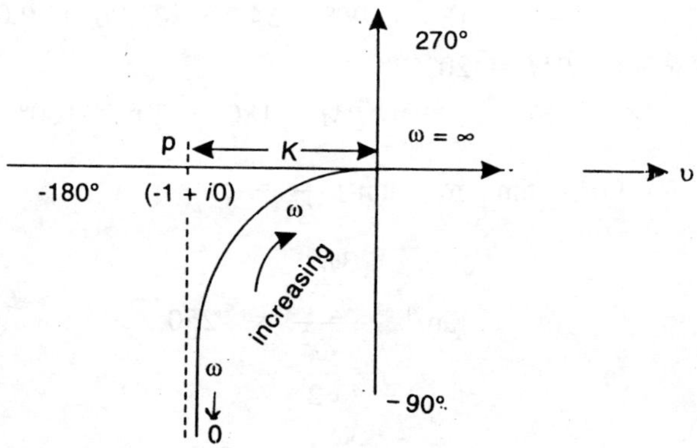

Fig. P. 8.17 (a)

The system would be stable if Nyquist contour drawn above does not encircle point $(-1 + j0)$. To satisfy this criterion, $k < 1$. If $k > 0$, then plot will be lie in RH-side and in this case also system could be stable.

Chapter 9

Miscellaneous Problems and their Solutions

Problem 9.1: Plot the magnitude of the signal,

$$x(t) = e^{j2t} + e^{j3t}$$

Solution:

Rewriting $x(t)$ as,

$$x(t) = e^{j2.5t}(e^{-j0.5t} + e^{j0.5t})$$

using Euler's identity,

$$x(t) = 2e^{j2.5t}\cos(0.5t)$$

and, $|x(t)| = 2\cos(0.5\ t)$

Hence, $|x(t)|$ referred as full-wave rectified sinusoid is drawn below.

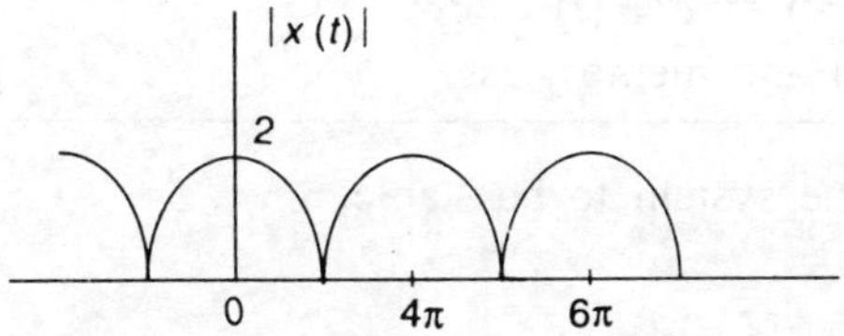

Fig. P. 9.1 (a)

Problem 9.2: Determine the fundamental period of the discrete-time signal

$$x[n] = e^{j\left(2\frac{\pi}{3}\right)^n} + e^{j\left(3\frac{\pi}{4}\right)^n}$$

Solution: The first exponential on the right-hand side of the given eqn has a fundamental period of 3 and the second exponential has fundamental period of 6. Then, for the entire signal $x[n]$ to repeat, each of the terms must go through an integral number of its own fundamental period. The smallest increment of n required is 24 and thus $n = 24$ is the fundamental period.

Problem 9.3: Find the step response of a discrete-time system with impulse response

$$h[n] = (-a)^n u[n], \text{ assuming } |a| < 1$$

Solution:

Step response

$$s[n] = h[n] * u[n] = \sum_{k=-\infty}^{\infty} h[k] \cdot u[n-k]$$

Now, since $u[n-k] = 0$ for $k > n$ and

$u[n-k] = 1$ for $k \le n$, we have

$$s[n] = \sum_{k=-\infty}^{n} h[k] = \sum_{k=-\infty}^{n} (-a)^k \cdot u[k] = \frac{1-(-a)^{n+1}}{1+a} \cdot u[n]$$

Problem 9.4: Determine the conditions on a such that the continuous-time system with impulse response

$$h(t) = e^{at} u(t)$$

is stable, causal and memoryless.

Solution: For the system to be stable,

$$\int_{-\infty}^{\infty} |h(\tau)|\, d\tau < \infty$$

or, $$\int_{-\infty}^{\infty} |e^{a\tau} u(\tau)|\, d\tau < \infty$$

or, $$\int_{-\infty}^{\infty} |e^{a\tau}|\, d\tau < \infty$$

This is possible when $a < 0$

Now, Since the impulse response is zero for $t < 0$ as $u(t) = 0$ for $t < 0$, memoryless as $h(t)$ is zero for $t < 0$.

Problem 9.5: Determine the difference equation corresponding to the block diagram of the system shown in the figure below.

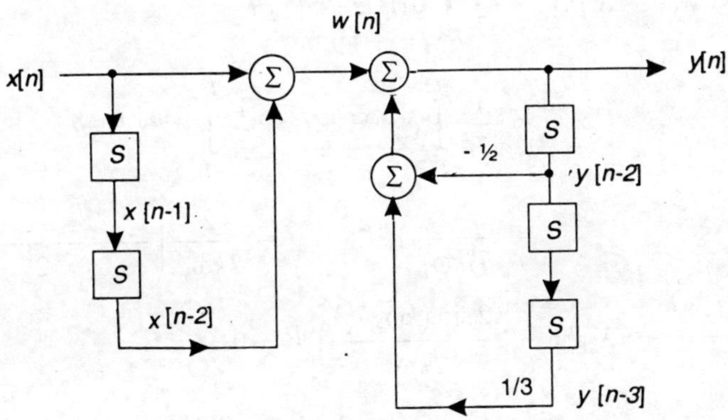

Fig. P. 9.5 (a)

Solution: The scalar multiplications and summations imply

$$\omega[n] = x[n] + 2x[n-2] \quad \ldots(1)$$

and, doing the same for output $y[n]$, we have

$$y[n] = \omega[n] + -\frac{1}{2}[y(n-1)] + \frac{1}{3}(y[n-3]) \quad \ldots(2)$$

From eqn (1) and eqn (2) we have

$$y[n] + \frac{1}{2}y[n-1] - \frac{1}{3}y[n-3] = x[n] + 2x[n-2]$$

Problem 9.6: Determine the fourier series representation for the square wave depicted below.

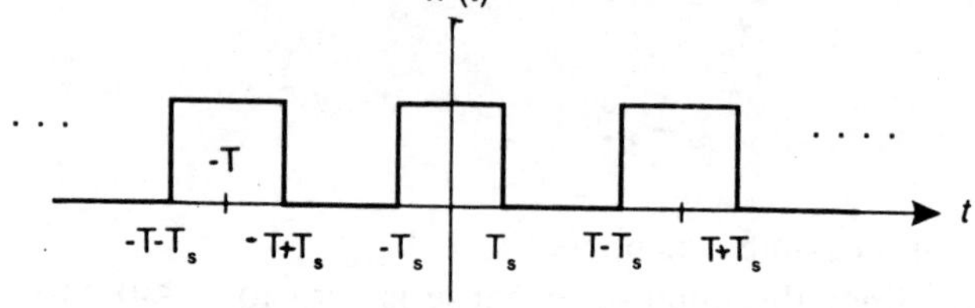

Fig. P. 9.6 (a)

Solution: The period is T, then $\omega_0 = \dfrac{2\pi}{T}$

Now, $$X[k] = \frac{1}{T}\int_{\langle T\rangle} x(t)\, e^{-jk\omega_o t}\cdot dt$$

$$= \frac{1}{T}\int_{-T/2}^{T/2} e^{-jk\omega_o t}\cdot dt = \frac{1}{T}\int_{-T_s}^{T_s} e^{-jk\omega_o t}\cdot dt$$

$$= \frac{-1}{Tjk\omega_o} e^{-jk\omega_o t}\Big|_{-T_s}^{T_s} = \frac{2}{Tk\omega_o}\left(\frac{e^{jk\omega_o T_s} - e^{-jk\omega_o T_s}}{2}\right)$$

$$X[k] = \frac{2\sin(k\omega_o T_s)}{Tk\omega_o},\ k \neq 0$$

For $$k = 0,\ X[0] = \frac{1}{T}\int_{-T_s}^{T_s} dt = \frac{2T_s}{T}$$

Now, using $\omega_0 = \dfrac{2\pi}{T}$, we have

$$X[k] = \frac{2\sin\left(k\dfrac{2\pi T_s}{T}\right)}{k2\pi}$$

Problem 9.7: Determine the Fourier series representation for the signal,

$$x(t) = 3\cos\left(\frac{\pi}{2}t + \frac{\pi}{4}\right)$$

Solution: The fundamental period of $x(t)$ is $T = 4$. Then

$$\omega_0 = \frac{2\pi}{T} = \frac{\pi}{2}$$

then, $$x(t) = \sum_{k=-\infty}^{\infty} X[k]\, e^{jk\omega_0 t} = \sum_{k=-\infty}^{\infty} X[k]\, e^{jk\left(\frac{\pi}{2}\right)t} \quad \ldots(1)$$

writing $x(t)$ as,

$$x(t) = 3\,\frac{e^{j\left(\frac{\pi}{2}\right)t+\frac{\pi}{4}} + e^{-\left[j\left(\frac{\pi}{2}\right)t+\frac{\pi}{4}\right]}}{2} = \frac{3}{2}e^{-j\frac{\pi}{4}}\cdot e^{-j\left(\frac{\pi}{3}\right)t} + \frac{3}{2}e^{j\frac{\pi}{4}}e^{j\left(\frac{\pi}{4}\right)t}$$

Hence, $$X[k] = \begin{cases} \frac{3}{2}e^{-j\frac{\pi}{4}}, & k = -1 \\ \frac{3}{2}e^{j\frac{\pi}{4}}, & k = 1 \\ 0, & \text{otherwise} \end{cases}$$

the magnitude and phase spectra are depicted in the Fig. 9.7.

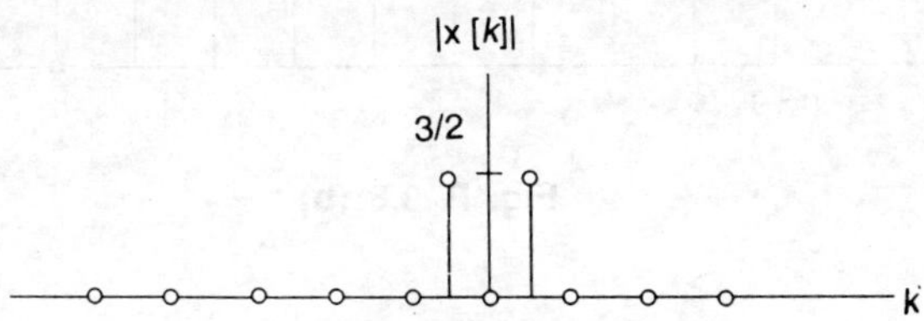

Fig. P. 9.7 (a)

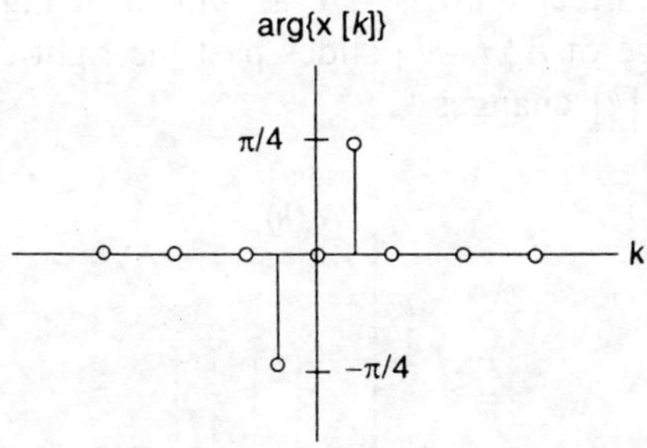

Fig. P. 9.7 (b)

Problem 9.8: A LTI system has impulse response given as,

$$h[n] = u[n] - u[n-10],$$

then determine the output of this system when the input is rectangular pulse defined as,

$$x[n] = u[n-2] - u[n-T]$$

Solution: The graph for $x[k]$ and $h[n-k]$ is depicted in Fig. 9.8 (a) and Fig. 9.8 (b) respectively.

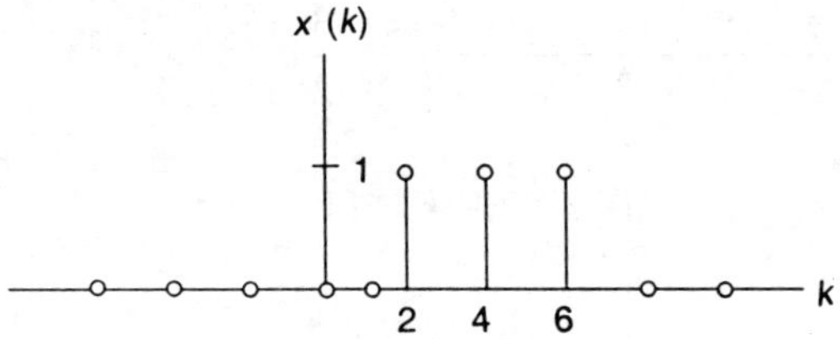

Fig. P. 9.8 (a)

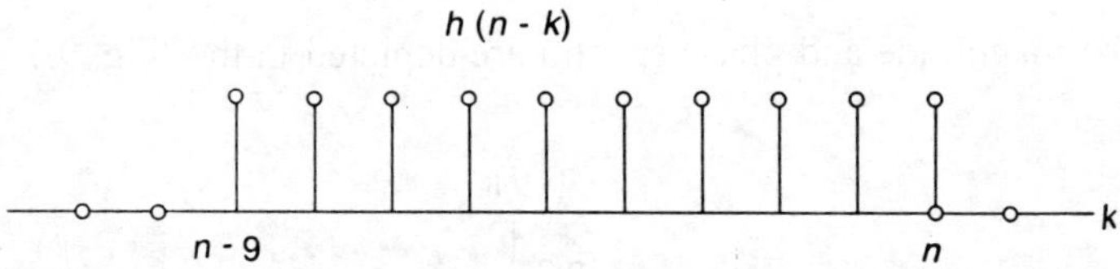

Fig. P. 9.8 (b)

For $\quad n \geq 2,$

$$\omega_n[k] = \begin{cases} 1, & 2 \leq k \leq n \\ 0, & \text{otherwise} \end{cases}$$

Which is correct until $n < 6$. as shown in Fig. 9.8 (c) when $n > 6$, the right edge of $h[n-k]$ slides past the right edge of $x[k]$ so that the form of $\omega_n[k]$ changes.

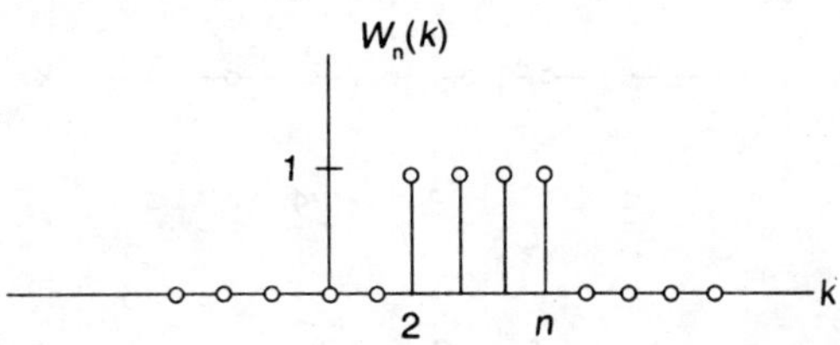

Fig. P. 9.8 (c)

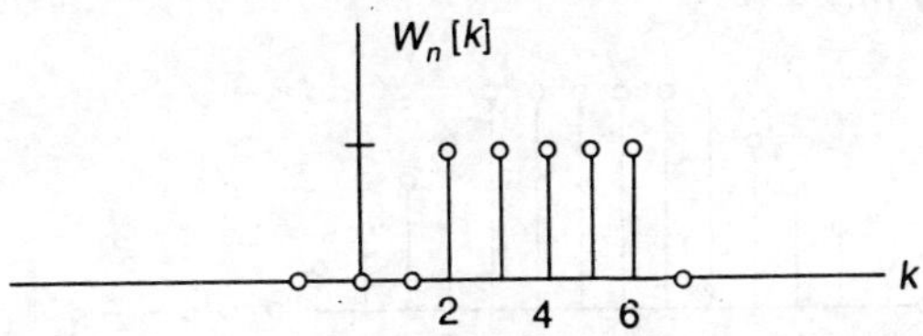

Fig. P. 9.8 (d)

For $\qquad n > 6,$

the $\omega_n[k]$ given as $\omega_n[k] = \begin{cases} 1, & 2 \le k \le 6 \\ 0, & \text{otherwise} \end{cases}$

which is shown in Fig. (d). This form is true until $n = 11$, because at that value of n the left edge of $h[n-k]$ slides past the left edge of $x[k]$. Hence third interval of time–shift is $6 < n \le 11$.

For, $\qquad n > 11$

$$\omega_n[k] = \begin{cases} 1, & h-9 \le k \le 6 \\ 0, & \text{otherwise} \end{cases}$$

as shown in Fig. 9.8 (e). This form of $\omega_n[k]$ is true until $n = 15$, since for $n > 15$, the left edge of $h[n-k]$ lies to the right of $x[k]$ and the functional form for $w_n[k]$ again changes, i.e., the fourth interval of time shift is $11 < n \le 15$

For $\qquad n > 15, \ \omega_n[k] = 0$

Now, For $2 \le n \le 6, \quad y[n] = \sum_{k=2}^{n} 1 = n - 1$

For, $n < 6 \le 11, \quad y[n] = \sum_{k=2}^{6} 1 = 5$

For $11 < n \le 15, \quad y[n] = \sum_{k=n-9}^{6} 1 = 16 - n$

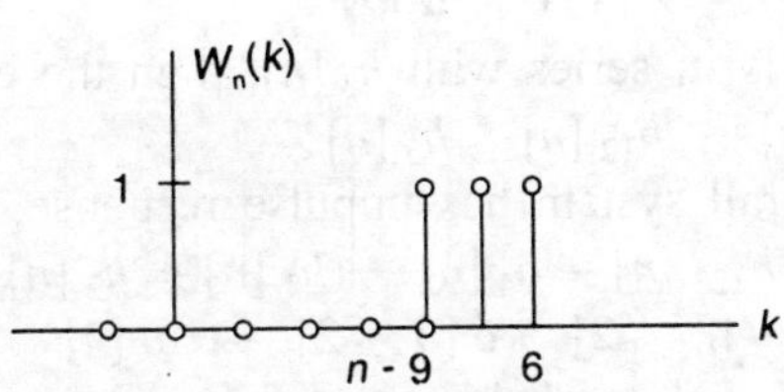

Fig. P. 9.8 (e)

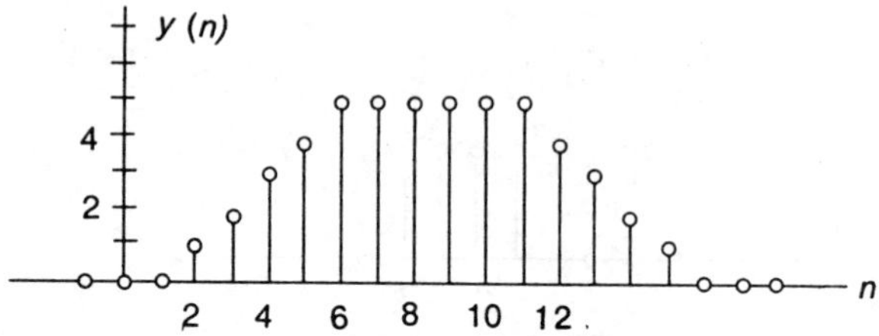

Fig. P. 9.8 (f)

(Fig. 9.8 (f) depicts the output $y[n]$)

For $\quad n > 15, y[n] = 0$

Problem 9.9: Consider the interconnection of LTI systems depicted in Fig. 9.9 (a) given below. The impulse response is given by

$h_1[n] = u[n] \qquad h_2[n] = u[n+2] - u[n]$

$h_3[n] = \alpha[n-2] \qquad h_4[n] = \alpha^n u[n]$

then, find the overall system, $h[n]$

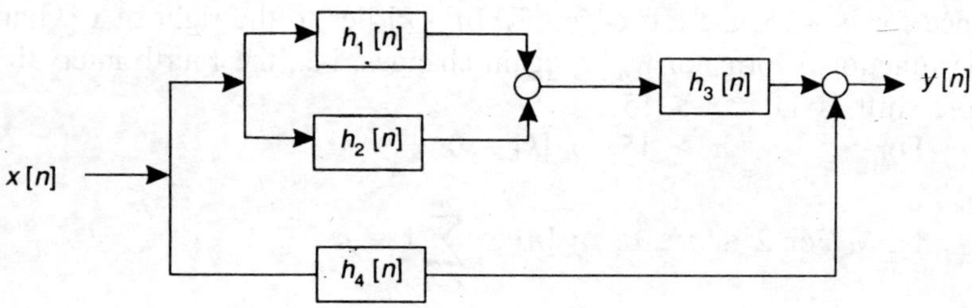

Fig. P. 9.9 (a)

Solution:

The equivalent of parallel combination of $h_1[n]$ and $h_2[n]$ is,

$$h_{12}[n] = h_1[n] + h_2[n]$$

Since $h_{12}[n]$ is in series with $h_3[n]$, then the equivalent will be

$$h_{123}[n] = h_{12}[n] * h_3[n]$$

Now, the overall system has impulse response,

$$\begin{aligned} h[n] &= h_{123}[n] - h_4[n] = (h_1[n] + h_2[n]) * h_3[n] - h_4[n] \\ &= u[n-2] * \delta[n-2] - \alpha^n u[n] \\ h[n] &= u[n] - \alpha^n u[n] = u[n][1 - \alpha^h] \end{aligned}$$

Problem 9.10: Find the difference equation for the block diagrams shown in the fig. given.

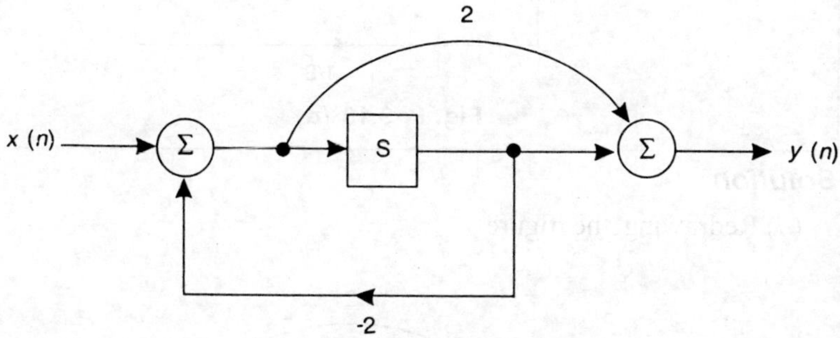

Fig. P. 9.10 (a)

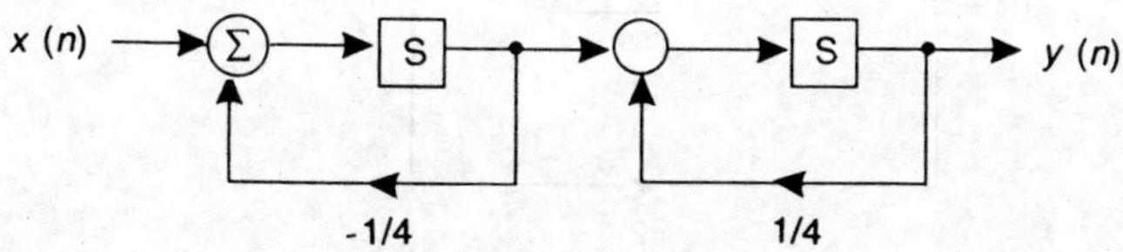

Fig. P. 9.10 (b)

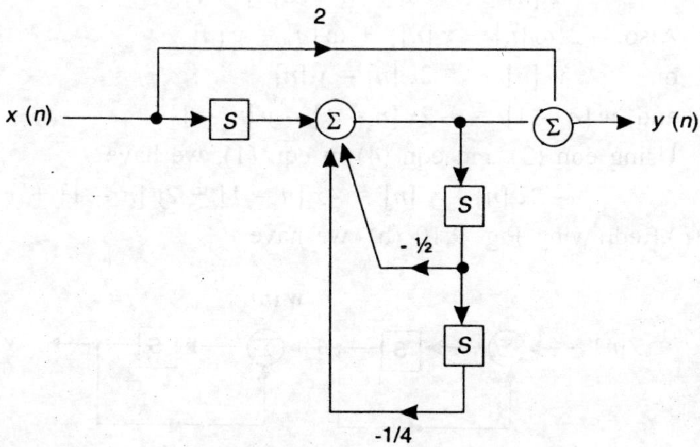

Fig. P. 9.10 (c)

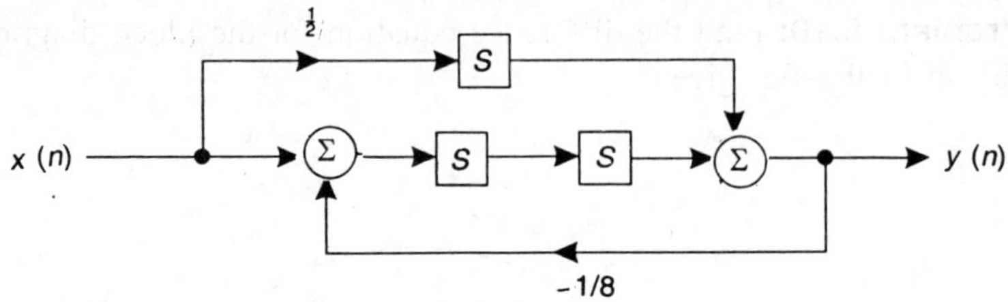

Fig. P. 9.10 (d)

Solution:

(*a*) Redrawing the figure

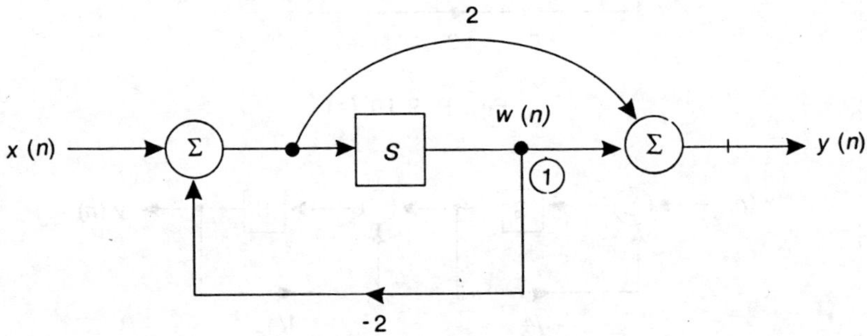

Fig. P. 9.10 (e)

For node (1), we have

$$\omega[n] = -2\omega[n^{-1}] + x[n] - 1] \quad ...(1)$$

Also, $-2\{\omega[n] + x[n]\} + \omega[n] = y[n]$ ...(2)

or, $w[n] = -2x[n] - y[n]$ ...(3)

and, $w[n-1] = -2x[n-1] - y[n-1]$ ...(4)

Using eqn (3) and eqn (4) in eqn (1), we have

$$-2x[n] - y[n] = 4x[n-1] + 2y[n-1] + x[n-1]$$

(*b*) Redrawing Fig. 9.10 (b), we have

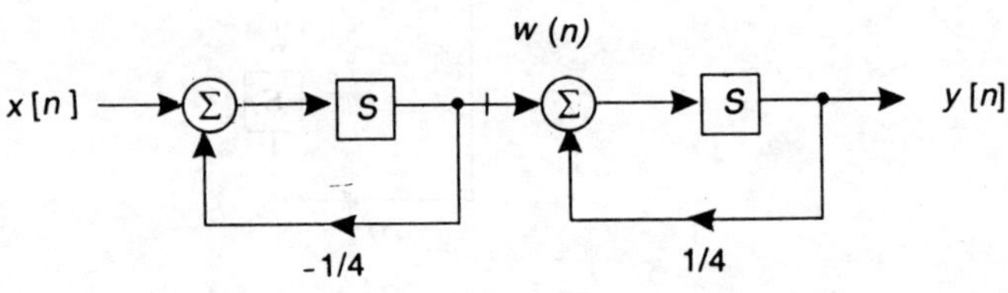

Fig. P. 9.10 (f)

For the first loop difference equation is,

$$w[n] = \left\{x(n-1) - \frac{1}{4}w(n-1)\right\} \quad \ldots(1)$$

In the second lo.. the difference equation is,

$$w[n-1] + \frac{1}{4}y[n-1] = y[n]$$

or, $$w[n-1] = y[n] - \frac{1}{4}y[n-1] \quad \ldots(2)$$

and $$w[n] = y[n+1] - \frac{1}{4}y[n] \quad \ldots(3)$$

Using eqn (2) and eqn (3) in eqn (1), we get

$$y[n+1] - \frac{1}{4}y[n] = x[n-1] - \frac{1}{4}y[n] + \frac{1}{16}\cdot y[n-1]$$

or, $$y[n+1] - \frac{1}{16}y[n-1] = x[n-1]$$

is the required difference equation.

(*c*) Redrawing Fig. 9.10 (C), we have

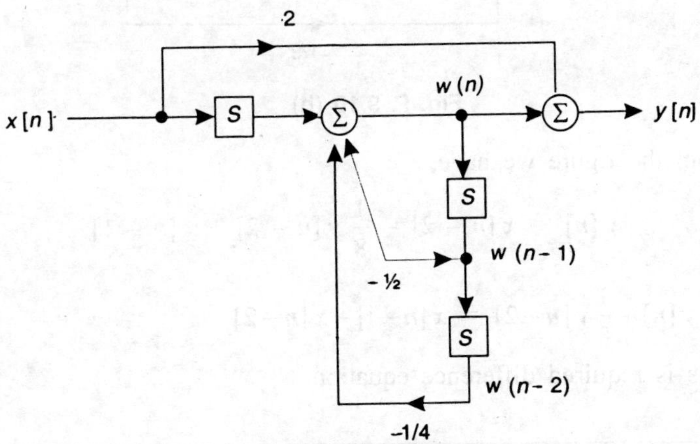

Fig. P. 9.10 (g)

From the figure, scalar multiplications and summations imply,

$$-\frac{1}{2}\omega[n-1] - \frac{1}{4}\omega[n-2] + x[n-1] = \omega[n] \quad \ldots(1)$$

Also $2x[n] + w[n] = y[n]$

or, $$w[n] = y[n] - 2x[n] \quad \ldots(2)$$

and thus,

$$w[n-1] = y[n-1] - 2x[n-1] \quad \ldots(3)$$
$$w[n-2] = y[n-2] - 2x[n-2] \quad \ldots(4)$$

Now, using eqn (2), eqn (3) and eqn (4), we have,

$$-\frac{1}{2}y[n-1] + x[n-1] - \frac{1}{4}y[n-2] + \frac{x}{2}[n-2] + x[n-1] =$$

$4[n] - 2x[n]$

$$y[n] + \frac{1}{2}y[n-1]\ \frac{1}{4}y[n-2] = 2x[n] + 2x[n-1] + \frac{x}{2}[n-2]$$

This is the required difference equation

(d) Redrawing Fig. (d), we have

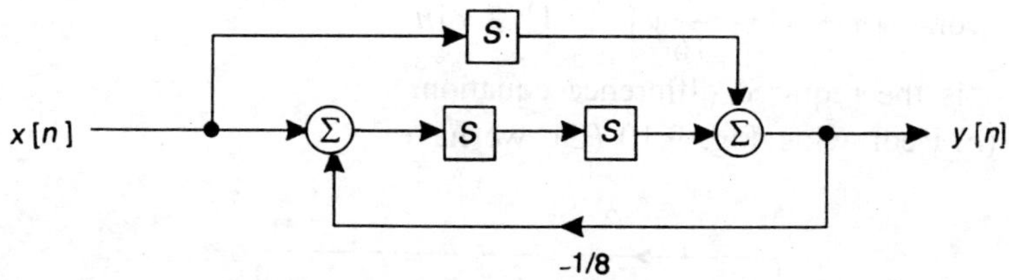

Fig. P. 9.10 (h)

From the figure we have,

$$y[n] = x[n-2] - \frac{1}{8}y[n-2] + x[n-1]$$

or $y[n] + \frac{1}{8}y[n-2] = x[n-1] + x[n-2]$

This is required difference equation

Problem 9.11: Let the input $x[n]$ to a LTI system H be given by

$$x[n] = \alpha^n \{u[n] - u[n-10]\}$$

and the impulse response of the system is given as

$$h[n] = \beta^n u[n] \text{ where } 0 < \beta < 1.$$

Find the output of this system.

Solution:

First the graph of $x[k]$ and $h[n-k]$ is drawn as shown below.

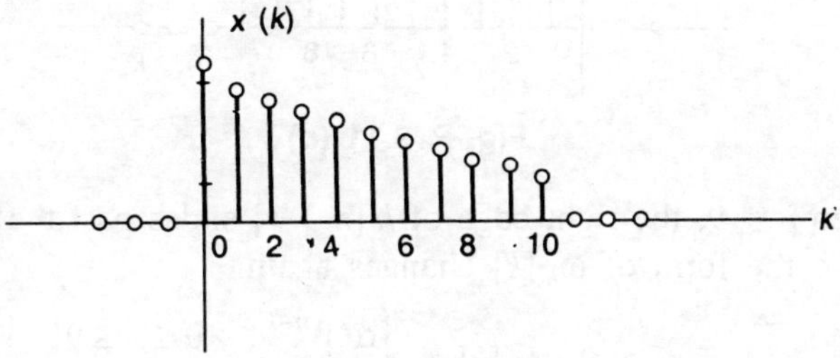

Fig. P. 9.11 (a)

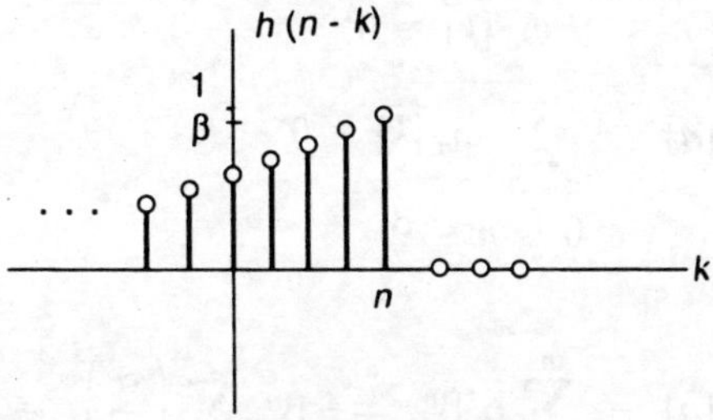

Fig. P. 9.11 (b)

From the figure, $x[k] = \begin{cases} \alpha^k, & 0 \le k \le 9 \\ 0, & \text{otherwise} \end{cases}$

Now, $h[n-k] = \begin{cases} \beta^{n-k}, & k \le n \\ 0, & \text{otherwise} \end{cases}$

Now, $\omega_n[k] = \begin{cases} \alpha^n \beta^{n-k}, & 0 \le k \le n \\ 0, & \text{otherwise} \end{cases}$

The above form for $\omega_n[n]$ is true when $0 \le n \le 9$ as depicted in Fig. (C)

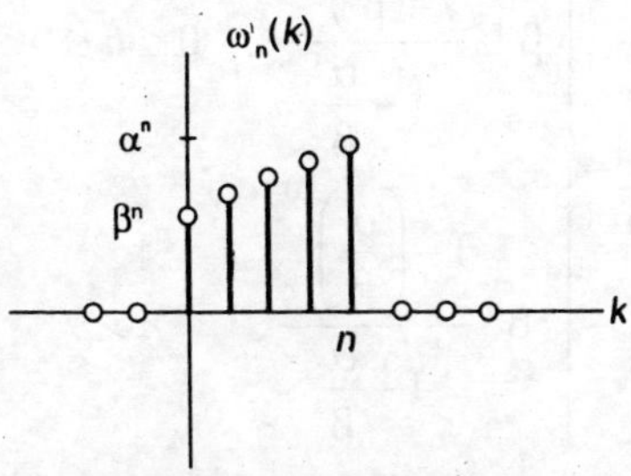

Fig. P. 9.11 (c)

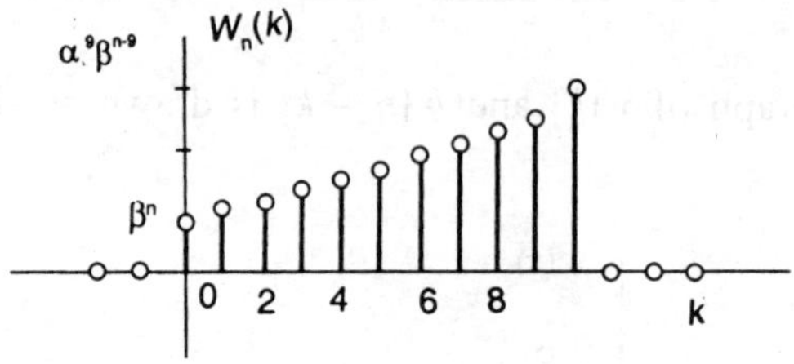

Fig. P. 9.11 (d)

When $n = 9$, the right edge of $h[n-k]$ slides past the right edge of $x[k]$ so the form of $\omega_n[k]$ changes again.

For $\quad n > 9,\ \omega_n[k] = \begin{cases} \alpha^k \beta^{n-k}, & 0 \le k \le 9 \\ 0, & \text{otherwise} \end{cases}$

Which is depicted in Fig. 9.11 (d)
Now, as for $n < 0$, $\omega_n[k] = 0$

$$y[n] = \sum_{k=-\infty}^{n} \omega_n[k] = 0$$

For, $\quad 0 \le n \le 9$

$$y[n] = \sum_{k=0}^{n} \alpha^k \beta^{n-k} = \beta^n \sum_{k=0}^{9} \left(\frac{\alpha}{\beta}\right)^k = \beta^n \frac{1-\left(\frac{\alpha}{\beta}\right)^{n+1}}{1-\frac{\alpha}{\beta}}$$

For, $\quad n \ge 10$,

$$y[n] = \sum_{k=0}^{9} \alpha^k \beta^{n-k} = \beta^n \sum_{k=0}^{9} \left(\frac{\alpha}{\beta}\right)^k = \beta^n \frac{1-\left(\frac{\alpha}{\beta}\right)^{10}}{1-\frac{\alpha}{\beta}}$$

$$\text{Thus,}\quad y[n] = \begin{cases} 0, & n < 0 \\ \beta^n \dfrac{1-\left(\frac{\alpha}{\beta}\right)^{n+1}}{1-\frac{\alpha}{\beta}}, & 0 \le n \le 9 \\ \beta^n \dfrac{1-\left(\frac{\alpha}{\beta}\right)^{10}}{1-\frac{\alpha}{\beta}}, & n > 9 \end{cases}$$

Problem 9.12: Find the inverse FT of

$$X(j\omega) = \begin{cases} 2\cos\omega, & |\omega| \le \pi \\ 0, & |\omega| > \pi \end{cases}$$

Solution: The FT representation,

$$x(t) = \frac{1}{2\pi}\int_{-\infty}^{\infty} X(j\omega)\cdot e^{j\omega t}\cdot d\omega = \frac{1}{2\pi}\int_{-\pi}^{\pi} 2\cos\omega \cdot e^{jwt}\cdot d\omega$$

$$= \frac{1}{2\pi}\int_{-\pi}^{\pi} \frac{\left(e^{j\omega}+e^{-j\omega}\right)}{2}\cdot e^{j\omega t}\cdot d\omega$$

$$= \frac{1}{2\pi}\int_{-\pi}^{\pi} e^{|\omega(t+1)}\cdot d\omega + \frac{1}{2\pi}\int_{-\pi}^{\pi} e^{j\omega(t-1)}\cdot dw$$

$$= \frac{1}{2\pi(t+1)}\left[e^{i\pi(t+1)} - e^{-i\pi(t+1)}\right] +$$

$$\frac{1}{j2\pi(t-1)}\left[e^{j\pi(t-1)} - e^{-j\pi(t-1)}\right]$$

$$= \frac{1}{\pi(t+1)}\sin[\pi(t+1)] + \frac{1}{\pi(t-1)}\sin[\pi(t-1)]$$

Problem 9.13: Determine the inverse FT of

$$X(j\omega) = \frac{-j\omega}{(j\omega)^2 + 3j\omega + 2}$$

Solution: Given,

$$X(j\omega) = \frac{-j\omega}{j\omega + 2 - 2(j\omega+1)}$$

$$X(j\omega) = \frac{j\omega + 2 - 2(j\omega+1)}{(j\omega+1)(j\omega+2)} = \frac{1}{j\omega+1} - \frac{2}{j\omega+2}$$

Using Linearity Property we have

$$X(j\omega) \xleftrightarrow{FT} x(t) = [e^{-t} - 2e^{-2t}]\,u(t)$$

Problem 9.14: Find the DTFT of

$$x[n] = \begin{cases} 2^n, & 0 \le n \le 9 \\ 0, & \text{otherwise} \end{cases}$$

Solution: $X(e^{j\Omega}) = \sum_{n=-\infty}^{\infty} x[n]\, e^{-j\Omega n} = \sum_{n=0}^{\infty} (2e^{-j\Omega})^n$

$$= \sum_{n=0}^{\infty} (2e^{-j\Omega})^n - \sum_{n=10}^{\infty} (2e^{-j\Omega})^n$$

$$= \frac{1}{1-2e^{-j\Omega}} - \frac{\left(2e^{-j\Omega}\right)^{10}}{1-2e^{-j\Omega}}$$

$$X(e^{j\Omega}) = \frac{1-2^{10}\cdot e^{-j10\Omega}}{1-2e^{-j\Omega}}$$

Problem 9.15: Find the inverse DTFT of
$X(e^{j\Omega}) = 2\cos(2\Omega)$

Solution: $x[n] = \frac{1}{2\pi}\int_{-\pi}^{\pi} X(e^{j\Omega}) \cdot e^{j\Omega n}\, d\Omega$

$$= \frac{1}{2\pi}\int_{-\pi}^{\pi} \frac{\left(e^{j2\Omega} + e^{-j2\Omega}\right)}{2} e^{j\Omega n}\, d\Omega$$

$$= \frac{1}{4\pi}\int_{-\pi}^{\pi} e^{j\Omega(n+2)}\, d\Omega + \frac{1}{4\pi}\int_{-\pi}^{\pi} e^{j\Omega(n-2)}\, d\Omega$$

$$= \frac{\left(e^{j(n+2)\pi} - e^{-j(n+2)\pi}\right)}{2\pi(n+2)} + \frac{1}{2\pi}\frac{\left(e^{j(n-2)\pi} - e^{-j\pi(n-2)}\right)}{n-2}$$

Problem 9.16: Find the FT of the time signal depicted in Fig. as shown below and as expressed by,

$$x(t) = \begin{cases} t, & |t| \leq 1 \\ 0, & |t| > 1 \end{cases}$$

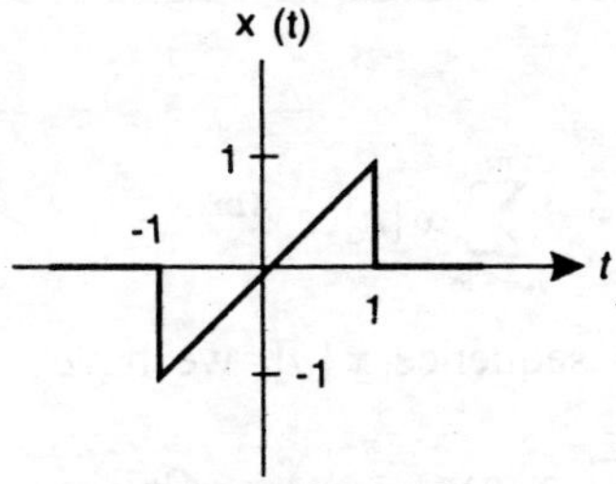

Fig. P. 9.16 (a)

Solution:

$$X(j\omega) = \int_{-\infty}^{\infty} x(t) \cdot e^{-j\omega t}\, dt = \int_{-1}^{1} te^{-j\omega t}\, dt$$

$$= \left[\frac{te^{-j\omega t}}{-j\omega}\right]_{-1}^{1} + \left[\frac{e^{-j\omega t}}{\omega^2}\right]_{-1}^{1} = \frac{je^{-j\omega}}{\omega} + \frac{je^{j\omega}}{\omega} + \frac{e^{-j\omega t}}{\omega^2} - \frac{e^{+j\omega t}}{\omega^2}$$

$$= \frac{j}{\omega} 2 \cos\omega + \frac{1}{\omega^2}(-2j \sin\omega) = j\frac{2}{\omega} \cos\omega - i\frac{2}{\omega^2} \sin\omega$$

Problem 9.17: Find the FT of $x(t) = te^{-at}u(t)$

Solution:

Since $-jtx(t) \xleftrightarrow{FT} \frac{d}{d\omega} X(j\omega)$

Now, $X(j\omega) = \int_{-\infty}^{\infty} e^{-at}u(t)\, e^{-jwt}\, d\omega = \int_{0}^{\infty} e^{-(a+j\omega)}\, d\omega = \frac{-1}{a+jw}$

Then $X'(j\omega) = j\frac{d}{d\omega}X(j\omega) = j\frac{jx-1}{(a+j\omega)^2} = \frac{1}{(a+j\omega)^2}$

Problem 9.18: Find the DTFT of $x[n] = 2(3)^n u[-n]$

Solution:

As, $X(e^{j\Omega}) = \sum_{n=-\infty}^{\infty} x[n]\, e^{-j\Omega n}$

Using the given sequence $x[n]$, we have

$$X(e^{j\Omega}) = \sum_{n=-\infty}^{\infty} 2\,(3)^n\, u[-n] \cdot e^{-i\Omega n}$$

$$= \sum_{n=-\infty}^{0} 2\,(3)^n \cdot e^{-j\Omega n} \text{ [since } u[-n] = 0 \text{ for } n < 0]$$

$$= 2 \cdot \sum_{n=-\infty}^{0} (3^{-1}e^{j\Omega})^{-n} = 2\sum_{n=0}^{\infty} (3^{-1}e^{j\Omega})^n = 2 \cdot \frac{1}{1-e\frac{j\Omega}{3}}$$

Problem 9.19: Find FT of $x(t) = e^{at}u(t)$ assuming $a > 0$.

Solution:

$$X(j\omega) = \int_{-\infty}^{\infty} e^{+at}\, u(-t) \cdot e^{-jwt} \cdot dt$$

Since $u(-t) = 0$ for $t > 0$, we have

$$X(j\omega) = \int_{-\infty}^{0} e^{+at} \cdot e^{-j\omega t} \cdot dt = \int_{-\infty}^{0} e^{at} \cdot e^{-j\omega t} \cdot dt$$

$$= \int_{-\infty}^{0} e^{(a-j\omega)t}\, dt = \left.\frac{e^{(a-j\omega)t}}{a-j\omega}\right|_{-\infty}^{0}$$

$$X(j\omega) = \frac{-1}{j\omega - a}$$

Problem 9.20: Let the input to a system with impulse response $h(t) = 2\, e^{-2t}\, u(t)$ be $x(t) = 3e^{-t}u(t)$. Then, find the output of the system $y(t)$.

Solution: The output

$$y(t) = h(t) * x(t)$$

then, $Y(j\omega) = H(j\omega) \cdot X(j\omega) = \dfrac{2}{j\omega + 2} \cdot \dfrac{3}{j\omega + 1} = \dfrac{6}{(j\omega + 2)(j\omega + 1)}$

$$Y(j\omega) = \frac{6}{j\omega + 1} - \frac{6}{j\omega + 2}$$

and, $\quad y(t) = [6e^{-t} - 6e^{-2t}]\, u(t)$

Problem 9.21: Show the DTFT of

$$x[n] = ne^{j\left(\frac{\pi}{8}\right)n} \alpha^{(n-3)}\, u[n-3]$$

Solution:

$$X'(e^{j\Omega}) = \sum_{n=-\infty}^{\infty} x[n]\, e^{-j\Omega n} = \sum_{n=-\infty}^{\infty} ne^{j\left(\frac{\pi}{8}\right)n} \alpha^{n-3}\, u[n-3] \qquad \ldots(1)$$

Using $\qquad -jnx[n] \xleftrightarrow{FT} \dfrac{d}{d\Omega} X(e^{j\Omega})$

or, $\qquad nx[n] \xleftrightarrow{FT} j\dfrac{d}{d\Omega} X(e^{j\Omega}) \qquad \ldots(2)$

and $\qquad x[n - n_o] \xleftrightarrow{DTFT} e^{-j\Omega n_o} X(e^{j\Omega}) \qquad \ldots(3)$

Now, $\quad X(e^{j\Omega}) = \displaystyle\sum_{n=-\infty}^{\infty} e^{j\left(\frac{\pi}{8}\right)n} \alpha^n u[n] \cdot e^{-j\Omega n}$

$$= \sum_{n=0}^{\infty} e^{j\left(\frac{\pi}{8}\right)n} \alpha^n \cdot e^{-j\Omega n}$$

$$= \sum_{n=0}^{\infty} \left(a \cdot e^{-j\Omega - \frac{\pi}{8}}\right)^n = \frac{1}{1 - \lambda e^{-j\left(\Omega - \frac{\pi}{8}\right)}}$$

Using eqn (3) in eqn (1), we have

$$X'(e^{j\Omega}) = \sum_{n=-\infty}^{\infty} n\left[e^{j\left(\frac{\pi}{8}\right)(n-3)} \cdot e^{+j\frac{3\pi}{8}} \alpha^{n-3} u[n-3]\right]$$

Using eqn (2),
then we get

$$X(e^{j\Omega}) = \frac{d}{d\Omega}\left\{\frac{e^{-3j\left(\Omega-\frac{\pi}{8}\right)}}{1-\alpha e^{-j\left(\Omega-\frac{\pi}{8}\right)}}\right\}$$

Problem 9.22: Determine the Fourier series representation for
$x(t) = 2\sin(2\pi t - 3) + \sin(6\pi t)$

Solution: The fundamental period of $x(t)$ is $T = 1$,

then $\omega_0 = \dfrac{2\pi}{T} = 2\pi$ and then $x(t)$ can be expressed as,

$$x(t) = \sum_{k=-\infty}^{\infty} X[k]\, e^{ik2\pi t} \quad \ldots(1)$$

Also,

$$x(t) = 2\frac{\left(e^{j(2\pi t-3)} - e^{-j(2\pi t-3)}\right)}{2j} + \frac{e^{j6\pi t} - e^{-j6\pi t}}{2j}$$

$$= \left(\frac{j}{2} \cdot e^{-j6\pi t}\right) + (je^{-2\pi t} \cdot e^{j3}) - (je^{j2\pi t} \cdot e^{-j3}) - \left(\frac{j}{2} \cdot e^{j6\pi t}\right) \quad \ldots(2)$$

Comparing eqn (2) with eqn (1), we have

$$x(t) \xleftrightarrow{Fs,\,2\pi} X[k] = \begin{cases} \frac{j}{2}, & k=-3 \\ je^{j3} & k=-1 \\ -je^{-j3} & k=1 \\ -\frac{j}{2} & k=3 \\ 0 & \text{otherwise} \end{cases}$$

Problem 9.23: Let the impulse response of a LTI system be $h(t) = e^{-2(t+1)}$ $y(t)$ if the input is $x(t) = e^{-|t|}$.

Solution: Sketch of $x(\tau)$ and $h(t-\tau)$ is shown in the figures,

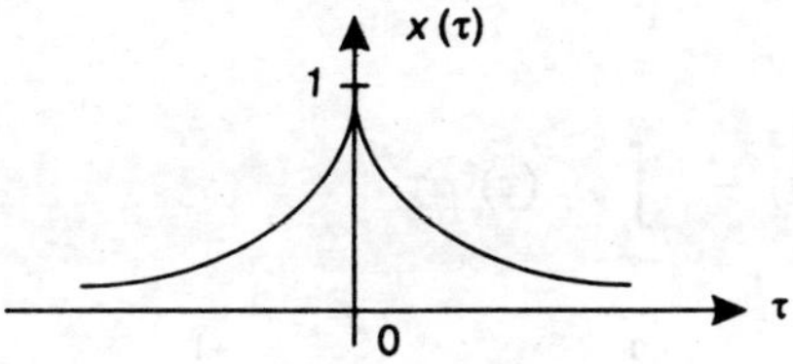

Fig. P. 9.23 (a)

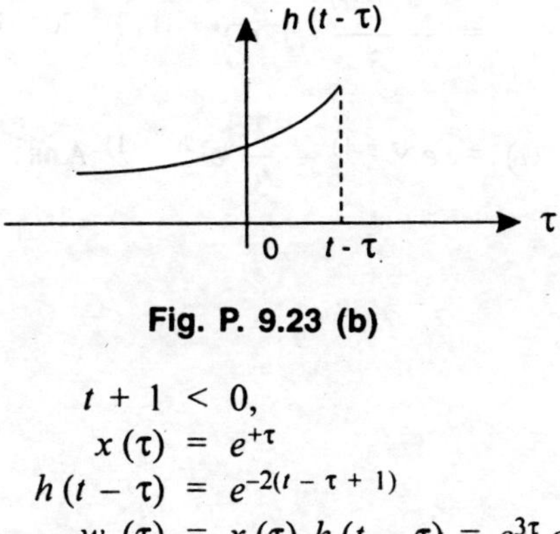

Fig. P. 9.23 (b)

For $\quad t + 1 < 0,$

$$x(\tau) = e^{+\tau}$$

and, $\quad h(t-\tau) = e^{-2(t-\tau+1)}$

then $\quad w_t(\tau) = x(\tau)\, h(t-\tau) = e^{3\tau}\, e^{-2(\tau+1)}$

and, $\quad y(t) = \int_{-\infty}^{\infty} w_t(\tau)\, d\tau = \int_{-\infty}^{(t+1)} e^{-2(\tau+1)} \cdot e^{3\tau}\, d\tau$

$$= \left[e^{-2(t+1)} \frac{\left(e^{3(t+1)}\right) - 0}{3} \right] = \frac{e^{(t+1)}}{3}, \; t + 1 < 0$$

and, $\quad y(t) = 0$, otherwise

Now, for $\quad t + 1 > 0,$

$$x(\tau) = \begin{cases} e^{\tau}, & -\infty < T \\ e^{-\tau}, & 0 \end{cases}$$

and $h(t-\tau) = e^{-2(t-\tau+1)}, -\infty < T < t+1$

Then, $\omega_t(\tau) = \begin{cases} e^{-2(t+1)e^{3\tau}}, & -\infty < \tau < t+1 \\ e^{-2(t+1)} \cdot e^{\tau}, & -\infty < \tau < 0 \\ 0, & 0 < \tau < t+1 \end{cases}$

Now, $$Y(t) = \int_{-\infty}^{\infty} \omega_t(\tau)\, d\tau$$

$$= \int_{-\infty}^{0} e^{-2(t+1)} \cdot e^{3\tau} + \int_{0}^{t+1} e^{-2(t+1)}\, e^{\tau}\, d\tau$$

$$= \frac{e^{-2(t+1)}}{3} + e^{-(t+1)} - e^{-2(t+1)}$$

$$y(t) = e^{-(t+1)} - \frac{2}{3}\, e^{-2(t+1)} \textbf{ Ans}$$

Reader's Notes

Reader's Notes

Reader's Notes

Reader's Notes

Reader's Notes

Reader's Notes